THE ART OF
ELECTROSTATIC
PRECIPITATION

THE ART OF ELECTROSTATIC PRECIPITATION

Jacob Katz, P.E.

DISTRIBUTED BY
SCHOLIUM
INTERNATIONAL, INC.
99 SEAVIEW BOULEVARD
PORT WASHINGTON, NEW YORK 11050-4610

1st Printing November 1979

2nd Printing January 1981

3rd Printing September 1989

Copyright© 1979 by Precipitator Technology, Inc.

All Rights Reserved

ISBN 0-9603986-2-7

Library of Congress Catalog Card Number 79-90504

Printed in the United States of America

To my wife, Estelle, and children, Owen, Linda, and Curtis, who have abided my many absences from home and allowed me to acquire experience with precipitation. They, in effect, are co-authors of this text.

Background of Author

Mr. Katz graduated with a B.S. in Electrical Engineering from the University of Pittsburgh in 1949. He was employed as a field test and service engineer for Research-Cottrell, Inc., from 1950 to 1956, working with the various industrial applications of precipitators. From 1956 to 1964, Mr. Katz was employed as a plant maintenance engineer and operating maintenance supervisor at the Duquesne Works of U.S. Steel Corp. He left U.S. Steel to receive a M.S. in Hygiene with a major in air pollution from the Graduate School of Public Health of the University of Pittsburgh in 1964.

Since 1964, Mr. Katz was first self-employed as an Air Pollution Consultant and Adjunct Instructor in Air Pollution at the University of Pittsburgh. He later formed and became President of Jacob Katz Associates, Inc., in 1970, which was later changed to Precipitator Technology, Inc., specializing in the field of electrostatic precipitators. He has authored and co-authored nine papers on air pollution, seven of which have to do with ESP's. Mr. Katz is a member of the Pennsylvania and National Societies of Professional Engineers, Association of Iron and Steel Engineers, and the Air Pollution Control Association (APCA).

Among APCA society positions, he has been Vice Chairman of Technical Council and Chairman of the TC-1 Committee concerned with collectors of particulate matter. Mr. Katz has also been active on ASME Power Test Code Committees for developing stack sampling methodology. He is registered in the State of Pennsylvania as a Professional Engineer.

Preface - 1st. Printing

Electrostatic precipitators have been used extensively for many years for the collection of particulate matter from industrial processes. Unfortunately, those who use the precipitator have not always understood it well because very little practical knowledge has been available to explain and augment the manufacturers' information.

This book has been written to help answer the many questions that arise about day to day operation and about the maintenance of performance levels in the precipitator. It is based primarily on my field work spanning nearly thirty years on this collector, including testing, evaluation, upgrading and internal inspections on hundreds of installations. Much of my background has given me contact with the utility, cement and iron and steel industries and how they use precipitators.

The precipitator is a complex but most interesting collector. More importantly, it is a highly effective means of achieving long-term solutions to the abatement of pollutant material. If this book removes part of the mystery that continues to confound the user of this equipment, it will have served its purpose.

I have included many of my own ideas and methods as well as ideas and methods already familiar to practioners' of precipitation to show how the collection efficiency of the precipitator can be improved. It is my hope that the information I discuss will provide focus and illustration on the general concepts of precipitation, concepts which can be made applicable to any particular precipitator. It amazes me how much performance margin actually exists in many precipitators and how the application of these fundamental concepts will often extract this margin.

I hope you will find the journey through these pages informative, useful and de-mystifying.

J. KATZ October 1979

Preface - 2nd Printing

The response to the initial distribution of this book has been gratifying. To improve its benefits to the practitioner of precipitation, I have included a supplement to the 2nd printing where I have clarified and enlarged on specific segments of the original text.

J. KATZ January 1981

Acknowledgements

This book, based on field experiences, could never have been possible without the help and contributions of the countless number of people I have worked with over the years, and I would like to thank them all. I would especially like to thank Bill Verrochi for his encouragement that gave me the confidence and enthusiasm to continue as a consultant during those early, difficult years. It was a pleasure to have been associated with John Dudeck, Tim McKenzie and Tony Catanese for a number of years, and I am appreciative of the help they gave me. The contributions of Dr. Harry White and Herb Hall to the field of precipitation in general, and to me in particular, directly and indirectly, will always be considered valuable.

For their help in the various stages of preparing this book, I am thankful to my wife, Estelle, for the typing of the drafts, to Bob Ramsey, Paul Dick and Bob Seiple of Cochran, Jones and Associates for their help on the figures, to Karen Hjelmervik for her help in trying to make some sense out of my English, and to Ed Siegel and his printing staff for help and suggestions during the final stages.

Table of Contents

LIST OF FIGURES

LIST OF TABLES

Chapter 1

Introduction

General Comments

The electrostatic precipitator uses electrical forces to capture either liquid or solid particulate matter from a flue gas system. We tend to classify the precipitator as a high-efficiency collector, comparable to the fabric baghouse or high-pressure venturi scrubber. As such, collection efficiencies of 99.5% plus are within a design range for most applications. A prime characteristic separating the various methods for high efficiency collection is that the precipitator concentrates its primary energy forces on the particle, rather than on the carrier gas stream. However, the gas stream or process characteristics will generally determine whether the particle will be easily collected or prove difficult to contain by electrical forces. In fact, comprehensive knowledge of the process is an important part of successful precipitation, and I intend to relate pertinent process factors in later chapters.

The basic physical design of the precipitator is relatively simple when compared to other forms of industrial equipment. The groups of parallel components often give a sense of security to the purchaser of the precipitator. All of this can be an illusion if certain safeguards are not implemented into proper design for the integrity of the individual components. I intend to point out areas of concern and suggest remedies for the critical equipment zones of the collector. There is no substitute for building maximum reliability into the designs of modern precipitators. Unfortunately, too many existing precipitators have not lived up to the collection performance expected for a variety of reasons. Many of these problems can be identified and corrected by judicial investigations. It may not be advantageous to correct the poor performance of an existing precipitator by the addition of more collector capacity. More discharge electrodes, more hoppers, more of any equipment that has to continually face internal gas conditions can add to maintenance costs and downtime. Field experience has emphasized the advantages of building greater collection efficiencies within reasonable physical sizes of precipitators.

The magnitude and character of the electrical forces will primarily determine the success or failure of the precipitator performance at each installation. Maintenance and operation personnel must learn to understand specific electrical signs if changes in the collector efficiency are to be explained and corrected.

I have observed precipitators operating much below their potential level for many years because basic concepts were not available to field personnel. Anyone would be disenchanted with an expensive collector if consistent and satisfactory performances did not appear possible. I believe the information to follow will provide the desirable approaches in design as well as pertinent knowledge and steps required to solve the many problems faced in the field.

The first four chapters include some background and fundamental concepts that should allow the reader to better understand the factors involved in the precipitation process. The next three chapters cover basic operating maintenance techniques that should identify many of the common problems that occur in precipitation. Chapters Eight through Thirteen will examine methods to evaluate and improve the performance of the collector. Chapters Fourteen and Fifteen submit some ideas and concepts for help at arriving at retrofit and new design. Chapter Sixteen will list references and suggestions for further reading.

There will be little space devoted to specific physical design or pictures of precipitator components. This information has been well covered in other books or papers. For the people who will use this book for a specific collector, I saw no need to describe the great variety of components that exist.

As you read through the chapters, the particles that are collected in the precipitator will be denoted as either particulate matter, dust, ash, or fume. Particulate matter is the valid term for representing this material, but the use of the common term "dust" is found frequently in the text.

Historical Aspects

Early practical installations of precipitators in the United States occurred in the early 1900's after the development of the mechanical high voltage rectifier by Dr. F. G. Cottrell. Smelter and cement industries were among the first to use this type of collector. Blast furnace and utility industries began using them in the 1920's. The electric power industry has become the greatest single user of precipitators for the collection of fly ash from pulverized coal-fired boilers.

Precipitators have been used either for the recovery of valuable material or for the abatement of pollutant matter. The collection efficiency design has grown from 85% in early days to the 92% to 98% range of the mid 1950's to present levels of 99.8% plus recovery.

Performance Comments

Each industrial process presents its own specific problems or application factors when precipitators are considered. Subtle changes in the process, of raw materials and equipment, can often produce a wide band of performance characteristics in a specific precipitator. These subtle changes probably produce the greatest number of installations which fail to meet expected collection efficiencies because of reduced electrical power input. A secondary contribution to sub-performance levels is a combination of design short-comings, including among others, gas distribution, gas sneakage, and insufficient sectionalization. A third group of factors that keeps the collector from performing consistently on a satisfactory basis is the balance of reliability, either in the evacuation of material from hoppers or in the failure of precipitator components. Each of these areas will be developed separately and in combination later.

I would like to enlarge on the above statements because they are pertinent to a practical understanding of precipitators.

1. With few exceptions, most precipitator designs are adequately sized for the conditions specified. This was true on the early designs of 94% as well as for the modern 99.5% levels. The attainment of the expected efficiency was predicated on the ability to utilize the available power supplies advantageously.

2. Gaseous and particulate matter characteristics generally control the voltage and current relationship that determine the approximate performance levels reached in a given precipitator.

3. When optimum high power inputs exist, design defects are generally masked, and while the collector may not be

performing at maximum efficiency, satisfactory levels are stil usually observed.

4. It is when optimum power input does not exist that weaknesses in design or other physical factors surface to help contribute to excessive emission losses.

5. Correction of all the secondary performance factors usually will not counteract the absence of high power input levels.

The above five (5) statements stress the key object of this book, that is, to explain in a practical way how industrial precipitators perform and what factors control their performance. There is no doubt that the process effects electrical characteristics, and this must be understood for successful precipitation to occur. But at the same time, other design factors must be placed in proper perspective according to their individual or combined effect on collection performance. It is often a difficult and complex task to understand and correct problems in precipitators because of the relationship between the variables. Misconceptions and the lack of practical information perpetuate the problems; users of the equipment correct one problem without an awareness of how they are effecting the rest of the process.

Most observations and information contained in this book are based on my field work, in many cases involving diagnostic studies in the field to correct problems. After all these years, I still have a healthy respect for the complexity of the precipitator, and the reasons for performance problems are not always clear. If I can offer one bit of advice to those entering the field, it would be not to make hasty judgements about what is wrong with a precipitator. Generally, the most visible cause of the trouble is not the culprit. Later examples will help clarify this statement. Above all, the effective use of internal inspections and valid records are important tools that can help maintain the performance levels of precipitators.

There are reasons for the successful performance of every precipitator as well as for the unsatisfactory units. Each precipitator may have distinct characteristics, but fundamental factors are common to all. These fundamental factors will be discussed in several different ways in the book.

Applications

The use of precipitators has been applied in all the basic as well as some exotic industries over the years. Collection of particulate matter in a dry type precipitator with flue gas temperatures between 250 to 700°F has been the most popular application. However, specific process characteristics will usually determine the design and type of precipitator utilized. There are process situations where the effectiveness of the electrical collector is questionable because the material is difficult to handle. Some typical industrial processes that have successfully employed the precipitator include:

Processes	Principle Material Collected
Utility	Fly Ash (SiO_2, Al_2O_3, Fe_2O_3)
Industrial Boiler Houses	Fly Ash
Oxygen Steelmaking Furnaces	Iron Oxide (Fe_3O_4)
Cement Kilns	Calcium Oxide, Silicon Oxide
Pulp and Paper	Sodium Sulfate

A large number of precipitators have also cleaned process gases from blast furnaces, sinter plants, open hearths, coke ovens, gypsum plants, catalytic cracking, smelters, sulfuric acid, and phosphoric acid plants. Other processes that use precipitators include electric arc furnaces, scarfing machines, incinerators, and the carbon black industry. Other precipitators have also recovered valuable metals in special situations.

The following comments describe the effectiveness of precipitators on the above processes:

Utility — Gas and fly ash variability cause potential deviations from design performance. Collection efficiencies are often related to the coal source. Generally, the precipitator works well here.

Industrial Boiler Houses — Similarly, a precipitator can be used here, but usually on a smaller scale. Process variation can sometimes make the gas flow rates cycle appreciably. The use of precipitators on industrial spreader stokers can be tough.

Oxygen Steelmaking Furnaces — A short and variable cycle makes for a difficult application. Difficulty with emission occurs at the start and finish of the heat. Moisture and gas temperature variations can cause corrosion troubles with the internal structure. The precipitator can work effectively if it is designed properly for this process. Electrical power input varies with the rapid process changes of the cycle.

Cement Kilns — Precipitators are highly efficient on the wet process kilns because of optimum gas conditions. Recent emphasis on fuel economy is presenting some problems at the lower flue gas temperatures. Precipitators used on the dry process kilns require some gas conditioning to be able to operate at satisfactory power input levels.

Pulp and Paper — Electrical collectors are generally used on the black liquor recovery furnace and are usually a successful application. Prime problems usually involve build-ups of material. Electrical characteristics are good.

Blast Furnace — Precipitators were commonly used for secondary cleaning of blast furnace top gas prior to its usage as fuel in stoves. Collection has been effective with the wet type precipitator.

Sinter Plants — This process became prominent within the last 30 years with precipitators cleaning the waste gases from the sinter strand. Raw material variables make this application difficult, but it can perform well if the gas conditions are controlled.

The effective application of precipitators for all the other industrial processes can be related to a satisfactory balance between electrical power input and the adverse physical or chemical characteristics of the material to be collected. If high power input and steady state conditions are present, success with precipitators is usually good. Without this set of conditions, the application is usually marginal and controlled by the process variables. More conservative sizing of the original precipitator is normally required to help overcome the inherent problems of the process.

Definitions and Physical Components

Some confusion has existed in past years regarding the terminology of the component parts of the precipitator. We have used the word "duct" and "gas passage" interchangeably or have commonly used the word "sections" to describe any portion of a precipitator. Recent attempts by committees of the Industrial Gas Cleaning Institute, Inc. and the TC-1 group of the Air Pollution Control Association have tried to bring some uniformity into precipitator terminology. The following descriptions should especially improve communication with regard to specification and bidding transactions.

Collecting Surfaces — The individually grounded components which make up the collecting system and which collectively provide the total area of the precipitator for the deposition of particulate.

Collecting Surface Area — The total flat projected area of collecting surface exposed to the electrostatic field (effective length x effective height x number of sides).

Effective Length — Total length of collecting surface measured in the direction of gas flow.

Effective Height — Total height of collecting surface measured from top to bottom.

Effective Width — Total number of gas passages multiplied by the center to center spacing of the collecting surfaces. (Disregard shape of collection surface.)

Effective Cross-Sectional Area — Effective width times effective height.

Gas Passage — Formed by two adjacent rows of collecting surfaces.

Discharge Electrode — The component which is installed in the high voltage system to provide the function of ionizing the gas and creating the electric field.

Collecting Surface Rapper — A device for imparting vibration or shock to the collecting surface to dislodge the deposited particulate.

Discharge Electrode Rapper — A device for imparting vibration or shock to the discharge electrodes in order to dislodge particulate accumulation.

Collection Efficiency — The weight of dust collected per unit time divided by the weight of dust entering the precipitator during the same unit time expressed in percentage. By its nature, this is not an instantaneous evaluation but a measure of the average performance that occurs during the sampling period.

Precipitator Gas Velocity — A figure obtained by dividing the gas flow rate through the precipitator by the effective cross-sectional area of the precipitator. (This is a descriptive quantity and not the actual velocity.)

Precipitator Arrangements

A convenient and logical method of comparing precipitators is the total collecting surface area and the amount of mechanical and electrical sectionalization. The basic terminology used to describe sectionalization follows and an illustration is shown on Figure 1—1.

Precipitator — A single precipitator is an arrangement of collecting surfaces and discharge electrodes contained within one independent housing.

Chamber — A gas-tight longitudinal subdivision of a precipitator (a precipitator without any internal dividing wall is a single chamber precipitator; a precipitator with a single internal dividing wall is a two chamber precipitator, etc.).

Bus Section — The smallest portion of the precipitator which can be independently de-energized.

High Voltage Power Supply — The power supply unit to produce the high voltage required for precipitation consisting of a transformer-rectifier combination and assorted controls. Numerous bus sections can be energized by one (1) power supply.

Field — A field of a precipitator is an arrangement of bus sections in the direction of gas flow that is energized by one (1) or more power supplies situated laterally across gas flow.

Description of Designs

A brief description of the physical design is included in this section although further details will be covered in later discussions. The three parts of any precipitator installation consist of:

1. Power Supply.

2. Collection Zone.

3. Material Removal Apparatus.

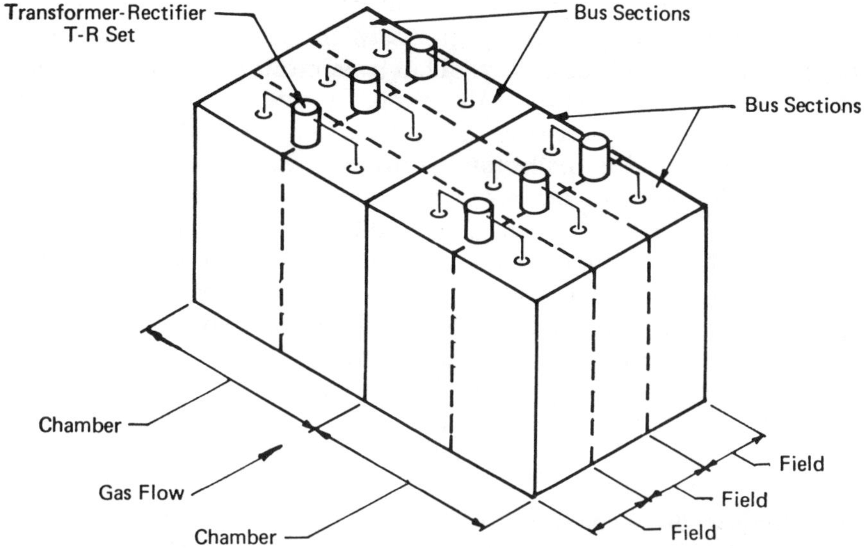

1 Precipitator showing 12 Bus Sections with 6 Power Supplies either F-W or Double Half-Wave.

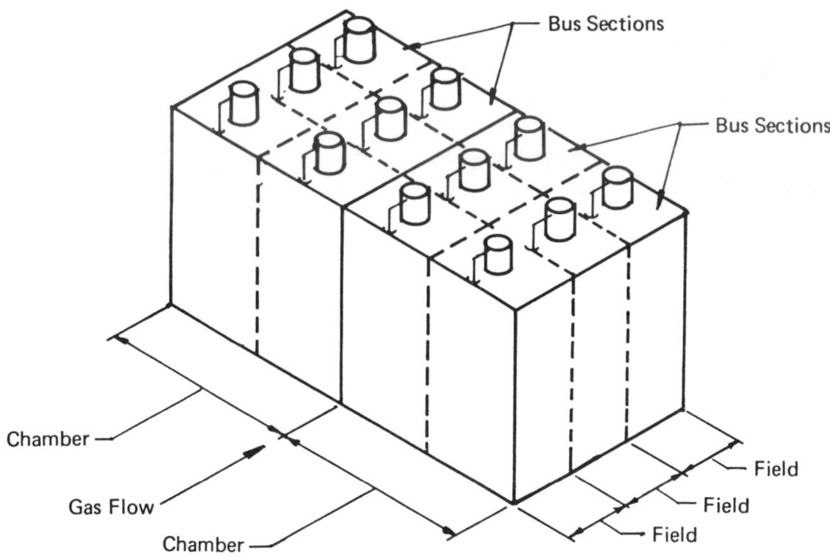

1 Precipitator, 12 Bus Sections, 12 Power Supplies Full-Wave.

Figure 1—1 TYPICAL PRECIPITATOR ARRANGEMENTS SHOWING TERMINOLOGY AND METHOD OF APPLYING POWER SUPPLIES.

The power supply usually consists of a single-phase high voltage transformer and bridge rectifier which is regulated by appropriate control circuitry. The precipitator transformer has developed over the years with special consideration being given to coping with the stresses and transients caused by electrical disturbances in the collector. Voltage rating of the primary winding of the transformer is nominally matched to the supply voltage and utilizes ratings from 400 to 480 volts in modern practice. Other voltage levels are used, but recent standardization often dictates the use of either auto-transformers or step-up transformers to match the nominal 480 volt system. The secondary winding rms voltage rating is commonly about 53,000 volts. Therefore, the turn ratio of the windings is generally found to revolve between 110 to 130 for most transformers used in precipitators employing 9 to 10 inch spacing. For gas passage spacing of 12 inches or more, greater secondary voltage ratings are required. The primary winding may or may not have taps to better match the voltage input.

The rectifier circuit may be connected either full-wave or double half-wave to the precipitator. (Figure 1—2.) Rectifier design has progressed from the early mechanical type through the kenotron vaccuum tube, selenium, to the modern practice of silicon rectifiers. Modern precipitator transformer-rectifiers that house both components in a common tank have proven reliable and relatively maintenance free. Failures may show up during initial operation, but this rate of failure should be less than 1% of the supplies installed.

The transformer-rectifier rating usually range from 15 to 130 KVA. Early ratings were 15 to 25 KVA with the recent tendency to apply 60 to 95 KVA units. The smaller size T-R sets will generally have greater impedance and stability under electrical disturbances. I will discuss sizing in greater detail later.

Part of the T-R design tends to have a built-in ballast resistance or reactance either in the primary or secondary circuits to help limit surges of current. On the early power supplies, this protective ballast was in the form of resistance grids, but choke coils and linear reactors have been used extensively in recent years.

An important part of the power supply is the type of control used to regulate the voltage level of the precipitator.

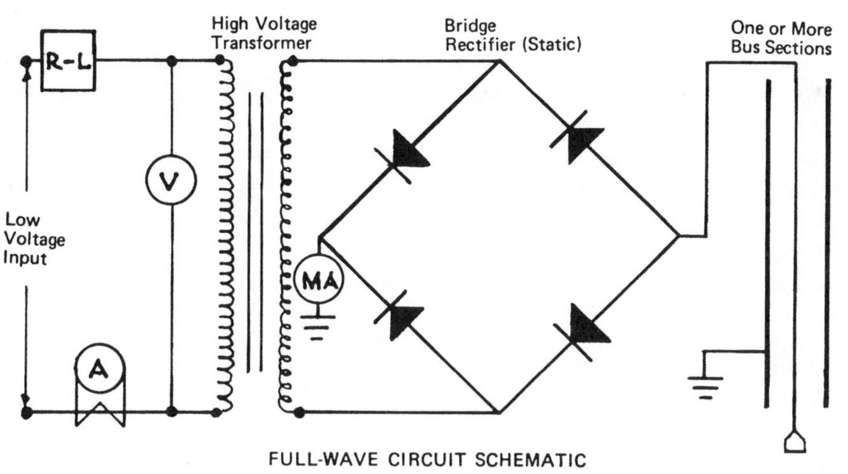

FULL-WAVE CIRCUIT SCHEMATIC

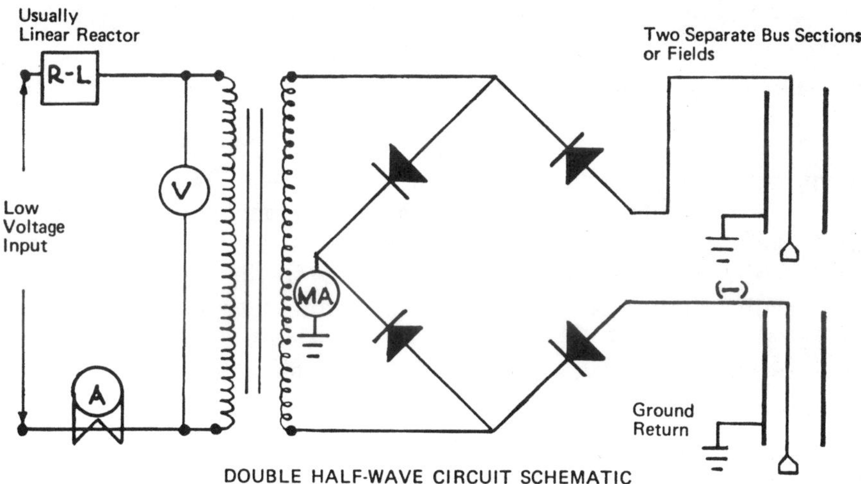

DOUBLE HALF-WAVE CIRCUIT SCHEMATIC

Figure 1—2 TWO POSSIBLE RECTIFIER CIRCUITS FOR AN ELECTROSTATIC PRECIPITATOR ARE A F-W OR DOUBLE H-W ENERGIZATION ARRANGEMENT.

With steady state process conditions and uniform electrical conditions, a manual mode of control would be sufficient. More often, variations in the process will cause changes in the electrical characteristics of the precipitator which can be best controlled with automatic circuitry.

The collection zone of the precipitator is where the action takes place and it assumes one of the two basic forms.

Parallel plate passages with discharge electrodes placed throughout the passage in direction of gas flow is called the dry-type plate precipitator. When structured with a parallel group of round or shaped pipes, each containing one centered discharge electrode, the precipitator is generally called a wet-type pipe design. In each case, the collecting surface is the positive grounded structure while the discharge electrode receives the negative high voltage input. (See Figure 1—3.)

Collection of particulate matter in a dry state from the flue gases of utility, basic oxygen furnaces, cement kilns, kraft recovery furnaces, cat-cracker, etc., is well suited for the dry-type plate precipitator. Mist substances and materials that are difficult to collect electrically are well suited for the wet-type pipe collector in which the collected material either flows down the inner furnace of the pipe or is washed down by a thin film of water. Applications such as coke oven gas detarring, sulphuric acid, and blast furnace gas have successfully utilized the pipe design.

Each manufacturer varies the design details of the precipitator, but basic concepts remain common to all. The dry-type precipitator is easily adapted to placing fields in a series arrangement with the horizontal direction of gas flow. The pipe type is normally a single field precipitator within one housing. Two or more housings of pipes in series can be used to attain greater collection capability. Since the wet surface in the pipe precipitator practically eliminates the loss of material once collected, series fields are required less in this application. There is renewed interest in the wet-type precipitator for difficult applications such as hot scarfers and sinter plants.

The dry-type plate structure has also been used in applications where water is used to flush the collecting surfaces for removal of material deposits instead of the rapper devices. Some difficulty was experienced years ago with this method, but that is not to say this method cannot be made to work satisfactorily.

Removal of the collected material from the collecting surfaces and discharge electrodes is only part of the dust removal problem for the precipitator. More difficulties arise in trying to successfully transfer the dislodged dust from the hopper to a remote disposal site than for probably any other phase of the collector. This will be discussed in greater detail in Chapter 5. The hoppers can include a pyramid, trough, or the flat bottom type which utilizes a drag-scraper to continuously push the material into a drop bin. The pyramid and trough hoppers nor-

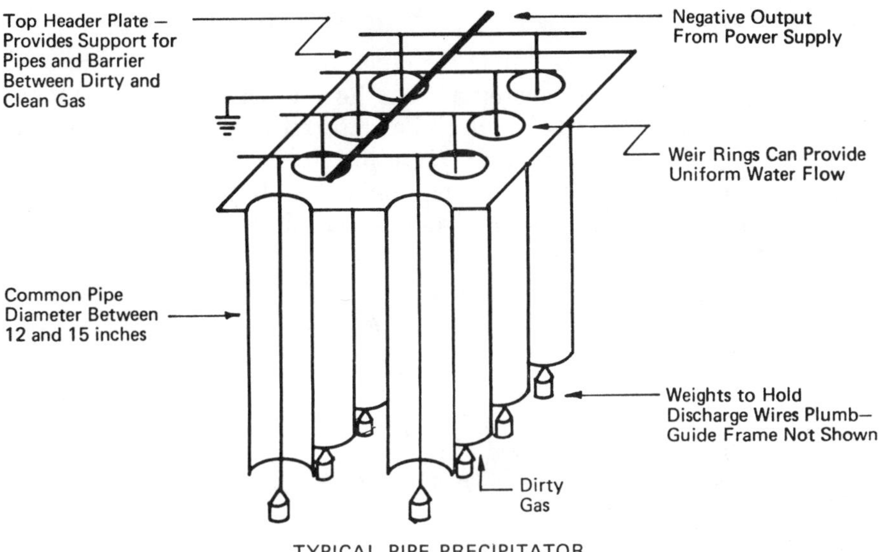

TYPICAL PIPE PRECIPITATOR

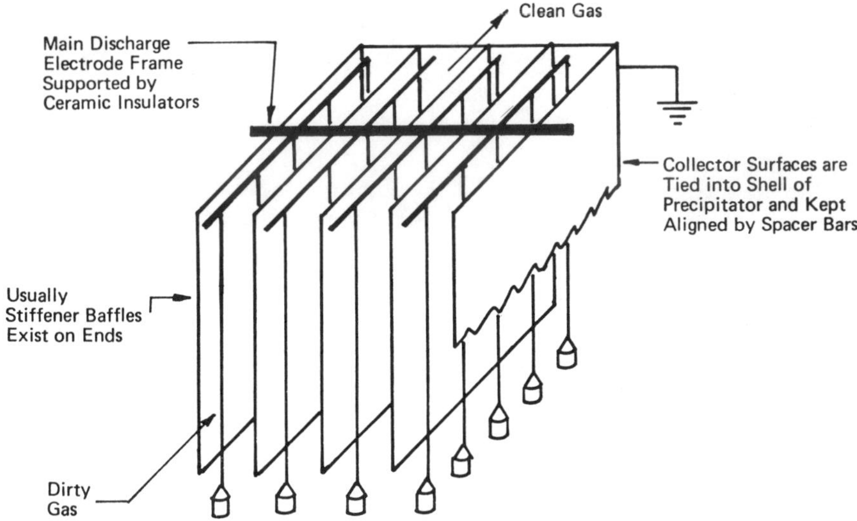

TYPICAL DRY-TYPE DUCT PRECIPITATOR

**Figure 1–3 THE COLLECTED MATERIAL IN A PIPE PRECIPITATOR IS
REMOVED BY LIQUID FLOW DOWN INNER SURFACE.
IN A DUCT PRECIPITATOR IT IS REMOVED BY RAPPING.**

mally use a variety of means for the evacuation of material including a vacuum system, a batch pressurized type, or a screw conveyor method.

The rapper has used many shapes and techniques to impart a blow to the electrode systems. Just how much impact or vibration is required depends on the characteristic of the material as well as the physical size and area of surface struck. Effective rapping of the collector surface is more important than the cleaning of the discharge electrode in most cases.

Chapter 2

Fundamentals

Introduction

This chapter discusses the fundamental concepts of the precipitator in a way that the reader should be better able to understand how the collector works. Even though this is important information, it will be possible to correct many of the problems covered in later chapters without this knowledge.

There are comprehensive documents available that cover the theoretical and mathematical treatment of the precipitator. These references can be found in Chapter 16 for use in further study. Nevertheless, I believe a need has existed for the basic concepts and relationships to be explained in a practical manner. I shall use a minimal amount of formulas, and utilize examples and discussion to help explain the relationships that do exist in the text.

Much of the information contained in this chapter will also be covered separately either in more detail or in a different way in later chapters. But the following sections should allow the reader to gain an overall viewpoint of the precipitation process. To accomplish this objective, I will briefly discuss what happens in the ionization zones, what the basic steps and makeup of the factors effecting the electrical characteristics are, and what constitutes the major design and evaluation concepts of the precipitator system.

Theory of Precipitation

Ionization Concepts

Whether the dry or wet-type of collector is used, the actual precipitation process occurs in the space between discharge electrode and the collector surface. It should be of interest to briefly discuss what tends to happen in the gas passage. This is an area still open to study and research, but high voltage phenonema has been well described in publications dating back to the 1920's.[1]* The corona discharge or the breakdown of gas at the discharge electrode can be considered the driving source

*Numbers in parentheses refer to the bibliography in Chapter 16.

and usually provides optimum activity at the breakdown point of the dielectric field which exists between the two electrode systems.

The process simply consists of charging dust particles through action of the corona discharge, and these particles pass through an electric field where they are attracted to the surfaces of opposite polarity.

There are five basic kinds of electrical design discharge that can occur in the gas spacing between two high voltage circuits:

1. The dark discharge is not visible, although chemical action occurs. The current increases nearly linearly with voltage.

2. The glow or corona is the most frequent initial discharge and is accompanied by a hissing noise. Current increases much faster than voltage.

3. The bush discharge usually falls midway between corona and a spark. It consists of a number of small sparks ending in space and originates in the surface irregularities or abnormalities of the conductors. Current increases with possibly a slight drop in voltage.

4. The spark is a common discharge that is unstable in that the increased current lowers the voltage appreciably. It is a sudden rush of localized electric current through the gas, and is only limited by the circuit power supply. The spark is the most complete form of gas "breakdown."

5. The arc is not ionization by collision, as are the preceding four categories, but is the result of other discharges. Current density is very high while the voltage decreases to a low value. This discharge is particularly unstable unless the current is limited by an external resistance in the electrical circuit.

In order for a current to flow across a gas space, ionization must be started and sustained by electron acceleration and collision with gas molecules, causing a cumulative effect.[2]

In the case of a negative wire in a cylinder, the following explanation describes the ionization principle of corona current: Electrons are released from the wire surface by a positive ion impact or photo-electric emission. These accelerating electrons in short movements near the wire surface generate new electrons and positive ions by molecular impact. The electrons quickly move away from the wire and attach to gas molecules to form negative ions or produce new collisions. Once the corona glow begins, ultraviolet radiation may produce additional photo-electrons at the wire surface. As the electron collision continues, the negative ions build a dense cloud throughout the space between the active region near the discharge electrode and the passive collector. The effect of this ion space charge is to help limit the ionizing field near the discharge wire and also to stabilize the corona discharge.[3]

This ion movement has two main charging effects on dust particles in the precipitator region existing between the electrodes:

1. Suspended dust particles are bombarded toward the passive anode under the force of a strong electric field.

2. Ion diffusion causes the negative ions to attach themselves to dust particles less than 0.2 micron in size and are by this method able to transport these smaller particles to the collecting surface.[4]

Basic Concepts of Precipitation

Probably the best way to gain an insight into the process of precipitation is to study a relationship generally known as the Deutsch-Anderson equation. This equation and adaptations of it are well covered in several books.[5,6] It describes the factors involved in the collection efficiency of the precipitator as shown in its simplest form:

$$\text{Collection Efficiency} \quad N = 1 - e^{-\frac{A}{V}W}$$

Where A = effective collecting electrode area of the precipitator. (sq ft)

V = gas flow rate through the precipitator. (acf per sec)

e = base of natural logarithim = 2.718

W = migration velocity. (ft per sec)

This equation has been used extensively in the above form in past years. Unfortunately, while the relationship is scientifically valid, there are a number of operating parameters that can cause the exponent to be in error by as much as a factor of two or more. I will discuss these factors later, but it is well to remember that the basic D—A equation can be used as an indicator or tool, but has limitations more often than not unless equated with some practical and empirical considerations by the designer. Values used can either be in the English or Metric systems. Conversion tables can be found in Chapter 16.

The exponent term W, known as the migration velocity, actually represents the speed of movement of the particle toward the collector surface under the influence of an electrical field. While we would consider it more an indicator than actual velocity, it does have a finite value that can be used for comparison purposes. This migration velocity is comprised of:

$$W = a \, \frac{E_o \, E_p}{2 \pi \theta}$$

a = particle radius, microns

E_o = strength of field in which particles are charged, stat-volts per cm. (represented by the peak voltage.)

E_p = strength of field in which particles are collected, stat-volts per cm (normally the field close to the collecting plates).

θ = viscosity or frictional resistance coefficient of the gas, poises.

The migration velocity, W, is quite sensitive to the voltage since the electrical field appears twice in the above equation. Therefore, the design objectives of the precipitator is to obtain maximum voltage fields with proper corona current flow for a maximum collection efficiency. Before we examine some of the values involved in the above relationship, the basic concepts can be described thusly:

1. When the area of the collecting surface is increased in relation to the gas flow rate or the quantity of waste gas passing through the collector per given unit of time, the collection performance will generally improve.

2. Conversely, for any given size precipitator, reduction of gas flow rate is beneficial, all other factors being equal.

3. An increase in the physical size of the individual particles passing through the precipitator is beneficial.

4. A decrease in the viscosity of the waste gases will generally increase the collection of particulate matter if all other factors are equal. Viscosity will decrease with a reduction of gas temperature.

5. A minor increase of the electric field in the precipitator can often materially improve collection performance. The voltage fields provide the driving force for movement of particles toward the collecting surfaces.

The meaning of migration velocity for a single particle of particulate matter, represented by an aerodynamic sphere, can be best illustrated by the following example:

a = 1 micron particle = 10^{-4} cm radius.
E_p = 4.8 kv per cm = 16 statvolts per cm
E_o = 6.0 kv per cm = 20 statvolts per cm
θ = 2.7 x 10^{-4} gas viscosity in poises at 500°F

Substitution in the migration equation:

$$W = \frac{a \; E_o \; E_p}{2 \; \pi \; \theta}$$

$$W = \frac{10^{-4} \times 20 \times 16}{2 \times 3.14 \times 2.7 \times 10^{-4}}$$

$$W = 18.8 \text{ cm per sec} = 0.62 \text{ fps}$$

If the discharge electrode is located 4 in from the collecting surface and the gas flow through the passage is 4 fps, the particle would reach the surface in approximately 2 ft of passage length.

Under similar dust and precipitator conditions but with the gas cooled to 300°F, the gas viscosity decreases to 2.3 X 10^{-4} poises:

$$W = \frac{10^{-4} \times 20 \times 16}{2 \times 3.14 \times 2.3 \times 10^{-4}}$$

$$W = 22.2 \text{ cm per sec} = 0.73 \text{ fps}$$

To determine the effect on the conditions of the first example by increasing the average and peak precipitator voltages by 20 per cent:

$$W = \frac{10^{-4} \times 24 \times 19.2}{2 \times 3.14 \times 2.7 \times 10^{-4}}$$

$$W = 27.2 \text{ cm per sec} = 0.9 \text{ fps}$$

The particle drift velocity to the collecting plate was thus increased 45 per cent when the precipitator voltages were increased another 20 per cent.

Consider the first example again, except with an average particle size of 5 microns:

$$W = \frac{5 \times 10^{-4} \times 20 \times 16}{2 \times 3.14 \times 2.7 \times 10^{-4}}$$

W = 95 cm per sec = 3.1 fps

Thus, a 5 micron particle should be able to reach the collecting surface in about 0.43 ft of passage length considering a 4 fps gas velocity.

A precipitator collection efficiency can be calculated for the 1 micron example by placing the value of 0.62 fps into the efficiency equation:

$$\eta = 1 - e^{-\frac{0.62\,A}{V}}$$

Consider the efficiency for one gas passage:

Collecting surface A = 20 ft high x 12 ft long x 2 sides = 480 sq ft.

Gas passage flow rate V = 20 ft high x 0.667 ft wide (8 in. spacing) x 4 fps gas velocity equals 53.4 cfs.

Substituting for A and V in equation

$$\eta = 1 - e^{-\frac{480 \times 0.62}{53.4}}$$

$$\eta = 1 - 0.0037$$

$$\eta = 0.996 \text{ or } 99.6 \text{ percent collection efficiency}$$

Disregarding all other factors, the results of this calculation would indicate that of 250 particles of one (1) micron radius entering the gas passage, only one (1) would exit the 12 ft long section.

The purpose of the above examples was not to provide design criteria tools, but rather to illustrate concepts. Variability of the process and particle characteristics among other factors make the job of designing precipitators exceedingly difficult.

Electrical Energization Concepts

In Chapter 1, I emphasized that optimum levels of power input to the precipitator is of utmost importance to achieve acceptable collection efficiencies within practical physical sizes. A brief summary of ionization principles and how voltage levels effect the final efficiency has already been covered in this chapter. I will continue with a brief description of the components that provide the method for energizing the precipitator in addition to the factors that effect the levels of power input.

A simple review of the precipitator energization would include the following steps:

1. A low voltage supply is transformed into a suitable high voltage — alternating current.

2. This high voltage AC is rectified into a pulsating D.C.

3. Connection of this rectified supply is made to an internal discharge electrode system that is insulated from ground potential by insulators. This part of the precipitator is now at a negative high potential and is the source of the corona energization.

4. The high potential on the negative discharge electrode (which could be a wire with a nominal diameter of 0.10 inch) causes a localized corona breakdown of the gas surrounding the electrode.

5. This corona sheath supplies the flow of electrons (either separately or more generally attached to gas molecules) through the gas space toward the passive electrode or collector surface — which is at ground or positive potential. It is this electrical charging of particles by negative ion attachment that provides the transport mechanism for the particulate to move through the gas stream to the collecting surfaces under the influence of the electric field. The distance between the two electrode systems is usually 4 to 6 inches.

6. The grounded electrode, which also acts as the primary depositor of material, returns the total precipitator current to the rectifier and transformer so as to complete the electrical circuit. Refer to Figure 1—2.

An industrial precipitator will comprise numerous parallel paths for the above phenomena. There are countless designs and differences in hardware, some of which will be described later.

The precipitator electrode system of discharge and collector can now be considered analogous to the two plates of a capacitor in which the plates are of opposite electrical polarity and the space between serves as the dielectric.

A prime object of the precipitator is to increase the voltage difference between the electrode systems in order to maintain a high charging condition. As with the capacitor, a limiting factor is an electrical breakdown of the dielectric or gas space (called the spark-over) which then becomes a localized path of high current flow at one point in the electrode system.

If precipitation is considered to be effective at high voltage differences between the negative and grounded electrodes, then the occurrence of a spark counteracts this performance criterion since it causes an immediate, short-term collapse of the precipitator electric field. Yet we strive for some spark-over at high voltage levels as an indicator that the maximum amount of power is being placed into the precipitator section.

Why a minimal amount of spark-over is acceptable is related to the number of voltage pulses that occur every minute. If 60 cycle full-wave is considered, then there are 120 rectified pulses per second or 7200 pulses per minute. Minimal spark-over usually means that between 5 to 100 sparks occur per minute for each power suppy. Since this is a small fraction of the energization during any given minute, it is usually more advantageous to operate at a higher power level even though some electrical breakdown does occur. But I will say more about this later.

High levels of voltage and useful corona power in the precipitator, all other conditions being equal, are synonymous with high collection efficiencies. Figure 2—1 shows a typical performance curve of the effect on efficiency by changes in the peak voltage of a precipitator. This simple curve can only represent one (1) situation because each precipitator will have its own characteristic curve based on many factors. The important point to remember is that small changes can produce substantial changes in power, and hence in the efficiency of the collector. This is especially true at the lower levels of power input. It is therefore important to understand the factors that effect the electrical characteristics of the precipitator.

Main Factors Effecting Electrical Characteristics

In this section, I intend to continue the development of relationships that better explain the factors effecting the electrical characteristics of the precipitator. The relationships will be enlarged upon in later chapters, but consider the following sections as simple building blocks which can help aid you in the understanding of the precipitation process.

Optimum power input to the precipitator varies among processes and even changes on a minute to minute basis for certain applications. There are seven (7) basic factors that I believe directly effect the electrical characteristics. These are:

1. Design of power supply.
2. Physical design of precipitator.
3. Design of electrode system.
4. Characteristics of gas stream.
5. Effect of process changes.
6. Characteristics of particulate.
7. Maintenance factors.

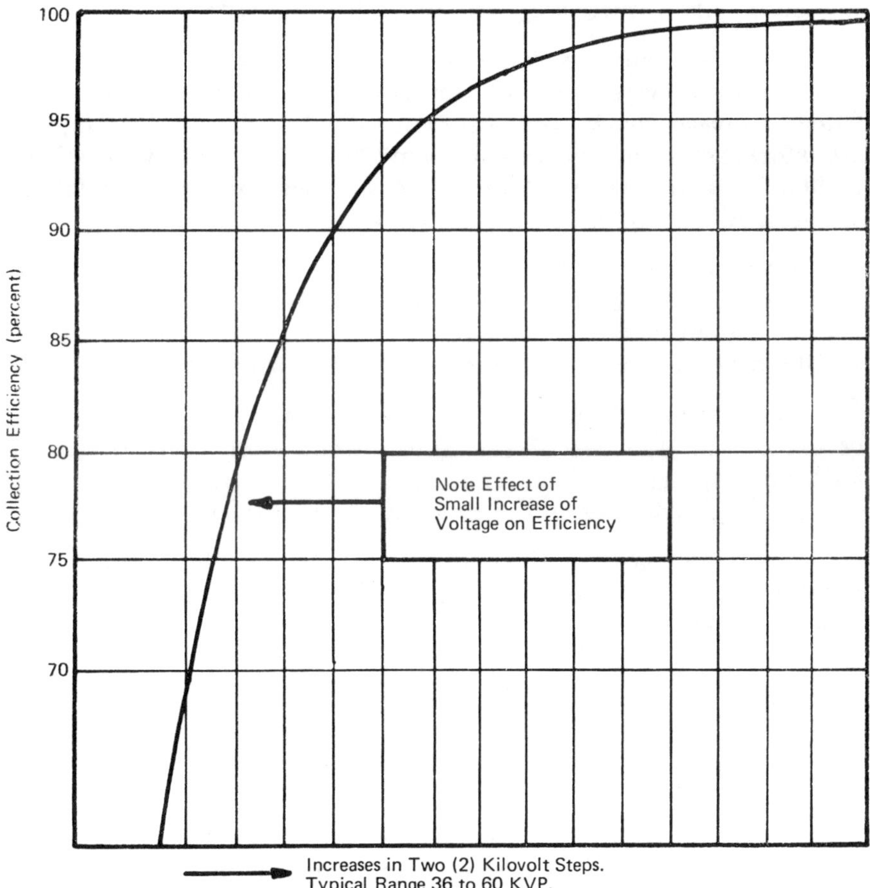

Figure 2–1 A TYPICAL ELECTROSTATIC PRECIPITATOR PEAK VOLTAGE VERSUS DUST COLLECTION EFFICIENCY CURVE SHOWS HOW EFFICIENCY INCREASES WITH VOLTAGE.

In this chapter, the factors will be discussed individually. The above factors often combine synergistically to effect the performance of the precipitator.

Design of Power Supply

The power supply consists of three basic parts:

1. High voltage transformer.

2. Rectifier.

3. Control, metering and protection circuitry.

The high voltage transformer is especially designed for precipitator service with the ability to withstand winding stress when a severe spark-over occurs inside the collector. Ratings of transformers normally range from 15 to 95 KVA with 45,000 to 55,000 secondary winding voltages and secondary output load currents of 250 to 1500 milliamps DC.

Modern rectifiers are of the silicon type and are usually mounted inside the transformer tank on new installations. Conversion kits are available to modify earlier installations containing rectifiers of the vacuum tube or mechanical type.

One recent improvement in the control of power input has occurred with the use of the Silicon-Controlled Rectifier (SCR) automatic voltage control. This circuit can minimize the effect of electrical disturbances in the precipitator by the speed of response.

Important parts of the control circuitry are the meters which monitor the variations in the electrical power input. The most commonly used meters are:

1. Voltage across the primary winding of the transformer in AC volts.

2. Current in the primary winding of the transformer in AC amps.

3. Ground return current from the precipitator in DC amps and usually designated as milliamps in the smaller power supplies.

Two (2) other meters sometimes found on control panels are:

4. Precipitator voltage through an appropriate voltage divider and noted as average DC kilovolts.

5. Sparkmeter which integrates the electrical breakdowns in the precipitator as average sparks per minute.

The primary voltage and current are reflections of the actual precipitator power input by the transformation ratios and accounting for voltage losses through the rectifier system.

In the absence of a sparkmeter, the sharp fluctuations of the other meter needles can be used to estimate the severity of the internal sparking. This will be discussed in the Resistivity Chapter.

The power supply must be matched correctly for the precipitator section or service expected, or several difficulties can arise:

1. The impedance of the power supply, including a ballast resistance or reactor in the primary winding of the transformer, may not be sufficient to dampen the severity of electrical breakdowns in the precipitator. This condition is especially likely if the power supply rating is much larger than the actual operating level.

2. If the physical size of the precipitator is too large compared to the size of the power supply, then lower than desirable precipitator voltages may exist because the current rating of the supply becomes the limiting factor.

3. The gas and particulate matter conditions can drastically alter the voltage-current relationship and produce lower voltage fields than expected because a small power supply becomes current limited.

Physical Design of Precipitator

Several physical design factors will effect the voltage-current relationship of the precipitator. They are:

1. The distance between the high voltage discharge electrode and the grounded collecting electrode will help determine the total voltage required across the spacing. Higher voltage and lower current levels tend to occur for greater spacings.

2. Another major factor that determines the ratio of voltage to current is the total area of the collection surface combined with the total length of wire electrode per precipitator field.

3. The number of precipitator fields in series will effect the voltage-current relationships of each field primarily related to the quantity and characteristics of the particulate. This will be covered later.

Along with the physical size of the precipitator, the actual construction of the individual electrodes will help determine electrical readings.

Design of the Electrode System

Electrode design will effect the voltage-current relationship in two (2) major ways:

1. Diameter or sharpness of the corona emitting points or edges of the high voltage electrode will determine the starting voltage at which corona will initiate. Therefore, for a given voltage, the smaller diameter wire may show more current. This is especially true at low voltage levels.

2. Any irregularities of the collector surface opposite the discharge electrodes might produce spark-over at a reduced voltage.

Characteristics of the Gas Stream

There are an infinite number of changes in the composition of a gas stream that can alter electrical characteristics of a precipitator. Some gases tend to elevate voltages and depress currents. Temperature of the gas will have an effect on the voltage-current relationship as will the velocity of the gas through the precipitator passages to a lesser degree.

To determine the effect these differences mean in the actual collection efficiency of a precipitator, a complex analysis of all the many variables will usually be required. I have some doubt that this analysis can ever be accurate. As a general rule, the higher the voltage reading, the more performance benefits will generally exist for any given current. As we shall see later, this does not always hold true for every rise in precipitator voltage.

Effect of Process Changes

As covered earlier, each process can have its own peculiar characteristics that will effect the electrical readings as observed on the precipitator meters. Uniformity of the process is of major importance for optimum performance unless the precipitator is specifically designed for a variable mode of operation.

The average precipitator can be sensitive to process changes in the following ways:

1. Changes to gas temperature (effect on density).

2. Changes in gas pressure (effect on density).

3. Changes in gas flow rate.

4. Changes in gaseous composition.

5. Changes in particulate chemical characteristics.

6. Changes in particulate concentration or loading.

7. Changes in the size distribution of the particulate.

8. Changes in the electrical conducting characteristics of the particulate.

It is difficult to separate the effect of one process change on another. If the rate of process change is rapid, the readings can change almost instantaneously. On the other hand, rapid changes of temperature may not be seen readily on the meters because of the heat sink effect of the precipitator. Some changes in the process will cause large variations of voltage-current readings, while others will cause subtle effects.

There are two major factors that are vital to the performance of the precipitator when related to the effects of process changes on electrical characteristics:

1. The power input may be altered by the process change so that previously collected material begins to erode from the collecting surface. I have termed the optimum combination of voltage and current in the precipitator as "Effective Holding Power." If by a change such as gas temperature, this "Effective Holding Power" decreases, then emission losses will increase by the reentrainment of particulate matter into the gas stream.

2. Process change can alter the spark-over voltage which will allow a change in the level of power input to the precipitator. If, for example, the moisture content of the gas stream is reduced for a specific installation, spark-over can occur at a much lower voltage.

The major reason that the reduction of water vapor can drastically alter the meter readings is usually connected to the changes in the surface electrical characteristic of the dust by the change in humidity.

Effect of Characteristics of Particulate

The precipitator is primarily concerned with applying work on a particle in a gas stream and forcing it toward a collector electrode by means of electrical forces. Whether this can be accomplished easily or not depends to a great extent on how the particulate reacts to this electrical activity.

The classification of this characteristic of the particle is simply related to its ability to conduct or resist the passage of electric current. This ability is not critical for the individual particle as it drifts in the gas stream, but becomes important after it deposits on the collecting surface.

If the grounded collector acts as a receptor for the precipitator corona current, then a layer made up of particles that resist a passage of electrons will cause a resistive sheath to form. This layer will tend to reduce current flow (consider this to be an increase of dielectric between the electrodes) which can actually raise the indicated voltage on the meter. This is often observed after a fairly conductive layer of material had previously existed on the collecting surface.

As the material layer on the collector surface becomes more resistive to a passage of current, initiation of spark-over occurs. In reality, another thin, capacitor-like arrangement has been superimposed on the collector surface. Therefore, the meter voltage reading becomes indicative of two voltage drops, that of the space between the discharge electrode and layer of material and another of the layer itself. The voltage drop across this layer is the prime cause of small localized break-downs which can then erupt into a complete electrical break-down of the space from the discharge electrode to collecting surface. This condition of sporadic spark-over usually results in a substantially lower power input of the precipitator in order to control at reasonable spark levels.

The term used to describe the ability of the corona current to flow across the layer of particles is called resistivity. Measurements are sometimes made to help determine the resistivity value of the particulate, usually expressed in ohm — centimeters. The theories and ramifications of this important subject will be covered in a later chapter. But it is well to know that the occurrence of spark-over caused by high resistivity can be overcome by methods of process modification or gas conditioning.

The size distribution of the particulate matter can have a bearing on electrical readings. For example, iron oxide fume from a basic oxygen vessel contains a predominance of sub-micron particles that will react like a space-charge in a vacuum tube. This can actually impede the flow of precipitator current and thereby elevate the voltage potential across the space. This condition can become serious enough to completely nullify the precipitator process dependent on electrode geometry and the concentration level of the sub-micron particles.

Maintenance Effects on Electrical Readings

The factors that have been discussed earlier did not take into consideration the maintenance variables that can complicate the analysis of the voltage-current observations. Build-up of material on electrode surfaces, electrical leakage over insulator surfaces, misalignment of the electrode systems can combine or react separately to determine the voltage level of a specific field. How the control circuitry of the power supply is set-up can effect the electrical stability of the field. In later maintenance and trouble shooting sections, I intend to point out how these variables are detected and to suggest corrective steps that can be taken.

System and Design Parameters

There are a number of parameters that are used to categorize precipitators both in the design and evaluation stage. I would like to discuss a group of parameters that I use in the initial evaluation of existing precipitators. While these indicators form only part of a group used in the field, they are considered the most pertinent and useful for the reader.

The important point to remember is that the original design parameters of the precipitator may not represent actual field conditions. I shall also briefly stress how to weigh the effect of these parameters on actual performances.

Gas Flow Rate

An actual gas flow rate, or the quantity of flue gas entering the precipitator, is an important measurement, and one that often deviates from the original design criterion. This value is used to determine several other parameters so I shall later describe how to gain confidence in this number. I also should point out that the quality of the flow is as important as the magnitude. For example, a gas flow rate containing a large percentage of inleakage air (not part of the actual formation of gas in the process) can be less desirable than greater quantities of undiluted gas flow.

Precipitator Gas Velocity

The parameter of precipitator gas velocity is a common indicator, but its meaning is often questionable unless knowledge of the gas distribution pattern at the inlet face of the precipitator is well known. As such, it is a term that should be used carefully in comparing different precipitator units or even the same collector under varying flow rates. Reference is made to the terminology of Chapter I:

$$\text{Gas Vel} = \frac{\text{Gas Flow Rate in Acfm}}{\text{Effective Cross Sectional Area}}$$

The effective cross sectional area denotes the full height of the collector surface including the top and bottom reinforcing bars times the number of gas passages times the width of one gas passage — without regard to any baffle or vertical obstruction on the collector. While this gas velocity value theoretically allows one to compare precipitators, one can see that the variation of structure can modify this velocity for its effect on the precipitator.

Another question arises in the use of this velocity value because the cross-sectional area of the inlet flange is less than the effective area of the collector proper. This situation and the use of deflecting baffles in walkways impose some difficulty in arriving at a true number.

Disregarding all the inherent problems, let's examine a typical collector with the following design:

Gas flow rate   — 525,000 acfm

Collector plate height   — 30 ft

Width of gas passage   — 9″ ϕ to ϕ

Number of gas passages — 78

Therefore:

$$\text{prec. vel} = \frac{525,000}{30 \times 9/12 \times 78 \times 60 \text{ sec/min}}$$

prec. vel = 5 ft/sec

Design velocities of 6 to 8 ft/sec were common up to about 1970, but 4.0 to 5.0 ft/sec is a reasonable range under modern efficiency goals. Below 4 ft/sec other factors evolve that tend to nullify the benefits of reduced flow. Low level gas flow means large sized collectors. One danger lies in the change toward ultra-high collector plates to obtain the lower levels of precipitator velocity. Gas distribution and maintenance problems become likely with these designs.

Treatment Time

The term treatment-time is often referred to as residence time. This parameter denotes the length of theoretical time a particle is exposed to the electrical field within the precipitator. It should be used with caution, but nevertheless, does serve as an indicator for the sake of comparison.

$$\frac{\text{Treatment}}{\text{Time}} \underset{\text{in seconds}}{=} \frac{\text{Effective length} - \text{ft}}{\text{precip. vel. ft/sec}}$$

Since effective length denotes length of collector surface in direction of gas flow, the above should be considered a net value. If walkways exist, a gross treatment time would result if the width of the walkways are considered.

Common time parameters for 99.5% collection designs range from 8 to 11 seconds. However, treatment times as low as 4 seconds can produce these same efficiencies under ideal power input conditions. Reasons for the existence of this wide band will be shown by examples in a later chapter.

Aspect Ratio

A term I find most important is the aspect ratio of the precipitator or the ratio of total length to the height of collector surface. This value provides an insight into what I call a "Margin Factor," or the ability to overcome detrimental effects over the lifetime of the unit.

The net aspect ratio can be calculated for the collecting surfaces:

$$\text{A.R.} = \frac{\text{Effective Length}}{\text{Effective Height}}$$

The gross ratio would also include an allowance for the width of any walkway or other space between the leading edge of the inlet field to the trailing edge of the outlet field. While this dead space does not contain charging ability, the benefits from terminal settling of particle agglomerates can be weighted into the overall performance of the collector.

Ideally, it would be nice to have a chamber many times larger in length than the height of the collector surface. Carried to an extreme, the design of the precipitator can serve as a partial settling chamber with collection efficiencies of 20 to 40% for specific materials. Of course, the cost would be prohibitive. Figure 2—2 shows various forms of Aspect Ratios.

As recently as the early 1970's, ratios of 1.0 were quite common. Higher design efficiencies occurred with the design of a greater number of gas passages or increased collector height to realize the reduced precipitator velocities. The tendency for more fields has forced the change to Aspect Ratios of 1.5 to 2.0. If this type of design is also coupled with low velocities, then the cost can become magnified in terms of capital investment and operating costs of the collector system.

Number of Fields in Series

Along with the Aspect Ratio, the number of fields available in the precipitator is another input to the "Margin Factor." Since the fields are located in series, it often allows the power supply connected to each field to operate at levels commensurate with the characteristics of the particulate matter at that location in the precipitator. The value of this arrangement is dependent on the process or material to be collected.

I have observed extreme positions on the philosophy of fields that ranged from the existence of a single field to ten (10) fields in the total length of the collector. Three (3) to four (4) fields were used extensively in the early 1970's, but five (5) to seven (7) fields are being used in the late 1970's. Confidence to control the process and the quality and reproducibility of raw materials usually determine the final number of fields chosen.

Number of Power Supplies

From the definition of a field, the number of power supplies has to at least equal the number of fields or be greater in some

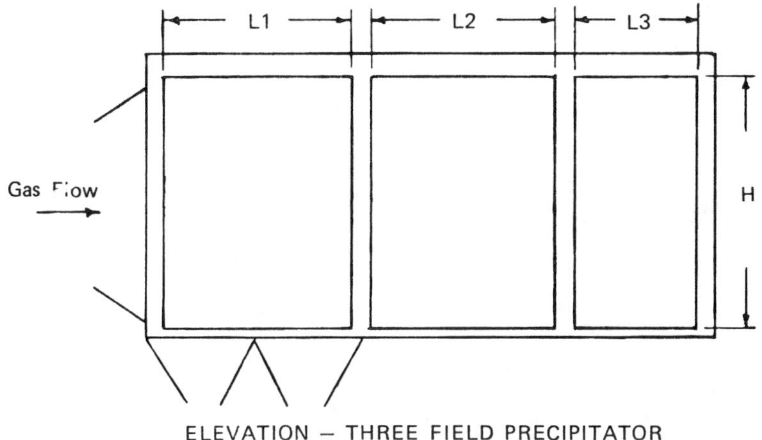

$$\text{Net Aspect Ratio} = \frac{L1 + L2 + L3}{H}$$

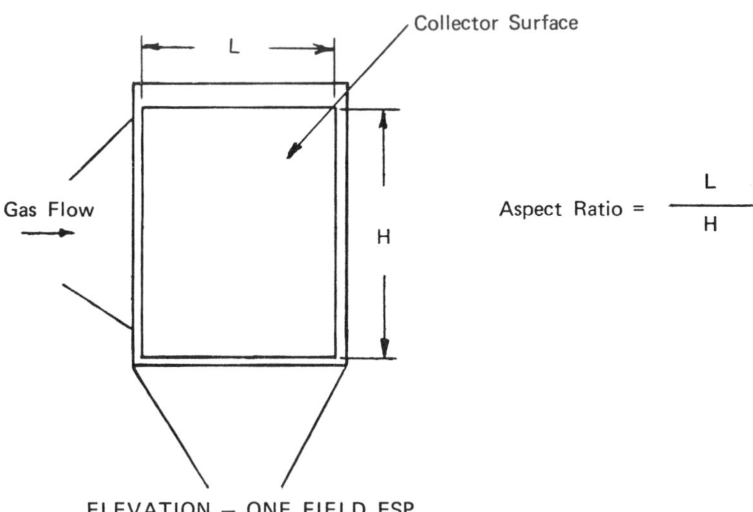

$$\text{Aspect Ratio} = \frac{L}{H}$$

Figure 2–2 REPRESENTATIONS OF THE ASPECT RATIO OF THE ELECTROSTATIC PRECIPITATOR

multiple depending on how many parallel bus sections are energized. It follows that the greater number of power supplies for any given precipitator will mean a smaller sub-division of the collector will be effected by localized electrical disturbances.

On balance, it is desirable to have a greater number of supplies, but the advantage can be nullified by the poor sizing of the transformer in relation to its location in the collector. I will say more about this later in the Case History and Design chapters.

Power Input

The power input to the precipitator is a parameter that is used in several ways. We are generally interested in the voltage-current levels after transformation into a pulsating d-c high voltage form for use in the precipitator.

Earlier descriptions showed that the levels of the voltage and current values are often controlled by the characteristics of the process, which effects the dielectric of the space between the discharge electrode and collector surface.

The characteristics of the power supply itself can modify waveforms and directly effect voltage and current values. For purposes of this section, I am only concerned with arriving at numbers that are useful and valid for analysis.

If secondary meters exist on the control panel, the precipitator power can be obtained by direct readings.

$$\frac{\text{Precip. Power}}{\text{in Watts}} = \text{Volts x amps}$$

Secondary voltages are usually read in average dc kilovolts through the use of a divider and multiplied by 1000 to obtain volts. The secondary current may have to be converted from milliamps to amps by multiplying by a factor of 10^{-3}. With some designs, the secondary current is measured in each leg of the rectifier bridge so that the total current is obtained by adding the readings of both meters.

If meters only exist in the primary circuit of the high voltage transformer, the meter will represent rms A-C values, and multiplication of these readings will provide a volt-ampere measurement. However, this value must be corrected by the losses in the transformer and rectifier to arrive at a useful power relationship in watts. The design and operating level of each power supply will cause some variations in the transformation of primary volt-amperes to actual secondary watts, but 60 to 70% can be considered a nominal range. I generally use a value of 65% to obtain a working number in lieu of actual secondary readings.

The peak voltage of the precipitator can be calculated by the following relationship:

$$\text{Peak Voltage} = \left(\text{Pri. AC Volts} \times \text{Transformer Turns Ratio} \times 1.41\right) - \text{Rect. Voltage Drop}$$

The rectifier drop will vary depending on the type of rectifier and operating level of the supply. For practical purposes, the following drops can be used for conservative estimates of peak voltage.

Mechanical Rectifier	30 Kilovolts
Vacuum Tube Rectifier	7
Silicon Diode Rectifier	4

Watts/1000 ACFM

When the total corona or precipitator power and gas flow rate is known, a useful density parameter is obtained for design and evaluation purposes.

$$\text{Watts/1000 ACFM} = \frac{\text{Total Watts Input}}{\text{Total ACFM} \times 10^{-3}}$$

A large range exists for this parameter and its meaning is sometimes clouded. Modifications in the process can produce the greatest change in this number.

Early fly ash designs in the 1940-1950's considered 100 watts/1000 acfm as quite satisfactory for the efficiencies that then existed. Designs of 200 to 300 watts/1000 acfm were common for most applications in the late 1960's and early 1970's for design efficiencies of 99.5%. Recent designs call for 800 to 900 watts/1000 acfm as the number of fields increase to attain ultra-high collection performances.

I use this number primarily to compare performance levels of a specific collector under different levels of power. I have observed high efficiencies with 170 watts/1000 acfm and low performance at 350 watts/1000 acfm, all on the same collector. Therefore, the number has to be used carefully.

Electrode Area per Power Supply

The designer will assign some relationship of electrode sur-face area to each power supply. Of course, the original design of corona current per square foot of collector surface or linear foot of discharge electrode can be drastically different than that observed in the field.

The object is to maximize the corona power commensurate with optimum voltage fields. Usually, the power supply does not operate up to its potential in the first half of the precipitator. The supplies in the latter half of the collector are often seen limited by the current rating of the set. The key point to be made is that the actual operating level of each power supply must be compared against its rating. Twice the electrode area can be connected to certain power supplies while additional trans-former-rectifier sets may be required in other zones of the collector. I will say more on this phenomena later.

Specific Collector Area (SCA)

A parameter that has found wide use is the ratio of the col-lecting surface of the precipitator relative to the gas flow rate expressed in thousands. The importance of this term is that it represents the A/V relationship in the exponent of the Deutsch-Anderson equation.

$$SCA = \frac{\text{Total Collecting Surface (sq ft)}}{\text{Gas Flow Rate (acfm x } 10^{-3})}$$

As with the other parameters covered, this term must be weighed carefully, but it does indicate that more margin is available as the number becomes larger. Modern conservative designs call for a SCA of 350 to 450 for 99.5% efficiencies, but these performances have been achieved at much less values. How-ever, much greater SCA's have also been required on applications of extreme resistivity.

Power Density

Another term often used is the corona power in watts ex-pressed per sq ft of collecting surface. Values of 1 to 2 watts per sq ft are quite common, but the range can be extreme, depending on the process.

Migration Velocity

As I described earlier, the migration velocity parameter represents the collectability of the particle within the confines of a specific collector. The assumption of a low W in the original design will denote a conservative size of precipitator. This parameter can be used for comparison purposes but has limited significance for the field man. When a collector is not performing at the expected or desired level, an understanding of the fundamental factors and their effect on precipitators will determine the eventual migration velocity rate. The next chapter will discuss these factors.

Chapter 3

Factors and Effects

Introduction

The title of this book refers to the "Art of Precipitation." After years of studying the field performances of this collector relative to design data, I firmly believe that scientific input has definite limitations for allowing someone to understand all the ramifications of the electrostatic process. While I do not scoff at the recent attempts of computer studies to provide a synergestic answer devised from a set of variables, I doubt that this method by itself can ever overcome the complexity of the field situation.

Earlier discussions covered some of the fundamentals and relationships that were intended to provide elementary background. It was noted that while the Deutsch-Anderson equation basically describes the precipitator process, designers have had to inject correction factors for the conditions existing in the field. In many cases, because the data base is considered poor, recent designs are sized very conservatively. Yet, some precipitators still fail to meet performance goals, and if they do, often have an exceedingly difficult time maintaining that performance.

I believe one of our problems is in the poor distribution of precipitator knowledge that no computer can rectify. Misconceptions are widespread for the designer as well as for the user of this equipment. Even though the following factors are discussed separately, I intend to show later an intertwining of these factors in field situations. This chapter is considered fundamental to a true understanding of the precipitator process. Particular examples of many of the concepts covered will be shown in later chapters.

Gas Flow Rate

While the true measurement of the gas flow rate commands an important place in specifications and performance tests, in practice, it is often less critical than other factors. This comment is especially true with the larger designs that exist today. Even with the 6 ft/sec or more velocity designs, it is often the quality of gas that weighs most importantly.

If the gas flow rate is reduced in certain processes, the

resulting change in gas temperature could adversely effect the electrical characteristics of the precipitator. Moreover, it is possible to have a 10% overload of gas flow and have an improved performance result in certain fly ash applications when compared to lower gas flows. The above example is to stress that over-emphasis on the levels of gas flow might make you miss the one main ingredient of good precipitation . . . effect on the power input characteristics. So while the Deutsch-Anderson relationship always indicates an improvement with reduced gas flow — be careful.

Gas velocities below 3.5 to 4.0 ft/sec through the precipitator often produce less collection improvement than expected while velocities above 8 ft/sec can distort performances because of the reentrainment of material. The optimum flow through the collector must also be weighed against the Aspect Ratio. In other words, the precipitator has a better chance of handling higher than design gas flow rates with Aspect Ratios of 1.5 than with those of 1.0.

Whether or not the relationship of higher gas flow rate to reduced efficiency becomes critical is dependent in large part on the characteristics of the particle. Certainly, large porous particles such as combustible grit found in fly ash applications will be sensitive to increased velocities. On the other hand, fine-sized particulate matter that tends to agglomerate in the deposited layer of the collecting surface will resist easy reentrainment into the gas stream. With low levels of power input and low Aspect Ratios, high gas flow rates can often be observed in reduced performances.

Another factor of the gas flow rate on the precipitator performance is as much its variability over short term periods as its magnitude. Variations of 5 to 10% in the gas flow rate will either produce little change in performance or cause distinct reentrainment puffs dependent on the electrical characteristics. For example, a 10% change in the excess air of a fly ash boiler can either result in a slight rise or decay of precipitator voltage. This change will generally produce a short transient period until equilibrium returns to the system. Therefore, uniformity of the gas flow rate is of prime importance.

Probably the effect of more or less gas flow on the distribution pattern of the gas as it enters the collector is as important as any other input.

Gas Flow Distribution

I am concerned about gas distribution problems from the standpoint of velocity, temperature and concentration of material as well as particle size. If one area of precipitation has become worse in recent years, it is in gas distribution. The trend toward larger collectors has meant a greater difficulty in transferring the gas leaving the inlet nozzle to an acceptable pattern at the face of the precipitator. Granted, optimum gas distribution is not as critical in the larger units (between 4-5 ft/sec) with all fields serviceable, but the margin can be quickly lost with outages of equipment.

Probably one fallacy in gas distribution is placing too much emphasis on the results of model studies. The model cannot forsee the fall-out of material during periods of reduced operation that will often distort the actual flow pattern.

Another problem that occurs in distribution is that a distorted pattern of gas flow entering the inlet plenum of the collector often tends to magnify this bias as the gas expands in the plenum. A common fault lies with the poor design of the turning vanes at the elbow or entry of the inlet plenum. Vanes can allow gas flow to short-circuit the separate channels if they are not extended sufficiently on the trailing edge.

A simple concept that escapes many designs is that the gas tends to continue in the direction pointed until it strikes an obstacle. Often the first obstacle met is a perforated plate with 40% to 50% open area. If the vectors of gas flow strike this perforated plate at some angle, a deflection occurs that tends to crowd the gas into small areas of the precipitator.

A common design for years employed a single or dual system of perforated plates at the immediate face of the collector. Unfortunately, the pressure drop of these diffuser devices in the low velocity zone of the plenum is usually insufficient to correct gas maldistributions.

Although major problems in gas distribution are occurring with the large designs, it is the older group of precipitators in the 7 to 8 ft/sec range that cannot stand a poor pattern. Poor gas distribution can produce spark-over at reduced voltages in the inlet field, which of course, compounds the loss in performance caused by zones of high flow.

Poor gas distribution can take many forms and will be covered in a later chapter. Even though it is not identified separately in the D-A equation, poor gas distribution is one of the main reasons performance expectations are not met in the field.

Gas Temperature

The level of gas temperature in the precipitator opens up many areas of interest, especially the effect on the viscosity of the gas stream. But the major effects of temperature lie in the modification of the electrical characteristics and the reactions of the particles as they deposit on surfaces. The effect on metal corrosion by changes in flue gas temperature will be covered separately.

Practically all of the particulate matter handled in precipitators will go through a wide spectrum of electrical characteristics for the temperature range of 200°F to 750°F. Much of this has to do with condensation effects and surface leakage at the lower range and conductivity changes in the bulk material at the higher temperatures. The true effects at any given temperature will depend on the moisture level and chemical composition of the particles. Of greater interest would be whether the precipitator is operating in critical temperature zones for that particulate material. For example:

1. High sulfur coal for pulverized coal-fired precipitators would be critical in the 250 to 280°F zone.

2. Lower sulfur coal for this same precipitator might find its most critical zone between 310 to 360°F.

3. Cement precipitators might find its most critical range in the 350 to 400°F.

Within these critical zones, the variation in electrical readings may occur with as little as 10 to 15°F movements in the process gases. I have observed some fly ash installations where a 15°F change has meant a 3 to 4 fold increase in emissions.

The ability to modify flue gas temperatures within critical zones is as important to successful precipitator performance as any other design feature. As with variations in gas flow, short term variations in flue gas temperature should be controlled in order to minimize losses from the collector. In fact, it is usually better to operate at a less than optimum uniform temperature rather than experience variations. The heat sink effect of the internal structure will tend to mask effects of the temperature cycle if it is less than 10 minutes in duration.

Another key point is the relation of the average gas temperature to its parts within the precipitator. In large collectors especially, it is possible to have temperature gradients of 50 to 60°F between individual chambers. If this occurs within the critical temperature zones, it is possible for the poor operating chambers to control the performance level of the overall system. Localized temperature deviations as might occur under conditions of severe air inleakage can control the power input of a field.

Resistivity

Since I have stressed that high levels of power input is the name of the game in precipitation, a characterization of this ability to reach effective voltage fields is most important. As pointed out earlier, the term resistivity is a measurement of the ability to pass the flow of electric current through the layer of collected material. This subject will be covered in greater detail in a later chapter, but I would like to briefly discuss several key factors.

1. Do not emphasize the measurement of resistivity as an important decision-making number. The value of 2×10^{10} ohm-cm or any other number should be considered only another tool in the analysis of an installation. The value itself can be quite misleading.

2. It is extremely difficult to obtain a representative dust sample from the gas stream and simulate the method of deposition on the actual collecting surface.

3. The design of the electrodes and the spacing in the precipitator may have much bearing on whether any given resistivity level will actually produce adverse electrical conditions.

4. Variations in resistivity can occur within short time periods as well as occur in different portions of the precipitator system at any given time. Identification of these problems are usually of much greater importance than the identification of numbers.

5. Since the precipitator is the best measure of resistivity, the ability to modify several of the key process inputs and observe their effects on electrical characteristics is a worth-while endeavor. More on these techniques in the Case History Chapter.

6. It is usually better to operate at a slightly higher resistivity range if precipitators are to be kept within reasonable sizes. Higher resistivity with the evidence of random spark-over

that occurs at relatively high power levels is usually an optimum operating level.

Properties of Gas Composition

The normal constituents of the gas, for all practical purposes in precipitators, provide sufficient capability for the ionization process. The dielectric space between the discharge electrode and collecting surface must contain sufficient gaseous molecules to promote electron and ion activity. Most process gas streams have varying mixtures of H_2O, O_2, CO_2 that are relatively active in the high voltage field.[5] Extreme levels of one of these gases over the other may tend to distort voltage-current relationships. But I have not observed major gas sensitivity with the normal basic process precipitator.

High levels of SO_2 will tend to show low voltage — high current relationships. High levels of H_2O tend to elevate voltage fields slightly. However, gaseous condensation effects on the surface of particles in the gas stream will also determine voltage-current levels as we shall see later. The key point here is not to be overly concerned with the effect of the major gases on electrical performance as with the subtle effects of some of these gases on the electrical characteristics of the particle.

Properties of the Particle

The complexity of the precipitator can be best related to the wide range of material it is called upon to collect. There are an infinite number of particle characteristics that can effect precipitation, including size, shape and chemical constituents. Fortunately, the precipitator can also nullify or mask a wide spectrum of material characteristics by operating at optimum gas conditions with a proper size collector. I will briefly look at some key points.

1. While the Deutsch-Anderson equation indicates that the finer-sized particles are more difficult to collect, too much emphasis is placed on this fact for most of the industrial applications. I have never been overly concerned with a particle size analysis for its small size fraction, especially in basic industries, when determining whether a precipitator can or cannot meet its performance goal. Of course, large percentages of fine particulate (less than 2 microns in diameter) will tend to aggravate the removal of material from electrode surfaces. Some effects on

resistivity can occur because of the large surface area and packing characteristics of the fine material. The chemical make-up is usually more important than the size alone. The key point to remember here is that all modern precipitators should have sufficient capacity margin to overcome the levels of fine particles usually found in the process effluent. Only when low power levels exist will difficulties begin to occur, and later chapters will show how these conditions are overcome.

2. Most of the fine-sized particles in a flue gas stream (under the 2 micron level) usually occur through condensation mechanisms. This usually involves particles that have been formed in spherical shapes by high temperature existing in the combustion zones of the process. The fine-sized material will sometimes tend to agglomerate or grow into multi-particle formations while moving in the gas stream. Very fine-particulate matter can also attach to the larger sized particles. Even the particles that remain in discrete form when entering the precipitator, once deposited, will tend to attach and form agglomerates. In other words, the effect of particle sizing on collection performance becomes a complex variable to analyze.

3. The production of particulate matter in each of the industrial processes conform to certain laws of particle sizing. Usually the large sized particles will be coarse and irregular in shape. Of interest with a spectrum of particle sizes in a specific process is the possible chemical-make-up of each size segment. I normally would group these segments in terms of chemical constitients as follows:

Below 2 microns in diameter

2 to 10 microns

10 to 100 microns

over 100 microns

It is quite possible to have several different populations of particles within the same effluent. I will cover some of these characteristics in the chapter on Process Factors. As the particles increase in size, surface porosity generally increases.

4. If a process has a wide spectrum of particle sizing, there is a distinct possibility of particle segregation entering the precipitator. For example, large particles will tend to concentrate along turning vane surfaces. The configuration of the duct or flue system ahead of the precipitator can play an important part in the segregation of the particle size groups.

Particulate Build-Up on Electrodes

Evaluating the effects of deposits on discharge and collecting electrode surfaces is a prevalent problem for personnel involved with precipitators. While I shall include more details in a later chapter, I would like to stress some key points at this time.

Discharge Electrode Build-up

While the negative electrode is the source of energization, it too can develop a coating of particulate. It is not unusual to observe a build-up of 1 to 3″ in diameter on groups of wires, for example, and also at different sections of the electrode. Large wire build-ups may or may not be critical on the electrical characteristics dependent on the nature of the coating. For example, an irregular build-up that forms the shape of beads over the total length of wire will not basically alter the meter readings or precipitator performance. However, if an uniform build-up of fine particulate coats the electrode so that it simulates a wire of larger diameter, then higher voltages may be required to initiate corona current. This condition can lead to a voltage sensitive precipitator in which spark-over occurs, because the voltage gradient is too great for the physical spacing. The barbed wire and punched plate type provide a corona discharge by use of physical points, thus reducing the effect of material build-up on the electrical characteristics. However, it has not been my experience for build-ups on discharge electrodes to be a prime source of defective precipitators.

Collector Surface Build-up

Particulate build-up on the collecting surface can provide substantial effects on electrical readings. A thin resistive layer can cause a spark-sensitive precipitator. If not resistive, material build-ups of 1 to 2″ thickness can often occur without an electrical breakdown of the remaining space.

It is practically impossible to operate with collector electrodes in a completely clean condition. Normally ⅛ to ½″ build-ups will be found on the collecting surface. Unfortunately, the higher resistive material will tend to have better cohesive forces holding the particles together on the surface of the collector. The case where thick build-ups occur can also produce changes in the voltage-current readings. Even though the voltage gradient may be similar to that of the wider spaces, the increase in gas velocity through the narrowed passage will decrease the overall precipitator efficiency. The indicated voltage will generally be less for the narrowed passages in an affected field.

In summary, the point to remember is that normal precipitators will operate with varying degrees of electrode build-up. Whether this build-up is meaningful can often be judged by comparing electrical readings of adjacent fields or cells. With a successful precipitator, build-ups will generally show some uniformity. When observations indicate great differences in build-up from top to bottom and side to side in specific fields, this might indicate that poor distribution of gas may exist. See the Operation and Maintenance Chapter 5.

Rappers and Reentrainment

Nowhere in the original Deutsch-Anderson equation is an allowance made for the losses that occur in transferring the collected material from electrode to hoppers. The interplay between the electrical forces holding the material on the collecting surface and the rapping device attempting to remove it provides a real challenge for effective precipitation. But this challenge does not merit the priority some people have placed on high rapping forces. This statement is valid as long as the rapper mechanism is sufficient to impart at least 10 to 25 G's to the support structure holding a group of collector plates. With many process conditions, even a sub-standard rapper system will not effect performance adversely. But when the build-up on the collecting surface reaches over $3/4''$, it is prudent to assess whether the rapper system is sufficient. I place great emphasis on the reliability of the rapper and control circuitry components. This has become more important as collection efficiency levels have increased.

The effect of rapping on precipitator performance is whether puffing losses are observed or measured, since this can denote a significant reentrainment of material from the collector surfaces caused by the rapper operation. While the reentrainment puff is usually a mechanical occurrence, the operation of the rapping device can sometimes effect the electrical characteristics at the same time, thus aggravating the magnitude of the problem. The vibration of the high voltage frames could produce an electrical disturbance dependent on the structural integrity of the discharge electrode system.

The rapping operation (remember this takes in all impact or vibrator type of devices) becomes less of a performance factor when other aspects of the system are functioning poorly. Another way of looking at it is that the chance for the rapping operation by itself to convert a poor performance of a precipitator to an acceptable one is slim. I will list some points that, while appearing elementary, should furnish some key guidelines for your understanding of the rapping problem.

1. With low power inputs such as might exist with high resistivity problems, the percentage of rapping loss relative to the total emission losses should be less than 30%. When rapper reentrainment is a dominant part of the losses, then either gas distribution or highly conductive material must be suspect.

2. The existence of observed puffs at the discharge of the precipitator is not a phenomena of only the outlet fields, but indicative of reentrainment that is occurring throughout the collector. Usually puffs noted for the outlet fields will mean much larger reentrainment from the inlet fields.

3. The size and type of particles, as well as process gas characteristics, will usually have a greater effect on reentrainment than the specific design factors of the rapper system.

4. It does not pay to expend time and energy on modifications of the rapper system (if a reentrainment problem exists) before corrections of other problems.

5. Excessive rapping intensity in order to attain sufficient cleanliness of collector surfaces to overcome the effects of high resistivity has not been successful in my experience. Actually, excessive rapping can sometimes worsen electrical characteristics in highly resistive cases by keeping the surface of the layer unduly irregular.

6. Modification of the frequency or cycle time of rapping will generally produce more benefits than changes in the intensity level.

7. The impact rap appears much better for the collector surface while the vibrator is desirable for the wire discharge electrode systems.

Electrode Design

I have often been asked whether one manufacturer provides a better precipitator internal design than the others. Certainly there are differences between manufacturers and between the wire and rigid frame designs, but in all candor, the design of the internal box is much less important than the characteristics of the dust and gas to be handled. For example, the worst design will do the job with good gas conditions and the best design will fail to meet performance with poor gas conditions.

Another way to look at it is that if the precipitator design is physically sufficient, I have never encountered an installation whose internal problems could not be overcome to attain performance goals. But make no mistake, poorly designed electrodes do cause difficulties with critical units and especially can generate maintenance problems.

The main reason that the design of the internal components plays such an important role in precipitation is that the sharp edges, or places where the electrical field is distorted, will often produce a spark-over at reduced voltage levels. Any protrusion or contour on the collecting surface will provide a greater sensitivity for spark-over than even a kink or bend in the discharge electrode. Any stiffener used on the collecting surface structure must be rounded sufficiently to reduce the stress concentration of the field. This becomes very important with high resistivity particles.

Probably the greatest cause of wire electrode failures in recent years has occurred opposite the upper and lower termination of the collector plate, when the diameter of the wire has not been effectively increased at this critical area. Even though the plate termination is contoured correctly, the corona field will still cause a major flux distortion to occur at this discontinuity of the collecting surface. While this may not cause problems in certain installations, the use of a well-designed shrouded wire is mandatory in most situations to keep localized spark-overs from occurring at these critical points.

I observed a number of cement precipitators that did not perform at anticipated levels because localized spark-over conditions occurred between the discharge electrode and a rounded support used for the collector elements. However, this case, as well as others involving spark-over areas, must be identified early and measures implemented to reduce this localized activity. See later chapters for the analysis and corrective steps of these problems.

Power Supply Characteristics

As precipitators have grown in size so have the power supplies grown in Kva ratings. This trend to larger transformer-rectifier capacities has injected some difficulties in stability, and yes, even in the performance of the precipitator if a gross mis-match occurs between the size of the power supply and the field to be energized.

It is well to understand a basic concept of precipitation that each field of any installation will only effectively absorb the amount of power that the existing gas, dust, and internal structure integrity allows. Therefore, the actual voltage-current requirements of a precipitator field may be drastically different than shown by the full load rating of the power supply.

The spark-over is the limiting point of power input if it occurs before the current or voltage limit is reached for any given supply. When the spark rate increases, at some point the disturbance of the field can diminish performance even though the indicated power readings have not appreciably changed. As additional power is forced into the precipitator field, a lower average voltage will result while the current flow will actually rise, often giving a sense of higher power input. This is the classic case often observed in the field when someone attempts to attain the current rating of the power supply. I have observed several cases where continuous break-downs occurred, and at the same time the panel meters indicated an absence of spark-over at full load current. Remember, the corona current does its most effective job when it is distributed among all the discharge electrodes within a precipitator field. But the localized spark-over actually funnels all the available current for that field through one path, rather than by the ionization process through countless thousands of paths.

Now most of the above conditions that have occurred in the past was basically man-made either when a manual mode was used, or with the early power supplies that did not have auto-matic control circuitry. The use of automatic circuitry utilizes a transient signal initiated by the spark-over to effectively control the power levels of modern supplies. Each manufacturer may vary the degree or complexity of these modern systems, but the object is to alter the input voltage to the primary winding of the high voltage transformer by controlling the gate of the silicon diode rectifiers in the low voltage circuit.

While I made an earlier point that by disregarding spark-over levels can be bad with the use of manual control, the auto-matic circuit can also produce erroneous actions in some cases. The basic fault usually lies in the fact that the control will over-correct or lack circuit stiffness so that the average precipitator field over a short period will be actually less than optimum. If the automatic circuitry is overly complex and contains un-reliable components, the net result can be less than expected.

There are certain processes, such as the basic oxygen furnace in steel making, where the automatic control will be called upon to vary power input by a factor of 10 within a 20 minute cycle. But in steady state applications, the automatic circuit will either vary power within a narrow band or only be called upon to react during emergency process disruptions. Each installation must be weighed as to how the automatic features perform in relation to all the other factors. For example, with a well designed precipitator, the response of the automatic controls can be much less critical than would be required if internal defects exist. All in all, I would like to list a number of characteristics that you should be alert to:

1. The automatic circuitry will usually perform better when the power supply is operating at 50% or more of its current rating.

2. The severity of the spark-over should determine the response of the circuit. With low power levels, the system should be tight with little down time of energization, and a fast ramp return to full voltage. On the other hand, if a power arc occurs because of a severe disturbance, it would be desirable for the automatic circuit to extinguish at least one power cycle before re-energizing the field.

3. After a decay of voltage, I usually like to see a return to full voltage much faster than one (1) second. This can be observed visually by the movement of the panel needles. Of course an oscilloscope would show this rate of recovery response quite clearly. In other words, the stiffer automatic response should almost simulate a manual control operation while providing help in suppressing the severity of the spark-over.

4. Be alert to the restrike or repetitive spark-over that lasts over a 3-4 cycle duration. This can be the result of an overshoot in the automatic circuit, but I would also be concerned about internal difficulties. Operation with manual control at a low spark-rate will usually determine whether the trouble is with the power supply or caused by internal problems.

5. The use of a properly-sized linear reactor in the primary circuit of the high voltage transformer can stabilize and diminish

the severity of spark-overs for many installations. Most manufacturers are now applying these ballast reactors that tend to increase conduction time of the precipitator current, while modifying the wave form from a peaky or fast rate of rise to a moderate slope. In reality, the impedance of the reactor provides system inertia, which helps draw the wave-form out toward a sine-wave pattern whenever the power supply is operating below full-load. The reactor will tend to reduce the peak voltage slightly while bringing the average voltage up and might produce a 5 to 15% increase in power input for certain applications. I will cover some of these applications in the Case History section.

Mr. Herb Hall was instrumental, many years ago, in promoting the use of linear reactors in conjunction with the proper design of the power supply. Actual sizing of the reactors will depend on the maximum voltage required for a desired current level, but a typical size is shown for a range of power supplies often found in use on the nominal 480 volt circuits:

Rated Secondary D.C. Current-ma	Size of Reactor Millihenries
250	13 ± 10%
500	7.5 ± 10%
750	4.5 ± 10%
1000	3.3 ± 10%
1500	2.2 ± 10%

6. The linear reactor will provide a reactive voltage drop dependent on the current flow through the primary winding. This voltage drop must be kept small enough so that the desired voltage can be attained at the primary winding of the high voltage transformer. I will say more about this in later chapters.

7. Probably even more important than the reactor is the match of the power supply to the load demand of the field. Not only is it important to operate above 50% of the power supply rating for overall stability, but I tend to observe better spark-over characteristics with the smaller sized, higher impedance supplies as long as they are operated near rating. This is a case where bigger is not better.

In summary for this section, I would stress being careful to allow the power supply to analyze what is happening in the precipitator rather than provide an artificial condition. Moreover, a precipitator that is sparking in a normal manner at some reduced power level, may be performing better than a collector that shows greater power input without spark-over. For example, resistivity conditions that produce some spark-over is usually desirable in fly ash applications when reentrainment losses can be high with the more conductive ash.

Full-Wave and Half-Wave Energization

Whether to use full-wave or double half-wave energization for precipitators is still one of the most frequently asked questions. This is the case where one (1) power supply can either energize two (2) fields or bus sections separately as shown in Figure 1—2 of Chapter 1, or supply common energy to the same precipitator area as noted by the arrangement in the upper portion of the figure. The full-wave connection is generally made with a wire jumper across the two output bushings of the transformer tank, or by a switching arrangement. Figure 3—1 shows the difference in the voltage waveforms between the two methods of energiza-

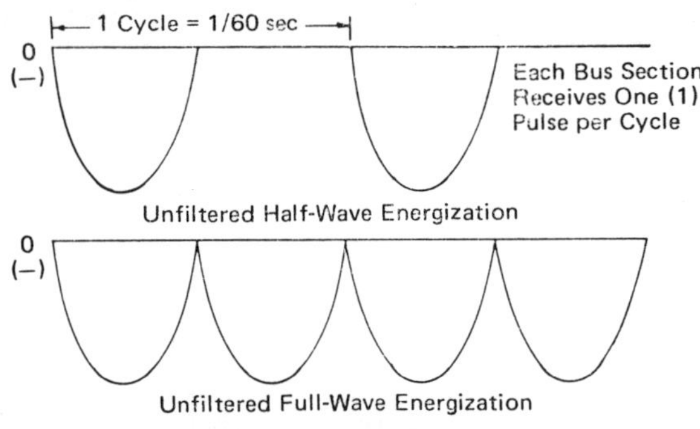

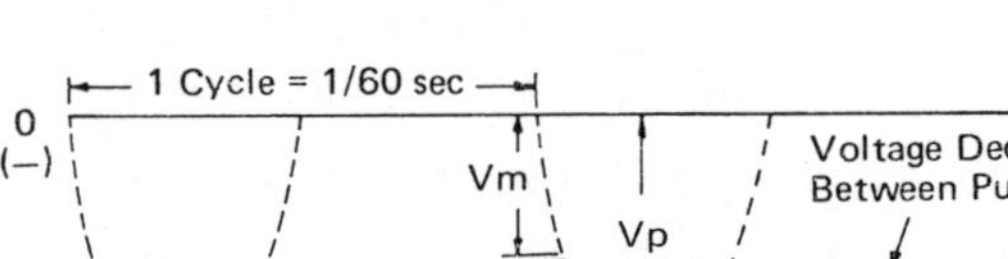

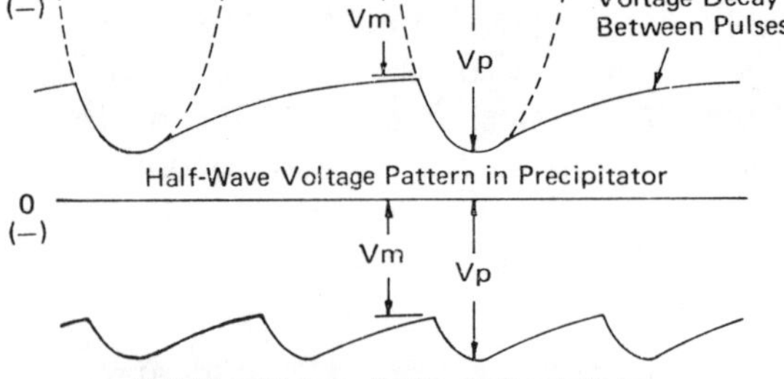

FIGURE 3—1

tion. Figure 3—2 also shows the precipitator current waveforms for the F-W and H-W cases. Basically, the two bus sections split the current flow of each pulse from the rectifier in the case of F-W , while each bus receives a separate pulse on the alternate half-cycles of the double H-W arrangement.

Years ago, I often obtained efficiency tests to substantiate if either F-W or H-W had any effect on performance. Some minor benefits were realized with H-W in a couple of high resistivity cases, but the differences were hard to distinguish in most of the installations. Where some help occurred, it was primarily due to isolated alignment difficulties in the one bus section, or by gas temperature variations between the sections. With high resistivity, the panel meter readings showed very little differences, while on occasion, the readings did indicate improvement on F-W with more conductive material. For example, a fly ash installation, 270°F, 270,000 acfm gas flow rate, and about 2% sulfur coal, gave the following comparison on an outlet field:

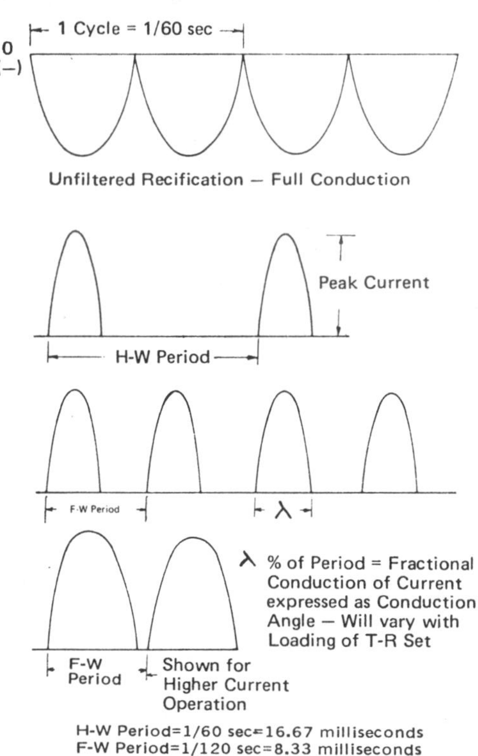

FIGURE 3—2

**Representations of Precipitator Current Waveforms
for F-W and H-W Energization.**

	Trans. Pri Volts	Sec. DC ma	Sec. Peak Voltage (KV)	Sec. Min. Voltage (KV)	Power Input KW
F-W	223	150	34	26	4.5
H-W	235	125	37	22	3.7

The above readings were obtained on a 15 KVA, 250 ma rated power supply. Control settings were adjusted for a higher power level and which gave the following readings:

F-W	243	200	36	26	6.2
H-W	262	175	40	22	5.4

Note that the minimum voltage remained the same while the power was increased. These voltages were measured by voltage divider and oscilloscope.

The characteristics of the F-W and H-W energization can be briefly summarized:

1. Full-wave will result in a lower precipitator peak voltage while the minimum field voltage will be generally higher, but the average secondary voltage will be similar to that of H-W.

2. F-W will tend to be better applied on outlet fields, or on more conductive ash where low minimum field voltages might aggravate reentrainment problems.

3. Where there is a tendency for sustained arc-overs to occur, because of electrode defects, the H-W might be better able to break the arc-path with the longer dead time between pulses.

4. With the larger power supplies of recent times, the F-W appears better able to provide stability when operating at low power levels. This will generally occur with the higher resistivity cases.

5. The H-W benefits for the large power supplies does not turn out as well as expected, since there can be some reflection into one bus from a disturbance of the other section.

6. Where spark-over may be sensitive to the peak voltage, some help can be obtained by reverting to the F-W energization.

Sectionalization

Another factor not given prominance in the analysis of precipitator performance is the effect of sectionalization, or the number of power supplies per installation. While basic theory may indicate that one (1) 100 KVA supply will provide the same end result as two (2) 50 KVA supplies for the same surface area (utilizing split high-voltage frames), in practice, the two smaller supplies will generally provide improved performance.

Just how many power supplies (or the surface area per supply) are required must be weighed for each process or installation. A higher number of fields in series will provide maximum benefits for a sharp change in particle concentration and character. Differences in gas characteristics, usually based on temperature stradification, might dictate the use of more power supplies in parallel coupled with smaller bus sections.

While sectionalization can often provide the best-match-up between power supply and precipitator area, the greatest benefits accrue in maintenance flexibility and performance reliability. Consider a two (2) field precipitator versus a four (4) field precipitator. Even though the overall collection efficiency may be the same for both units, the calculated efficiency for each field is as follows:

	2 Fields		4 Fields	Cumulative eff.
Eff. Inlet	92%	Inlet 1	76%	76%
		2	71 %	93%
Eff. Outlet	88%	Outlet 1	66%	97.6%
		2	59%	99%
Overall Eff.	99%		99%	

The above efficiencies are based on exponential laws that has each field collecting the residual material exiting the preceeding zone at some specified efficiency.

You will note that the efficiencies of each series field will decrease from inlet to outlet as the concentration of material is reduced. It can be seen that the loss of one field will produce a nominal performance of 90% for the two field precipitator while a loss of one supply in the other unit will still allow an efficiency of about 97% to exist. I will discuss more of this phenomenon in the design sections.

Operation and Maintenance Factors

Many success stories of the high collection performance of precipitators are well documented. But to many users, a constant battle is waged to maintain these performance levels. A major reason for this situation lies in the basic design of components for the overall system that produce sensitivity for break-downs. While I discuss many design concepts in this book, a concerted effort must be made by the user to understand all the inputs to the potential problems of maintenance.

Obviously, it is exceedingly difficult to predict where some of the maintenance troubles may occur, but there are eight (8) key areas that can be emphasized.

1. Raw material and operation forecasts — original design.

2. Design concepts.

3. Construction phase.

4. Initial check and training.

5. Personnel assignment.

6. Control of process.

7. Record keeping.

8. The actual maintenance program.

Raw Material and Operation Forecasts

This may be the toughest, but yet the most important phase to help minimize future problems. The user has to forecast possible raw material or operation changes that can effect the precipitator. The manufacturer's representative must know what to ask for in terms of the parameters. For example, with a wet process kiln that is currently operating with a 500°F back-end gas temperature, additional chains may be contemplated that will reduce the gas temperature to 400°F. This future course may alter the design of rapper and hopper evacuation equipment as well as the size of the precipitator.

Design Concepts

Just how much is placed into a design to overcome operating and maintenance problems must be discussed as much in detail as the emphasis on performance levels. Simple concepts, such as the number of bolts on access doors, or the number and location of walkways can be quite important for inspection activity.

While each option or change may add additional cost to the initial installation, it may prove far less costly for the long term.

Construction Phase

The best precipitator design can be adversely affected in the fabricating and erection phases. Just how the quality of welding is controlled, or the shaping of the component is finally accomplished in the shop could have a significant effect on the final operating characteristics of the precipitator.

How the material is handled on the job site can be important to prevent a tendency to distort long electrode elements. Weather protection is of primary concern if long storage time is required.

There are advantages for the user to assign an inspector during the actual erection phase. Cross checks of the actual erection procedures are important.

Check-Out and Training

The manufacturer will usually send a service man during the last days of the precipitator construction period for the final check-out of electrode alignment and electrical circuits. The user representative should use this opportunity to observe techniques and various measurements used in the analyses of the circuits. On-site experience has advantages compared to later training. It is up to the manufacturer service man to provide pertinent methods and answer all questions methodically.

Whoever is assigned by the user to accompany the manufacturer's service man should carefully record all information in a start-up manual that would be available for plant use. In addition, the manufacturer will usually offer a training course over and above the normal operating and maintenance manual.

Personnel Assignment

Just who to assign to the precipitator system should be given much thought. The value of the initial check-out and contacts with the manufacturer can be lost if the user representative is moved to another assignment.

A person who can be assigned long term to oversee the precipitator system and monitor the process as to how it affects the collector is probably the best investment a company can make. Recent years have shown the advantage of close supervision for the large precipitator installations. It takes a special person not only willing to spend hours in dusty, uncomfortable working conditions when necessary, but willing to take the responsibility to assure successful conditions to the precipitator.

Control of Process

The weakness in recent years in coping with precipitator problems, both from a maintenance and performance standpoint, is the lack of knowledge and understanding of the effects of the process on the conditions that can occur in the collector.

Of special importance is the effect of start-up and shut-down procedures of the process on present and future troubles of the precipitator. Gas temperature and moisture levels can affect hopper evacuation conditions. Seasonal and daily ambient temperature variations must be considered. Raw material variations should be closely monitored. Batch operations will impose the potential for hopper difficulties.

Record Keeping

Minimal efforts for recording difficulties and failure of equipment will often produce benefits. While we can all agree that it is too much to expect an installation to be 100% free of break-downs, repetitive failures should not be considered normal. For example, the pattern of where wire electrodes fail can point out the potential causes. Periodic recording of the electrical characteristics of the precipitator fields can identify internal difficulties or changes in the process. I recommend that several key process factors be logged with the precipitator readings such as production, gas temperature entering collector, gas moisture, and air flow levels.

Actual Maintenance Program

Sincere efforts in each of the above categories will go far to help minimize the situation where problems continually exist with the precipitator. The first year is critical in this overall program. Internal inspections by trained personnel should form the basis of ascertaining future trouble spots. Emphasis must be placed on periodic check-outs and cleaning of certain apparatus. The key point — do not continue to live with repetitive problems.

There is no substitute for the building of maximum reliability into modern precipitator designs. Attempts to build sheer size into the collector in order to overcome a failure rate factor may be a dangerous path. More electrodes, rappers, hoppers or more of any equipment that has continually to face flue gas conditions is fraught with danger — if there are weak spots in this equipment.

The economic advantages to be gained by improvements in the design for reliability vary, depending on the specific process characteristics, and even on the attitudes of personnel at certain plants. Therefore, the information required to overcome potential trouble spots should be freely discussed in the early negotiation and specification stage for the purchasing of a precipitator.

Chapter 4
Resistivity

Introduction

Much emphasis has already been placed on the fact that effective precipitation coincides with the occurrence of optimum amounts of electrical power input in the corona process. While power input is sometimes limited by structure or individual component defects, the performance of limited power installations occurs under conditions of excessive-electrical resistivity of the collected material.

Earlier coverage briefly described the scope of resistivity concepts, but more information will now be given to allay some of the misconceptions and provide you with a working knowledge of the subject. The subject is of greater importance with fly ash applications, but basic resistivity situations can occur at times with all cases of the dry precipitator collection of particulate matter.

All finely divided particles that are generated in the basic industrial processes have critical temperature zones which can effect the electrical operation of the precipitator. The chemical composition of the bulk of dust particles contain common constituents even if the make-up varies somewhat in weight fractions. Given similar gas conditions, it might be hard to differentiate electrically whether it was fly ash, cement dust, or iron oxide being handled in the collector. Of course, that is a simplistic statement because in practice there are an infinite number of process conditions where differences in raw materials can alter the electrical characteristics of the particles.

Fortunately high moisture contents in the flue gas stream (such as found in wet process cement applications or other applications where spray water is used to cool the gases) will usually nullify the subtle chemical particle composition and provide ample power inputs. You can call water vapor a primary conditioning agent that will control resistivity problems if the quantity of water used is effectively matched to the gas temperature levels of the flue gas entering the collector.

When moisture levels in the flue gas are low — usually below 10% by volume — the chemical make-up of the particle becomes a dominant factor in controlling electrical characteristics.

I would like to stress the point that any given value of resistivity could mean little in the bottom line of a precipitator — that is, the ability to meet performance or a specific regulation. It is the match-up of all the variables discussed in this book that will generally determine whether the collector will be satisfactory, even at reduced power. Higher resistivity conditions can actually provide improved adhesion characteristics of the particles.

It is difficult for me to specify various resistivity levels as denoting good or bad operation. I have observed factors of 100 between one job to another as to where spark-over or conductivity becomes a problem when related to a specific resistivity curve. There are general guidelines that have been widely dicussed by Dr. Harry White and others. I consider the zones to react as follows:

Comments	**Resistivity Range**
10^4 to 10^7 ohm-cm	Usually highly conductive material — hard to retain — low voltage fields present.
10^8 to 10^9 ohm-cm	Sensitive stage where lack of resistive characteristics can sometimes hurt — especially in fly ash cases.
10^{10} to 10^{11} ohm-cm	Appears to be the best range to shoot for — should show some spark-over in precipitator.
10^{12} to 10^{13} ohm-cm	Range usually associated with low sulfur coals — reduced power in all fields can exist.
Over 10^{13} ohm-cm	Not commonly observed in basic industries with normal moisture contents. Can produce severe electrical disburbances.

The description of spark-over as briefly covered in an earlier chapter was defined as an electrical breakdown through an isolated gas path between the negative and positive electrodes. The case of the threshhold resistive spark-over where the discharge streamers occur from the deposited dust layer is considered the start of back corona. This situation is not unlike that which occurs in atmospheric lightning when positive streamers from earth actually draw the localized stroke from the negatively charged clouds. But in the case of very high resistivity, a severe back-corona condition can occur characterized by greatly reduced voltage and high current densities without spark-over. While I have heard of this case, I have never observed it in the field. In this case, the surface layer of the collected material evidently produces sufficient positive ions to practically nullify the precipitation process. The case of high conductivity may also look like the classic case of severe back-corona, in that the indicated power patterns shows low voltage-high current characteristics.

One important concept is that each process will produce a particulate matter whose resistivity will usually decrease rapidly on the low temperature side of a peak, while decreasing at a lesser rate on the high temperature side. Figure 4—1 shows a typical plot of resistivity obtained by laboratory analysis for dust entering a cement precipitator. A typical fly ash from an eastern bituminous coal source with 2 to 3% sulfur would show a similar pattern at approximately 6% moisture by volume. In fact, some of the poorest performances that have occurred in coal-fired boilers in the last decade have been the result of over-conditioned fly ash that easily reentrains back into the gas system. These

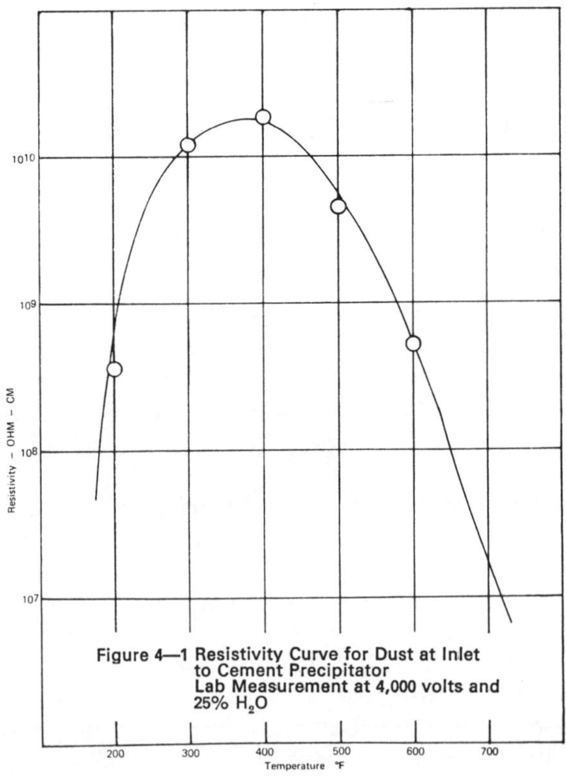

Figure 4—1 Resistivity Curve for Dust at Inlet to Cement Precipitator Lab Measurement at 4,000 volts and 25% H_2O

cases provide high power inputs in the precipitator without spark-over and usually occurs with high sulfur coal and flue gas temperatures below 280°F.

Measurements

I pointed out earlier that the measured value of resistivity is often suspect. This is true whether it is obtained by the in-situ or by laboratory techniques. This is not to lightly disgard the attempts to identify the electrical characteristics of the effluent material by a single number or pattern. The measurement itself can be considered another tool in precipitator analysis. But if a resistivity curve indicates the peak falls between 250 and 280°F for a specific fly ash sample, while the field obesrvations indicates that this zone provides a conductive range, then I should believe the field observations.

The basic problems in resistivity measurements involve the difficulty of achieving representative samples and to simulate the deposition of this material as it may occur on the actual precipitator surface. Some historical values are based on bulk samples obtained from hoppers. It is almost impossible to obtain a valid sample this way. A mass sample obtained through the traverse of the cross section of a flue may be actually more desirable than several single point measurements by the in-situ method. Just how the material segregates in the flue can determine the value of the in-situ measurement.

A common problem that may occur between the time of the measurement and what is observed later in the precipitator may be related to subtle changes in the process material, among other factors. On applications where little background exists, resistivity measurements can be an important tool, if not for finite values, then certainly for trends. However, established manufacturers should have sufficient experience to better judge the effects of the gas and chemical constituents on electrical properties than by any other technique.

The most difficult part of being able to pin-point how a precipitator will react over a long term period is in the prediction of the change in process and raw materials. I have observed drastic changes in resistivity in the same installation over a span of years. Coal analysis can vary greatly even from the same source. A precipitator can be faced with a poor ash condition after modifications to the heating surface of a boiler. I believe the following information will help identify some of the problems.

Effect of Chemical Composition

Fly ash characteristics vary through a wide band, based more on chemical constituents than any other factor. On most pulverized-fired applications using eastern bituminous coal, the moisture ranges between 5% and 8% by volume at the precipitator inlet. At this level, the absorption of SO_3 becomes a dominant conditioning factor and how much SO_3 is produced relates not only to the sulfur content of the coal, but to subtle mechanisms for converting SO_2 to SO_3 in the boiler system. It is possible to have as much SO_3 formation from a 1.5% S coal as from a 3% coal, dependent partially on levels of combustion air and the catalytic actions of iron and other components. Usually, with 1.0 to 1.2% or less sulfur contents in eastern coal, one can expect difficulty with electrical performance in the 300 to 400°F range.

As the sulfur range of eastern coal approaches 1%, there is also evidence to suggest that the pyritic portion of the sulfur content becomes a factor. Benefits appear to accrue if the pyritic percentage is greater than 50% of the sulfur entering the boiler. The effect of iron on converting SO_2 to SO_3 may be important here. In fact, higher Fe_2O_3 contents in the collected ash is a good sign. While the sodium content is less of a factor with eastern coals, values of sodium oxide (Na_2O) under 0.5% by weight may become a resistivity factor as the sulfur content reaches 1% or less. Much depends on the relationship of other factors.

The sub-bituminous and lignite coals of the western fields offer similar problems, but the lack of sulfur makes low sodium oxide and the higher basic components of calcium and magnesium much more of a critical factor.

I have observed subtle chemical effects in other processes such as iron oxide fume and cement dust, but no major electrical problems have occurred when sufficient conditioning moisture was available.

Effect of Moisture

Even fly ash electrical problems present with low sulfur or sodium conditions appears to abate as moisture is added into the flue gas. Improved electrical conditions are possible by the use of spray water and steam injection into fly ash ducts. The 1 to 1.5% increase in water content, in conjunction with the lower temperature, can provide a double benefit to power input. From what I have heard, the higher moisture western coals appear to have beneficial electrical effects over that of the lower moisture fuels.

Moisture contents above 25% by volume usually provide excellent electrical characteristics for iron fume applications in the 350 to 500°F range. I have never observed problems with cement applications with over 34% by volume of water vapor in the 450 to 600°F range. Somewhere in the 12 to 20% moisture range, one can expect a transient stage of rapid electrical changes in most process precipitators.

Effect of Particle Size

One of the subtle factors in resistivity is the effect of particle size on electrical characteristics. Where high moisture levels or optimum gas temperatures exist, I have not observed any poor collection results with fine particulate matter — of course, taking into consideration the higher performance requirements of the fine material. Of concern, is that the finer particles may allow the collector surface deposit to increase in depth by the greater cohesion action of the higher resistivity material.

High levels of very fine material can also produce greater problems with build-up on the high-voltage discharge electrodes, but it is the build-up of material on the collecting surfaces that will generally determine break-down voltages.

Another factor in the particle size distribution is the chemical make-up of the size fractions. Usually, the under 5 micron, and especially the sub-micron segment, will constitute the particles formed by the condensation mechanism, and will be more spherical with minimal surface porosity. While these fine particles have the greater surface area for a specific weight fraction, they also appear to take less uptake of gases — especially SO_3 for conditioning purposes.

The larger size fractions, over 30 microns in size, appear to provide benefits in the packing characteristic of the collecting surface layer to help reduce resistivity. After all, the irregular pattern of large and small particles provides a much less torturous path to ground. However, the levels of combustible, or coal particles that did not completely burn in the combustion zone, can provide both benefits and disadvantages. Combustible levels between 3 to 6% by weight of the inlet ash to the precipitator is considered desirable for electrical benefits in most fly ash applications. Great amounts of combustible, over 12%, can begin to effect the resistivity adversely.

Of interest, is the likely possibility that large amounts of combustible may provide a deterent to good electrical operation by absorbing too much of the available SO_3 in the flue gas. I have observed at least one case where unusual spark-over occurred on a high sulfur—high combustible situation. The combustible make-up will generally be found in the larger size fractions. A typical analysis of an inlet ash can show:

Size Fraction	% Combustible Range
Greater 149 microns	7 to 9%
Less 149 > 74	3 to 5
Less 74 > 44	1.5 to 2
Less 44 microns	0.8 to 1.5

It is generally believed that the existence of a sufficient number of combustible particles form parallel paths to ground for the corona current, even in low sulfur situations, and aid in the electrical stress relief of the dust layer.[7]

An interesting facet on the characteristics of fly ash is the influence of a mechanical collector placed before the precipitator can have on the power input and performance of the electrical collector. The resistivity characteristics of the fly ash can be usually altered to produce the following conditions:

1. If the ash is slightly resistive in the 300°F to 400°F range, a reduction of the large particles from the remaining ash segment entering the precipitator, could result in further deterioration of electrical conditions. A finer fly ash would have a tendency to pack on the collecting plates which may increase the sparking characteristics. Ash build-up on the discharge electrodes usually increases.

2. With higher sulfur coals, and gas temperatures in the 260 to 280°F range, an actual increase in resistivity would aid in the cohesiveness of the deposited ash layer on the collector. The finer ash would have less tendency to be reentrained in the gas stream. The higher ash resistivity would help optimize the precipitator voltage-current relationship under the high-conductivity condition.

3. If the carbon content of the fly ash is a major ash conditioner, removal of these particles can cause an excessive

sparking problem in the precipitator. This situation may occur when the sulfur content of the coal ranges between 1 to 2% and the flue gas temperature ranges between 300°F and 350°F.

I have observed several installations where a pre-cleaner collector of 75-80% efficiency reduced the performance of the precipitator appreciably from previous levels. As I noted before, the larger particles might have also adsorbed the bulk of the SO_3 gas. In other cases, precipitator units on adjacent boilers gave completely different electrical characteristics on the same coal when one of the units included a mechanical collector.

Effect of Gas Temperature

I have spent a good part of my field time exploring gas temperature effects on the electrical characteristics of fly ash. High sulfur coal with low exit gas temperatures can produce a precipitator sub-normal performance, just as low sulfur coal with higher gas temperatures can. An understanding of these characteristics can often explain a performance that varies with different coal compositions and seasonal gas temperatures in the critical range between 270°F and 320°F. Gas temperature changes of 10°F and 15°F in this critical range can substantially alter the electrical properties of fly ash from bituminous coal.

A paper I wrote in 1965 contained a table where the words "Poor," "Fair," and "Good" indicated the probability of expecting an optimum voltage-current relationship in each of the combinations shown.

Table 4—1

Sulfur content of coal %	Precipitator Gas Temperature 250-290°F	290-380°F	380-460°F
0-1	Good	Poor	Fair
1-2.5	Fair	Fair	Fair
2.5 and above	Poor	Good	Good

High ash conductivity can produce a high corona power situation in the precipitator, but the collecting performance may actually suffer because of the existence of low voltage fields and excessive ash reentrainment from the collecting plate. Charged particles will migrate across the electrical field with sufficient migration velocity but on contacting the collector surface, they rapidly lose their charge and easily reentrain into the gas stream.[8] This process can occur repeatedly and involve the same particles throughout the length of the precipitator, especially combustible grit, and thus overall performance is reduced. The above table can only be a rough approximation because of all the variables involved.

Effect of Gas Distribution

Poor gas distribution in the precipitator will normally not effect the resistivity problem, but might aggravate electrical disturbances in the inlet field. Of course, a very poor gas pattern will effect performance to a greater degree at the lower power levels. However, it has been my experience that a highly conductive condition will be effected more by poor gas distribution.

Probably more important than the effect of a poor velocity distribution on resistivity is the segregation of particles entering the precipitator inlet. If a gas stream is allowed to direct most of the large particles toward one portion of the inlet cross-sectional area, some electrical break-down may occur in the zones containing the bulk of fine particles.

Other gas distribution effects on resistivity could occur with different temperature zones at the entrance of the precipitator. While this condition was less prevelant with the smaller fly ash units, recent large designs make temperature variations of 50°F in different zones of the collector quite possible. Since 10 to 15°F can distort electrical characteristics in a specific installation, then these large temperature differentials can produce zones of varying performance. This situation becomes especially important when a single power supply encompasses two zones of different temperatures. It is likely to have the whole field controlled by the dominant electrical characteristic of one zone, and thereby result in an overall reduction of performance. I intend to cover more of this important condition in the Case History Chapter.

Air dilution is another factor that can effect resistivity in localized areas by in-leakage through insulator vents, or through openings in the shell structure or door seals. This condition is usually more critical with the high negative static applications. Even high moisture operations can be effected by this factor. Large amounts of dilution air used for cooling purposes or through leakage at air heaters and kiln seals can provide spark-over in a precipitator that should normally enjoy sufficient power inputs. Excessive air in-leakage at an air heater might produce some spark-over even though the gas temperature is reduced to what is considered a desirable low resistivity range.

Particle deposition on both the collecting surfaces and discharge electrodes is not normally uniform throughout the precipitator, and thus can effect the electrical characteristics in localized zones. Rapping will not provide uniform cleaning because of the harmonics and attenuation that occur in the metal structure. This lack of uniformity will often cause uneven patterns of corona production within the physical area controlled by individual power supplies. How important this situation becomes certainly depends on the overall design of the precipitator. I do not place too much emphasis on uneven dust build-ups except as a help to analyze velocity, temperature, or alignment problems. In other words, I normally expect to see some non-uniform patterns, and most precipitators will perform satisfactory under this condition. Perfectly clean surfaces observed during an inspection would concern me more.

The surface contours and shape of the deposition may be more important than the thickness. I have reduced spark-over in some high resistivity precipitators by increasing the time between rapping, and thereby minimize surface disturbances. Often this condition will take 5 to 10 minutes to take effect, and improved power will last for 15 to 20 minutes before the next rapping blow. The non-rap period provides time to smoothe over irregular edges of the dust layer caused by the effects of rapping.

Effect of Particle Concentrations

Where highly conditioned particulate matter exists, the levels of particle concentration will alter the voltage-current relationship from the inlet to outlet fields of the precipitator, but will not show much effect on resistivity characteristics. The inlet field may show some spark-over related more to elevated voltage level seeking out local stress points. But when resistivity increases, greater levels of dust concentrations will tend to aggravate the spark-over situation. As noted earlier, there are exceptions to this condition when the effects of particle composition, especially in the fine-size fraction, can produce reduced spark-over levels.

With low ash contents of coal below 8%, it is possible to work with reduced sulfur contents in the 1% range and still achieve reasonable power inputs. The major factor here would be the ash/sulfur ratio of about 8 to 1 for most eastern bituminous coals. Some difficulty to the electrical performance may exist if the ash chemistry is drastically changed when both ash and sulfur is reduced by the washing process.

Effect of Electrode Design

I indicated earlier that the variations observed in spark-over within a high resistivity range may have as much to do with internal discontinuities on the collector surface as for any other reason. Sharp edges, support or stiffener edge effects, surface holes or damage spots are some of the internal imperfections that will concentrate the build-up of current on resistive dust deposits. These discontinuities will allow spark-over to occur with microscopic dust layers. While I am more concerned with irregularities on the collector surface, sharp kinks in the discharge electrode can also sensitize a field.

Field Analysis of Resistivity Conditions

The resistivity conditions in some precipitators can vary from hour to hour, day to day, or even over longer periods. The problem for the user of this equipment is to be able to detect and understand this phenomenon with the tools available. I intend to point out some basic analysis concepts in this section with additional examples in a later chapter.

Since spark-over is an important monitor, the quantity, severity, voltage level, and location of this breakdown can be weighed carefully to ascertain the degree of the resistivity problem. Spark-over is detected by use of the control panel meters, sound and visual observations, and the use of an oscilloscope.

Meter Observations

The monitor meters on the control panel generally will provide the primary tool to determine if a resistivity problem exists. While the secondary side meters are most useful, the primary winding indications of voltage and current are the two meters most found on all controls, and I shall use these indicators in the following analysis. It does not matter what meters are really used, it is the relationship of voltage and current levels and their movements, that tell the story. I have even used the product of primary voltage and secondary current to analyze specific jobs and plotted this relationship of ficticious watts by calling it "FATTS." The precipitator spark-over can be noted by a transient flash in a detector circuit or sparkmeter, in some cases.

Even if a sparkmeter is available, the movement of the primary circuit meter needles can be used to indicate the severity of the spark-over. Some basic concepts are:

1. When a spark-over occurs on a modern power supply controlled at the threshhold level, both the voltage and current meter needles can dip sharply.

2. The degree of needle movement will usually depend on the ratio of the operating to rated current of the power supply.

3. With some automatic voltage controls and older power supplies, a spark-over will be seen by a sharp drop on the volt-meter with an equally sharp rise on the ammeter needle.

4. With a well controlled power supply at approximately 100 sparks per minute, the decay of the voltmeter may be 15 volts per fluctuation.

5. At the level of 100 sparks per minute, there should be a number of cycles where the meters come to rest before the fluctuations begin again.

6. Valid meter readings should be taken during the rest condition of point 5. Do not record the average of meter swings. Always record the high point of the voltmeter.

7. It is possible that the precipitator spark-over could occur on a practically continuous basis so that no meter fluctuation is seen. This condition will show an abnormally low voltage and high current.

The key point is that the needles should move sharply, not in any uniform frequency, with the meters reacting basically together. In a condition of resistivity when the leakage through the dust layer surface is controlling the break-down potential at relatively low power levels, the needle movements generally are of short duration and severity. When an internal defect is involved in a specific field, the voltmeter needle movement can show a much greater swing downward, of probably 40 to 60 volts, with a fast return to the normal base levels. This type of breakdown would constitute a power arc — the duration may last for more than one cycle — and is sometimes accompanied by a change in the noise level of some manufacturer controls. The occurrence of the arc can be superimposed on a resistivity condition, but is not considered a normal or satisfactory type break-down. I will discuss the arc-over more in the maintenance chapter. I should point out that the movement of the meter needles may be a better indication of spark-over than the sparkmeter itself. Secondly, there may actually be more low intensity sparks existing than indicated by the needle movements, but if you abide by a number of needle rests during any given minute, the degree of spark-over would be considered acceptable for most fields.

The detection of spark-over, or its absence, can be used to identify the severity of a resistivity problem by noting the voltage-current relationships and the patterns of readings between the fields of the precipitator. I shall use actual examples in fly ash applications to demonstrate the use of this technique. The size of the power supply and circuit design are somewhat pertinent, but for the sake of simplicity, I will discard all process and equipment information in this coverage. What you should obtain is an understanding of resistivity patterns and trends by utilizing the readings of the control panel meters.

Condition: High resistivity, full load current of T-R sets 80 amps. Four equal fields. 400 volt primary rating.

Field	Pri Volts	Pri Amps	Sparks/Min
Inlet	200	7	150
2	180	8	100
3	190	9	100
Outlet	190	9	100

Note that all four fields above showed similar characteristics. As the resistivity improves to the medium range, the readings can show:

Field	Pri Volts	Pri Amps	Sparks/Min
Inlet	260	17	100
2	240	24	100
3	270	44	20
Outlet	290	70	occ.

Note that the current has progressed in increasing steps as the spark-over is reduced in the direction of gas flow. As gas and dust conditions further improve to what I would consider optimum power characteristics, the readings can show:

Inlet	320	64	50
2	290	76	20
3	280	80	occ.
Outlet	270	80	none

Note that the voltage begins a stepping down action from inlet to outlet while spark-over decreases, but sparking is still present on three fields to some degree. Currents have increased appreciably on the front two fields. If this installation were faced with further gas conditioning that might produce a reentrainment problem, the readings could show:

Inlet	280	80	none
2	260	80	none
3	250	80	none
Outlet	248	80	none

Note the absence of spark-over and reduction of voltage for the greater currents. There tends to be a voltage flattening on the latter fields as material reentrainment has its effect. While you may not see evidence of a normal spark-over pattern, it will be possible to see meter movement during rapper operation. For example, the voltmeter meter may rise on the inlet field from 280 to 295 volts in a smoothe fashion, which on occasion, may produce a one to two second burst of sparks during the reentrainment period.

The above examples tend to show that the electrical patterns as noted by the voltage-current readings can indicate which resistivity range exists.

Some key points:

1. Very high resistivity conditions in the 10^{12} to 10^{13} ohm-cm range will generally show very low currents, especially with four fields or less. As the number of fields increases to 6 or more, the latter couple of fields may show a doubling of current, but never near the rating of the power supply. Spark-over is usually high for all fields.

2. With the resistivity in the 10^{10} to 10^{12} range, the power input tends to increase appreciably from inlet to outlet, especially with more than three fields. Spark-over should diminish in the outlet fields. This condition generally can provide good perform-

ances in high velocity applications at the nominal 99.0 to 99.5% level.

3. Resistivity in the 10^9 to 10^{10} range will generally provide good readings with high power and some spark-over in the front half of the precipitator. Usually the latter half will show current limit conditions because of the reduction of particles as well as the conditioned state. A sizable drop in the voltage is usually seen between consecutive fields from inlet to outlet.

4. The highly conditioned state shown will occur with possible resistivities in the 10^6 to 10^8 range or less.

All of the above resistivities could be off by a factor of 10 in either direction but the relationships will still allow a satisfactory analysis of the field conditions. There are, of course, exceptions to the above power trends for any given installation and some of the reasons for these exceptions will be shown in the Case History Chapter.

Sound Observation of Spark

Whether the spark-over noted on the meters is caused by resistivity or some localized condition can be often evaluated by the sound of the break-down. Resistivity-caused sparks will generally occur randomly throughout the field. If sparks are heard in the hopper area, side doors, and the top structure, it can often add confidence for the highly resistive state. Whether the spark, which should be detected by a short snapping sound, is heard, will often be determined by the degree of insulation and excessive background noise. I have often used a metal rod as a sound amplifier, and placed it between the ear and the shell of the precipitator. The use of sound is much less effective than the opportunity to visually observe the spark.

Visual Observation of Spark

The use of safety glass to observe various internal zones of the precipitator will generally provide an ideal method to detect the type and location of spark-over. Observation ports can be installed in special plates for the temporary replacement of existing doors, or actually installed in the existing doors. The use of side doors at walkways is quite satisfactory. Hopper doors have also been used as well as other parts of the hopper shell. The object is to detect whether the spark occurs randomly or appears consistently in one zone. These observations are most useful in the middle resistivity range where not all fields are sparking and the power patterns are questionable. While the localized breakdown will usually be of greater severity than the random type spark, much depends on the capacitance of the field and stiffness of the control system to differentiate between the two break-downs by meter observation alone.

With high concentrations of particulate matter in the inlet field, it is difficult to visually observe a large area, but usually the second field will have cleaned up sufficiently for viewing. While darkness is desirable, some extraneous outside light will not be detrimental to locating the direction and severity of the spark-over. With light excluded, it might be possible to detect small bluish streamers off various points on the collecting surface structure in a very high resistivity case. The spark may either be seen direct, or by reflection, and should consist of a bluish-white flash at different areas of the field. If one flash appears much brighter and consists of a reddish tone, it generally signifies a power arc. Even in a highly resistive situation where sparks are observed for several minutes, it may appear that while these breakdowns occur in the upper, middle, and lower zones of the field, some sparks tend to return to the same point, time and time again. The reason for this may lie in local discontinuities, but if the number of flashes at the repetitive points are less than 50% of the total observed, then this can still be considered a normal resistivity situation. The type and random nature of the spark-over, coupled with the meter readings, should add confidence to your analysis.

Oscilloscope Observations

Evidence of spark-over can be observed on an oscilloscope by a signal input either from the primary or secondary circuits. Transients caused by the spark-over can be detected by either the volt-meter circuits, primary current transformer or the ground return circuit through the D.C. current meter. Usually the D.C. meter will have sufficient resistance to provide a good trace, but if not, a low value of resistance can be placed in the circuit.

The scope will show the spike of the spark, in a well designed system, to rise about 30 to 40% above the envelope of the normal peak of the precipitator current waveform. Figure 4—2 shows a spark rate of about 16 sparks/minute, with 13 peaks counted over a 50 second period. Figure 4—3 shows a pick-up in breakdowns to about 80 sparks/minute. Figure 4—4 shows how the spark transient effects the voltage and current waveforms.

50 second span —
Low Spark Rate at 280 volt primary
and 200 ma secondary

Figure 4—2

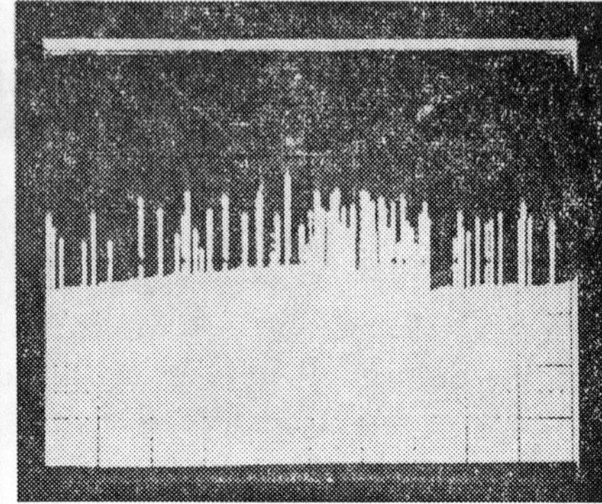

50 second span —
High Spark Rate at 300 volt primary
and 300 ma secondary

Figure 4—3

Oscillograms are shown of Precipitator Corona Current Envelopes at Two Different Spark Rates.

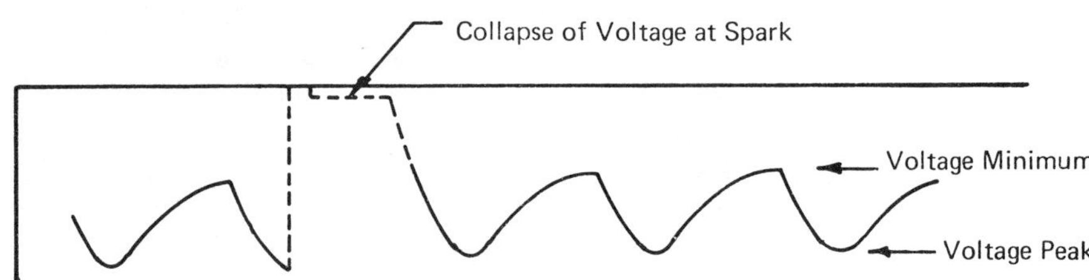

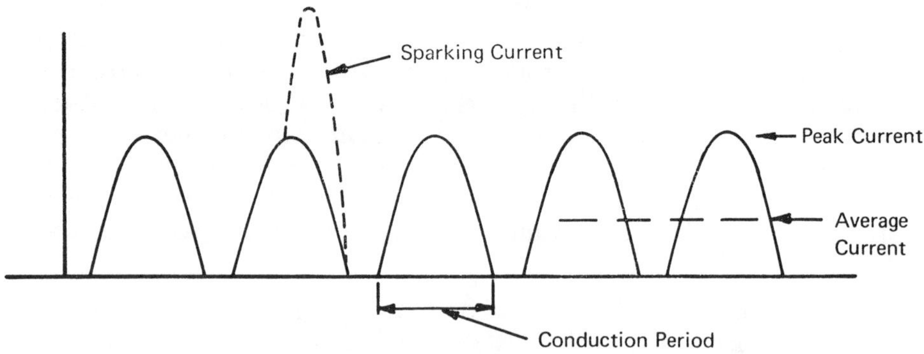

Figure 4—4 Waveforms of Precipitator Voltage and Corona Current Showing Effect of Spark-over.

76

Chapter 5
Operation and Maintenance

Introduction

This chapter intends to provide many practical items and advice for personnel who continually work with the precipitator. I will not stress all the differences between designs, nor will I stress how to correct all of the potential problems that can occur. All precipitators have common trouble areas, yet, all precipitator designs will perform well given certain operating conditions. However, it is the recurring trouble areas that nullify good performance and often cause costly production losses because of the outages required to correct these difficulties.

Many users have never realized how sensitive the precipitator can become to relatively minor changes in the process. They have never realized that the existence of thousands of component parts exposed to flue gas conditions is like being with a tiger in a cage — it is dangerous. This does not have to be. If the user attacks this collector like other complex equipment, with proper training and personnel assignment, and with proper records of pertinent operating and maintenance parameters, then these situations can be controlled. I made a strong point in Chapter 3 in stating that a good maintenance program can determine the success of a precipitator installation. The assignment of at least one knowledgeable full-time maintenance person on large installations is emphasized again. Continuity of inspections and monitoring of the variables by someone familiar with the concepts outlined in this book will often be the difference between a satisfactory or poor performance of the unit.

A good operating and maintenance group can often overcome equipment deficiencies if problems are recognized in time, evaluated properly, and if corrective measures are implemented on a routine basis. I have observed major expenditures required to overcome problems that would have been much less costly if detected and corrected earlier.

Safety Considerations

Before I get into some of the ways to evaluate maintenance activity, I wish to stress the necessity of having a healthy respect for the dangers of the precipitation system. Personnel involved with this equipment must be well trained in all the safety procedures, and high-voltage safeguards should be continually evaluated and upgraded to minimize the occurrence of accidents. There is no substitute for at least one well-trained person assigned to the precipitator for trouble shooting and to conduct internal inspections. Here are some of the key points in an effective program:

1. Double-check (prior to the initial start-up) that all controls are connected to the correct power supplies, and that identification is clearly marked. I always suggest that the identification system be well coordinated and familiar to personnel in the plant. For example, if everyone is used to talking in terms of east and west, then utilize these terms in the labeling process.

2. Make sure all key interlocks are initially lubricated with a light touch of graphite powder, that the keys work easily, and that the covers provide a tight dust seal. All locks should be activated on a periodic basis. When a key breaks or a lock fails, it sets the stage for having to by-pass this safety device even for a short period, a potentially dangerous situation. The fact that management places great emphasis on the care of this system helps assure the right attitude by all other personnel.

3. Although the key interlock has been a primary tool for safety, the use of grounding sticks or cables are mandatory. It is the practice of many plants to provide further safety control by disconnecting the high-voltage leads either at the T-R set or at the feed-through insulator. See Figure 5—1. It is a good idea

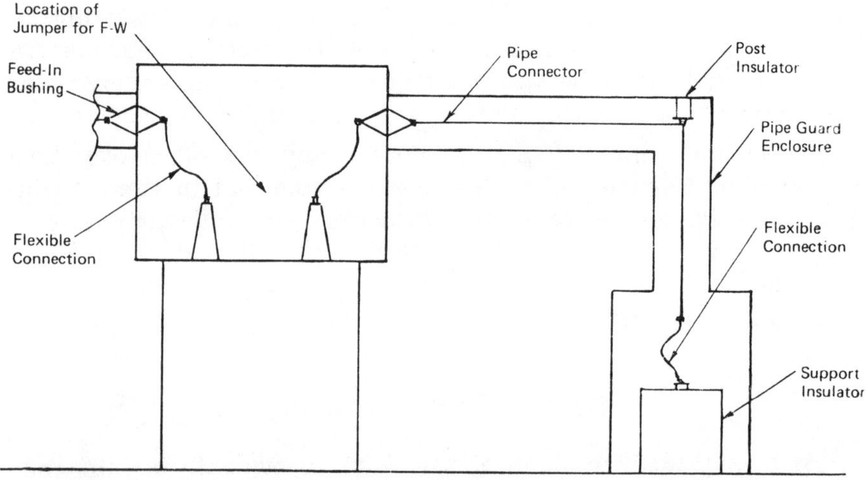

Figure 5—1 Layout of High Voltage System from T-R Set to Precipitator

to clamp the grounding cable to the high voltage structure for a specific field, rather than to rely on a hook device. The first contact of the high voltage components is usually made with a ground cable or chain connected to the end of at least a 3 foot long dry wooden or insulator probe. This type of grounding technique should be frequently checked for the continuity of the flexible metal connector to a solid ground; the usual method is through a bolt to the shell of the precipitator. If isolation and ground switches are utilized, either internally in the T-R tank or externally located between the power supply and support insulators, they should be periodically checked for positive contact to ground.

4. I highly recommend some positive method to determine when personnel are either inspecting or doing work within the precipitator. While this is less important with the smaller units, the large modern sizes provide many areas for someone to get hurt and have a difficult time communicating to the outside. A control room log and some identification at the access door are effective.

5. An area that must be carefully handled is the opening up of hopper doors, especially when personnel are not on solid footing. The concern should be for any excessive build-up that may be lodged above the door area. One method to ascertain any possible danger is to slightly open the door, and probe the area above the door with a 5 to 6 foot long wooden stick. Knowledge that the hopper was full prior to the shut down indicates a need for caution. The lower portion may have cleared, but it is still possible to have mounds of material poised in the upper portion of the hopper. I suggest that when hopper doors are left open, all hoppers be quickly checked for potential build-ups that may dislodge during some structure vibration, and possibly strike an unsuspecting person walking through the area. There should be special concern for this condition if either hot material or pyroforic material exists.

6. A suggestion which applies for access doors and hoppers is to provide sufficient hand-holds and step areas for personnel to safely get into or out of the openings. While these aids must be checked periodically to make sure the welds are secure, a well-placed step or hand-hold can prevent severe body twists or back-aches. There is always some concern that any obstruction in the hopper will cause build-up of materials to occur. This concern should be minimal with proper design. See Figure 5—2 for some suggestions. I also believe that special planks or ladders should

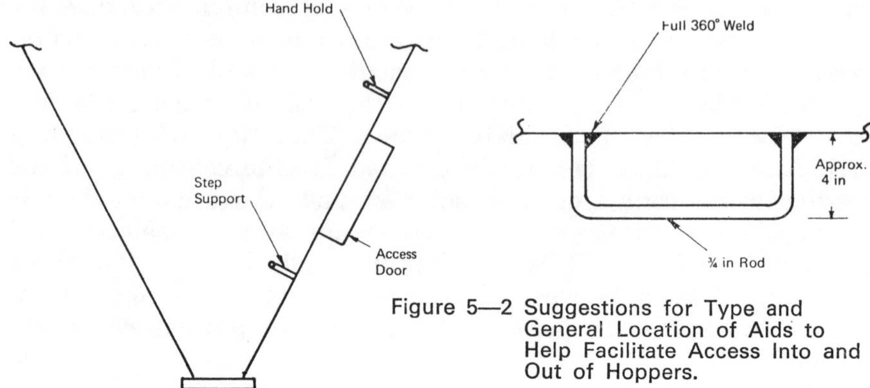

Figure 5—2 Suggestions for Type and General Location of Aids to Help Facilitate Access Into and Out of Hoppers.

be stored and only available for hopper work, to help assure safe and easy access to the components that can only be reached through the hopper zone.

7. Another suggestion, that is less important for the well-keyed installation, involves listening for the sound of corona before even using the ground stick. The corona sounds like a hive of active bees and is generally quite noticeable as an access door is opened. Secondary detection sense would be the strong pungent smell of ozone. There is no substitute for adherence to strict safety lock-out rules, but all aspects of this high voltage system should be known.

8. Make sure that all personnel are out of the flue areas and stack downstream of the precipitator before an electrical air load is obtained during a shutdown. This is also true of the immediate duct work upstream of the collector. Levels of ozone during these test periods can be appreciable.

9. Safety signs should be placed at every entry point of the system where high voltage can be contacted. Make sure that access doors in duct work that can allow personnel to come in contact with the precipitator proper is adequately protected. This is especially true of a duct at the outlet of a precipitator that does not have gas distribution plates blocking the electrodes.

Internal Inspection

The internal inspection of the precipitator is an important operating and maintenance tool that can provide many benefits if done thoroughly. This is where a person who is trained to do the job on a periodic basis can build-up a knowledge of observations and data that will help define causes of difficulties and even catch areas of impending trouble. The internal condition of the small precipitator with only one field may require this careful inspection more than the multi-field unit. Internal defects in any one field of the bigger collectors can often be detected electrically during operation, while it might be difficult in a small unit to ascertain whether a change in the voltage-current characteristics is due to the process or to an internal precipitator source.

Here are key inspection items:

1. A thorough inspection of the component parts and structure should be made during and at the conclusion of construction. It is especially important that a record be obtained of all distances between the internal electrode systems and the main support beams, for bench mark measurements in a clean, pre-operating state. See Figures 5—3 and 5—4 for some suggestions. The key effort should be to determine the plumbness or squareness of the two electrode systems to one another.

2. During the pre-operation inspection, emphasis should also be placed on locating any welded connections between electrode

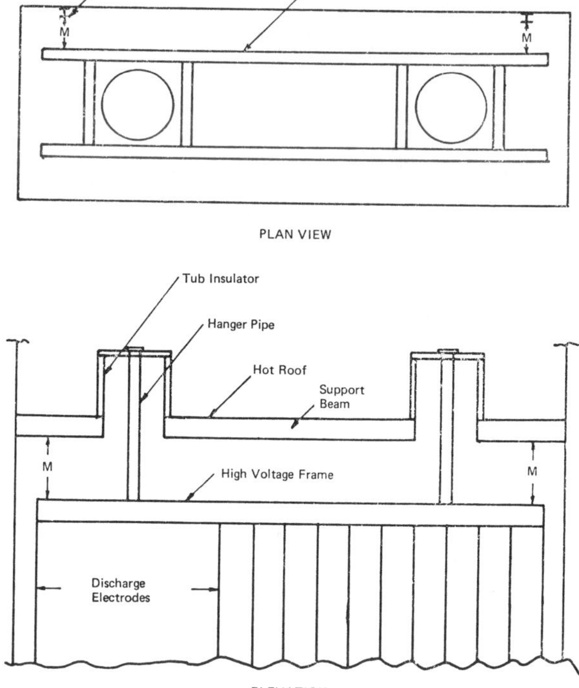

Figure 5—3 Alignment Check Points of High Voltage Frame as Noted by Letter M.

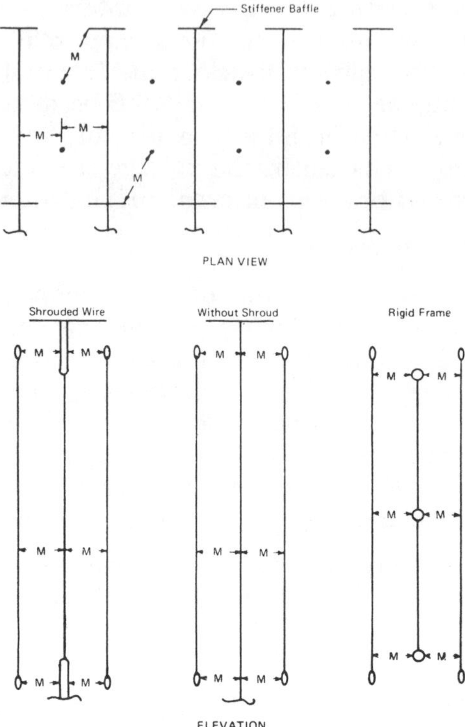

Figure 5—4 Points of Alignment Measurement as Designated by the Letter M.

systems and the shell, or any areas of construction damage on electrodes. For example, a splash of weld even on a rigid frame pipe mast may cause localized spark activity during operation. Even though the manufacturer's construction and service engineers make last minute inspections to catch these items, it is often difficult to detect all the imperfections during even one or two inspections. I have often crawled through some fields three or four times and found different things wrong each time. Note all items of concern even though corrections are not made. There are usually compromises on questionable defects, but if you have a record of where the imperfections are, this information may provide the answer to some localized break-down that may occur later. All imperfections will not cause difficulty in operation, and some judgment between the manufacturer and user must be reached after the options are discussed. There can be no compromise on integrity of alignment throughout the precipitator. It is especially important that expansion provisions for growth in the electrode systems be carefully checked.

The downward growth in most modern electrode systems may be 1½″ to 2″, but the relative growth of outer shell will not be the same. While it is possible to tie the lower portion of the collector plates to the side wall, sufficient slot clearance must be present between the collector and the tie-bar for the difference in expansion growth. It is especially important that a slight freedom exist between the individual collector plates and the common tie-bar used to maintain design clearances. I like to see at least ⅛″ side motion for each plate. Each manufacturer, or specific design, varies its techniques for this connection, but it does open up areas of concern for its translation in the field.

3. It is the first internal inspection after placing the precipitator in service that looms very important for future characteristics. If major structure defects are going to show up, the first several months will be ample time for this to occur. Growth problems or tendencies to shift high voltage frames after normal operating temperatures are reached should show up in this inspection. The first inspection should take special note of the following conditions:

a. Record any unusual build-ups of material and their specific locations. For example, note if the upper 3 feet of the high voltage electrode system is coated substantially more than the bottom 3 feet; or notice how the thickness of build-ups measures from side to side in the precipitator field. Within each gas passage, do you detect a thicker build-up on one side of the collector surface compared to the opposite side? These observations can help in ascertaining gas distribution or alignment problems. Recheck all the dimensions obtained during the pre-operation measurements. Record the thickness of material on electrode surfaces from inlet to outlet.

b. Pay special attention to any clean spots on the discharge electrodes, especially opposite any protrusion or termination of the upper and lower collector where the discharge electrode passes. This is true, even if shrouds exist on the wire electrodes. Look for any bowing or warpage of the collector surfaces or the rigid frames of the high voltage system. The key point of this careful inspection is to develop a base line for how the initial operation period effected dust patterns and the structural integrity of the precipitator.

c. Process information and precipitator readings should have been taken at least on a daily basis from the initial start-up to the first inspection. This information should be made part of the inspection record. Special inspection emphasis should be made in the fields where electrical characteristics appeared abnormal during operation. (See the later sections on electrical indicators and trouble-shooting.) This information can provide clues for the analysis of future difficulties, either by inside or outside company personnel. It is unfortunate when this type of history is not available to help explain changes in performance that may occur in the future.

4. The existence of heavy build-ups on the collecting surfaces should be of some concern during this first inspection. If the build-up is confined to one field or isolated area, that specific rapper must be suspect. If there is a general build-up throughout the precipitator, either the intensity level of the rapping system or the process characteristics must be suspect. If sparking existed before the shutdown of the fields, certainly the cohesive nature of the resistive material must be suspect. Of concern would be the effectiveness of the shut-down procedures employed.

The existence of heavy build-ups on the collecting surfaces which have been allowed to cool, build-ups of, say, greater than $\frac{1}{2}''$ thickness present some danger to the next operating period. The extra weight of hardened material on the collectors can further dampen the rapper shock or tremor. Therefore, a chain reaction caused by this extra weight can allow more build-up to form during the next operating period and further nullify the rapper system.

5. It is quite difficult to specify detailed guidelines as to what is a normal or abnormal dust pattern observed during an inspection. If the stack appearance and collection efficiency is satisfactory, I never argue with success. Unfortunately, success of one operating period may not coincide with performance success during a later period, and that is why a comparison of inspection results should be maintained. With some installations located in geographical zones that will experience ambient temperature variations of 80°F to 90°F between seasons, significant changes in build-up may occur. This condition can be observed more along the wall zones of the collector.

Alignment of Electrode Systems

The net effect on the electrical readings by the characteristics of the gas and particulate matter will often be contingent on the proper spacing or alignment between the electrodes. Meter readings may indicate a resistivity problem, while close spacing or even a specific electrode design may be causing the spark-sensitive precipitator. The comparison of similar electrical sections in a multi-unit precipitator is useful to identify internal defects. There is no substitute for careful measurements and inspections.

The degree of misalignment allowable in a precipitator is dependent on a number of factors. In highly resistive conditions, deviations of $\frac{1}{4}''$ in some discharge electrodes from the center of a passage could provide sensitive electrical breakdowns at slightly reduced voltages. Other installations can stand electrode misalignment of 1'' and more, and the effect on performance cannot be easily observed. Nevertheless, significant misalignment between the high voltage and collecting electrodes, whether with weighted wire or rigid frame, should be cause for alarm. This section will attempt to provide some practical guidelines in the detection and correction of these problems.

1. Some bowing of the electrode system can occur during the storage, handling and erection stages of the precipitator. If at all possible, any gross bends should be detected before the structure is secured. I would not be overly concerned with very gradual surface bends of $\frac{1}{4}''$ to $\frac{3}{8}''$ off center for the 30 foot high collector plate. It is often observed that the collector surface has a slight waving in the metal from top to bottom that might occur in the rolling-fabrication process. Again, this is usually acceptable with some designs. The key point is that clearances between the discharge electrode and each of the two collector surfaces of the gas passage be as uniformly centered as practical.

During the pre-operation inspection, I generally will accept all discharge wires if they are within $\frac{1}{4}''$ of center at the top and bottom terminations of the collector plates. Within the middle zones of the gas passage, deviations from center can be as much as $\frac{1}{2}''$ within a 9'' wide passage. These dimensions are also good to hold for wider gas passages. Every wire should be measured during the pre-operation period, if at all possible, even though this can take several days on the larger installations.

The above dimensions should be more restrictive through the center zone (that is, the zone of the passage located between 3' of the top and bottom terminations of the collector plate) if any cross bar or stiffener exists on the collector surface in a horizontal direction.

2. During the first inspection that may occur after about 2 to 3 months of full load operation, a repeat of these dimensions should be made. There should be minimal measurement differences from the prior set — especially at the top and bottom terminations. Be alert for horizontal and vertical bowing that may be either localized or widespread. The last collector surfaces along the outside precipitator walls are sometimes subject to deflection caused by a temperature gradient. See Figure 5—5 for examples of these types of defects.

If the observed areas of bowing coincides with fields that contained electrical disturbances much different than adjacent fields, then the manufacturer must be alerted. A slight bowing that still keeps the distance between discharge wire and collector at a 4″ clearance or more may still be considered satisfactory during this post-operation inspection. Make sure the alignment measurements take any material build-up into consideration.

A slight bowing of the collector surface may be corrected by applying heat to critical zones of the plate. Sometimes a small patch of heat at a stiffener rib will tend to draw the plate back

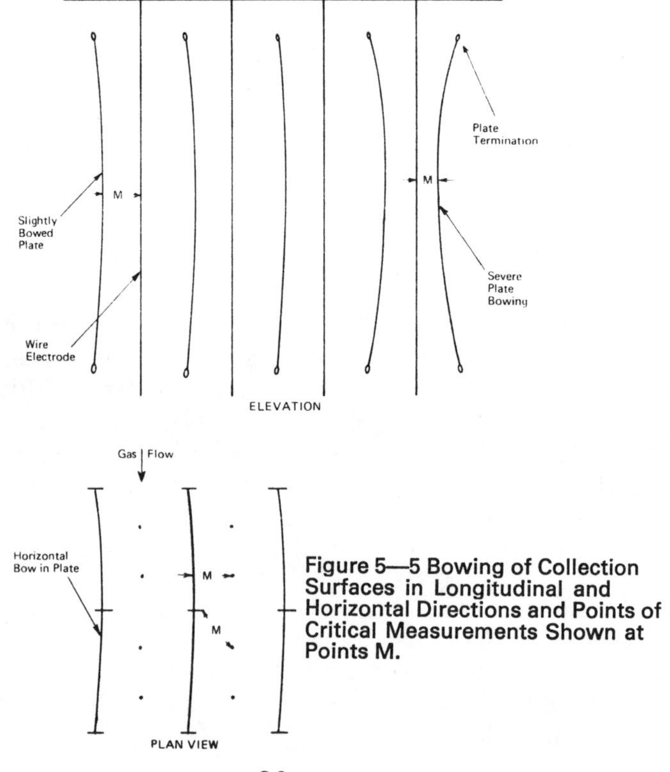

Figure 5—5 Bowing of Collection Surfaces in Longitudinal and Horizontal Directions and Points of Critical Measurements Shown at Points M.

into shape. When the vertical bowing is severe, some notching of the collector stiffeners may be required to help bring the plate back. This will usually require patching and is generally not desirable. Most of the manufacturers will have service or construction personnel trained in these field techniques. When vertical bowing does take place, there is a tendency for the plates to bow uniformly for many consecutive gas passages. If the bowing is under ½″ at the maximum deflection, it may be possible to shift the high voltage frame slightly in order to better center the discharge electrodes within the gas passage.

Some vertical bowing can be corrected if it occurs at the front and rear panels of the gas passage by the use of tie-bars that can maintain desired distances. These spacer bars, in the shape of flat bar or light gauge angles, should be loosely connected to each plate and then tied off to the wall in a manner to allow expansion. See Figure 5—6 for some suggestions.

Horizontal distortion and bowing at the bottom termination of the collector plates can be corrected by spacer bars at the points of maximum deflection, much like the tie-bars shown in

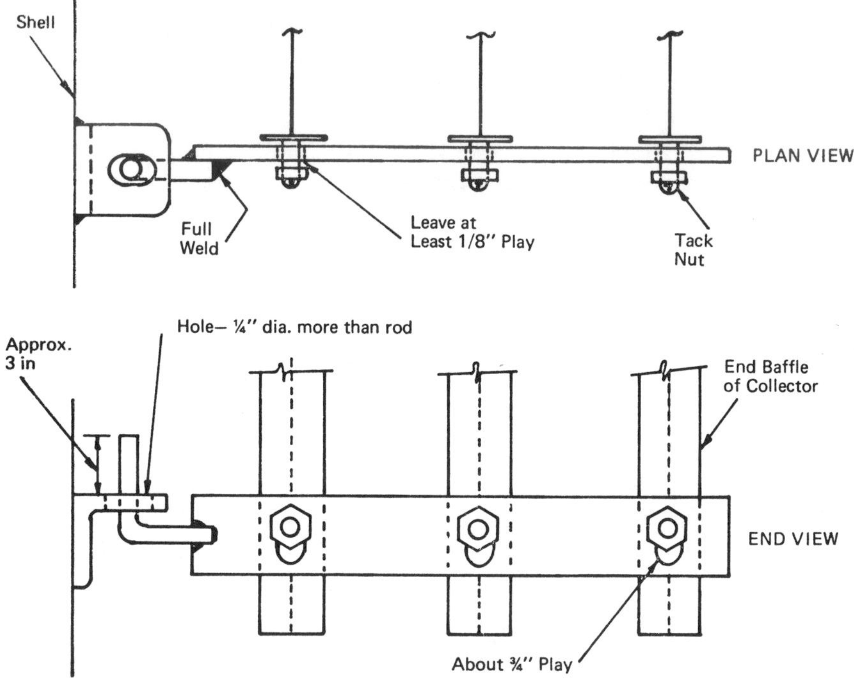

Figure 5—6 Suggestion for Spacer Bar Tie-Off to Shell of Precipitator

87

Figure 5—6. The rigid frame design would pose no problems with this spacer while the weighted wire installation may require the removal of one row of discharge electrodes so as to maintain electrical clearances.

3. A major cause of misalignment can occur from the build-up of hopper material when the lower internal structure is breached. Since hopper difficulties especially occur during initial operation, provide extra monitor services during these periods. The material can distort or bow the lower portions of the collector surfaces by initiating a heat-sink condition. The pressure of dust build-up can shift both the high voltage frames and collector surfaces even after the specific field is deenergized. Even the bottle weights can be lifted somewhat by this upward force and cause slack to occur in the wire electrodes. I have observed dust build-up as high as 2′ to 3′ within the gas passage. Anti-sway insulators, if present, can be shifted by a dust load which will tend to maintain the misalignment, even after the hopper build-up is removed during operation.

4. Misalignment can occur when support insulators are replaced for the high voltage frames. The height of the precipitator internal structure will determine how a slight error in the support adjustment will effect the clearances in the lower zone. The error in adjustment at one support insulator by $\frac{1}{16}''$ to $\frac{1}{8}''$ can mean reduced bottom clearances of $\frac{1}{2}''$ to $1''$ in 30′ high installations. I always suggest an internal alignment check after an insulator replacement. This is especially important with the stiff electrode designs that provide frame stability.

5. A subtle form of misalignment can occur during the very act of checking for alignment in the first place. Depending on the design, it becomes important to make sure electrode wires are not bent while personnel are checking the top structure. Another rigid frame design was observed to have misalignment between the two sets of electrodes because the spacer rods for the frames provided an easy step pad during inspections.

6. One of the main causes for misalignment occurs with substantial temperature gradients in the electrode structure. While this could occur during start-ups and shut-downs, the most frequent cause is an upset in the process or secondary combustion. In excessive heat cases, the gas distribution pattern will usually determine the area of greatest deformation. If high temperatures occur in the upper strata, the support frame for the discharge electrodes could droop, causing critical dimensions between wires

and plates. A fire-ball will normally attack the zones of mal-gas distribution in most precipitators. Any minor explosion will cause some damage to the collector surfaces depending on the location and the speed of pressure build-up. A major upset in the process should be noted and an internal inspection made at the first opportunity. Of course, the electrical readings would be prime indicators of serious internal damage after a high temperature explosion.

A question always arises if distortion could occur during the normal temperatures observed during operation, and yet show no deformation during the cold inspection. Although this condition occurs in rare situations, primarily attuned to the hot precipitators, I have never observed the effects of this condition in practice. Universally, when failures or damage to electrodes was noted during an internal inspection, the reason for these occurrences was also seen. Simply put, if a necked-down wire was detected, a misalignment or collector discontinuity was present.

7. Misalignment problems occurs frequently at the bottom zone of the precipitator where the pendulum effect will be more pronounced. It is for this reason that internal observations for misalignment at the bottom structure assumes more importance. This is even more important if a greater height of collector surface, a smaller length of field, or a higher number of gas passages exist for the two point suspension of the high voltage frame.

8. Some wire electrode designs use a rigid pipe that connects the upper high voltage frame to the bottle weight guide structure for maintenance of clearances at the bottom of the field. These pipes are either 1¼″ in diameter or formed in oblong shapes, and the large radius contour of the surface will generally allow clearances of 1½″ or less to the collector surface without electrical break-down. However, once spark-over occurs and causes irregularities in the pipe surface, then electrical break-downs become more frequent at reduced voltages until sizable holes in the pipe wall can occur. The main intent of these comments is to alert you to another cause of misalignment, the bowing of the rigid pipe due to temperature growth problems, or poor adjustment for plumbness at the top insulator support.

9. Along with the anti-sway stiff pipe design, the use of insulators mounted on the hopper walls to help maintain high voltage frame alignment opens up another area of concern. Unequal temperature expansions can cause difficulty if the frame

bridges two hoppers. Proper design for the growth in the insulator hardware is an important feature to check. The frame should also be allowed free growth downward without coming to rest on the insulator.

10. Another method for holding the bottom guide frame free of the weights has utilized wires fixed to both the upper and lower frames. Usually four (4) wires are used, which have produced slack on occasion, since it is quite difficult to maintain equal tautness by this method. As the expansion occurs due to operating heat, three of the four wires might take the weight of the lower frame, allowing the fourth wire to bow toward the collector surface. I have observed several failures with this type of arrangement. One remedy is to operate with a three wire support which minimizes the chances of slack in the odd wire. As a last resort, if the design of the bottles and guide rings are compatible, all the wires can be removed to allow the lower frame to rest on the weights.

Electrode Failures

Much controversy has surrounded the area of wire discharge electrode failures in the so-called American designed precipitators. I would like to clear up some of this background, and relate where and how most of the electrode surface failures occur, and what can be done about it. The 100% absence of electrode losses is too much to ask for, but routine internal failures is an abnormal condition and the reasons must be analyzed and corrected and not allowed to cause reduced performance and costly outages. I have observed precipitators that have not had any electrode failures for a period of years. Others have lost wires on a weekly basis. There are reasons for both situations. The following comments should bring this whole area of operation and maintenance into workable perspective.

1. Wire electrodes, and, for that matter, any corona emitting portion of a rigid frame, are not subject to repetitive replacement on a routine basis. During the initial operation period of a new installation, you might expect to see several failures based on handling or fabrication defects. After the initial shake-down period, high voltage discharge components should last for years.

2. The bulk of failures that have occurred after the initial operation period, and which account for about 90% of the total

failures within the last 10 years, is due to electrical erosion. This has occurred primarily at the top and bottom terminations of the collector plates when shrouds were not present on the wire electrodes. See Figure 5—7 for notes on valid shroud design. Whether this type of electrical erosion becomes a problem for an unshrouded wire depends on the integrity of alignment, the shape of the plate termination, or the resistivity of material and gas conditions. I have also seen an installation go from minimal electrode failures to several per week for unshrouded wires when voltage levels were raised by doubling the number of power supplies during a retrofit.

Another condition for the electrical erosion of a wire electrode to occur within the gas passage is when the tautness of wire is less than optimum. This normally allows excessive wire movement to occur during a spark, which then further elevates this motion into a hanging power arc between wire and plate or wire to a vertical stiffener fin. This type of arc will tend to damage the wire surface so that subsequent electrical break-downs will become easier.

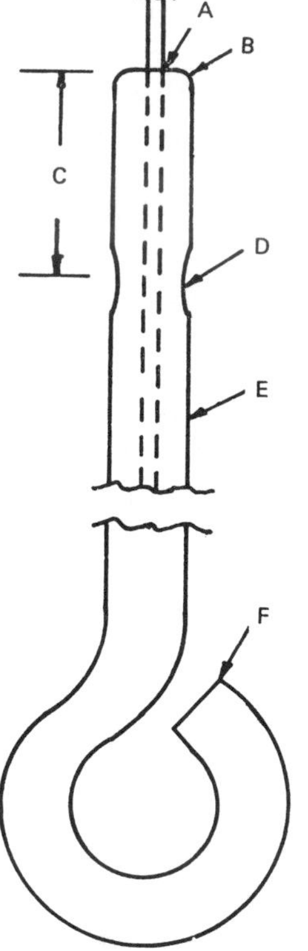

Note A— Contour inside edge of hole 1/16″ R.

Note B— Contour outside edge of shroud with 1/8″ R — with no tool marks.

Note C— Do not locate crimps opposite any overlap edge or protrusion on plate.

Note D— No sharp edge or ridge on crimp. Locate in direction of gas flow in high resistivity ESPs if wire cannot rotate.

Note E— Minimum shroud diameter is 3/8″ — although 1/2 to 5/8″ if preferable.

Note F— Contour not needed if end is outside of collector surface area.

Figure 5—7 Suggestions for Design of Metal Shroud Used on Wire Discharge Electrodes.

Any discontinuity on the collector surface that is positioned opposite a discharge electrode may draw a spark. Since each spark does emit some metal ions from the discharge electrode, a continuous or repetitive spark from the same point will eventually thin down the cross-sectional area until a necking-down occurs not much different than what occurs in the heating and drawing of glass tubing. These discontinuities might consist of a flat bar stiffener, an overlap of metal at the upper and lower terminations, or even clips used to tie panel sections of the collector surface together. How long a wire will last before this condition produces a failure statistic is based on the ballast impedance in the power supply as well as the capacitance of the field. The smaller size of the power supply and field will be advantageous toward reducing the severity of the wire damage. A quick acting automatic voltage control can also reduce the severity of damage. But the key is to locate these points of localized breakdown and either correct or remove the wire. That is one basic advantage of the weighted wire design — a certain number of wires can be completely removed from the gas passage without any measurable effect on performance. I strongly suggest not replacing wires that fail without investigation; the cause may still exist and begin the process of wire damage all over again. I have seen records of 4 to 8 wires replaced in the same slot before I had a chance to stop the practice. It is better to operate with 5% or more of the wires removed than to have those wires control the level of power input. Another point is not to have more than 10% of the wires removed in any one gas passage throughout the length of the precipitator.

Detection of electrode damage is usually first noted visually by differences in the cleanliness of the area compared to adjacent zones. Under closer observation, some pitting or irregularity can be seen. The use of your fingers to feel the contour of the electrode will usually disclose some change in thickness. A wire brush is often helpful to clean off surface scale that can give an erroneous conclusion. I have used micrometers on occasion to obtain a finite measurement of the erosion area.

Localized damage of the discharge electrode can also occur by an electrical path through material bridging the dielectric space as may exist during a hopper build-up. This usually happens when the protective circuitry does not contain an under-voltage relay used for the shut-down of the power supply, when the primary voltage decays below 125 to 150 volts for about a 10 second period. The actual trip-out level should depend on the average operating voltage. If the circuit is not deenergized and the current path is maintained for some time, the material will generally clinker and often fuse between the high voltage and collector components. Aside from the localized damage to the wire, the clinker will usually stay attached even after the hopper material has been removed.

3. The mechanical fracture or fatigue of the high voltage electrode probably accounts for another 3 to 5% of failures. This condition usually occurs in specific locations of the electrode, normally related to bends, at the interface of wire and shroud or the frame termination, or at an excessive vibration node.

Type of wire material will be important but probably less critical than its location to the rapper itself. If a shock rap is used, a break may occur at a point of electrode termination nearest the blow, such as might exist in a rigid frame. A 60 cycle electro-vibrator or air vibrator, normally used on the weighted-wire designs, might set excessive vibration on certain wires, relative to the rapper shaft, dependent on the harmonic nodes through the high voltage frame. In both cases, care should be taken to operate at minimal intensities. In the weighted-wire design, length of vibration should be no more than four (4) seconds in duration. As with the electrical erosion problem, I would not replace the electrode.

The mechanical failure in the discharge electrode will normally give a full cross-section break with the ends of the broken component showing evidence of a crystalline structure.

Other types of mechanical failures may occur through some rubbing action. This type of electrode erosion damage may take longer time periods to surface. If poor design is the cause, it might be widespread within a specific installation.

4. The third cause of discharge electrode damage would be corrosion oriented and encompass a variety of attacks. This type of damage is usually more confined to certain areas of the precipitator and often is characterized by extensive surface irregularities. I shall cover more on corrosion effects in a later chapter, but careful adherence to gas temperature control and proper material of design should make this problem of less concern with the discharge electrode. Corrosion will be much more critical for the collector surface because of the large surface area and the light gauge metal involved.

5. Collector surface difficulties with electrical erosion and other forms of damage has occurred with much less frequency

than with the discharge electrode. Often trouble with the one system will effect the other as in the case of localized arcing. Another danger lies in the excessive rapping of the collector surfaces, especially if coupled with difficult to remove build-ups that add substantial weight to the hanger devices. There have been occasions where distortion in the support hangers has occurred with higher rapping than the system can take, and which produced a misalignment leading to subsequent electrode failure.

The purpose of pointing out the many areas where difficulties may occur is not to cause undue concern, but to make you aware of danger areas. These areas can be discovered by careful internal inspections and record keeping that will isolate the areas of repetitive trouble. As you read the following sections, I believe you can obtain further understanding to help identify these areas. A precipitator, like many other pieces of equipment, will only perform with minimal outages and maintenance costs when the mystery clears about the causes and effects of breakdowns. While the power supply can be considered the brain and heart, the internal components are the body of the precipitator and this body must be kept healthy to allow the heart to do its job. I cannot cover all the possible difficulties that may occur nor all the various solutions, but it is important to know what you have initially and not allow troubles to mushroom into major problems.

Particulate Removal from Hoppers

Difficulties with the evacuation of material from hoppers can provide a continuous headache for plant personnel. The large-sized precipitator can have a good potential performance nullified by hopper troubles. This does not have to occur. Unfortunately, in the quest to produce high-performance units in order to meet stringent regulations, the emphasis on good hopper design was less than the concerns with SCA or power input. Hopper difficulties are often fought for years because key modifications are not applied, especially to correct the underlying causes. I believe this lack of commitment to eliminate hopper troubles has produced a sad reputation for precipitators. The problems are many, the solutions often hard to come by, but do not hesitate to expend time and money to obtain a satisfactory performance of the hopper system.

Process characteristics, the type of material being handled, and moisture and gas temperature levels are important inputs in understanding the reasons for hopper problems. Without a reliable evacuation system from the hopper flange to the storage area, the burden on the hopper components may be too much. However, there are a variety of other inputs that are continually at work to threaten successful material evacuation. Based on many personal field experiences, here are the major factors.

Insufficient Heat and Insulation

Lack of proper insulation and heat input, especially at the bottom apex of the hopper has probably accounted for most of the troubles experienced. Regardless of the gas and dust characteristics, the ability to keep the wall surface temperature of the lower hopper no less than 250°F is most important. A combination of inputs are required:

1. Insulation material should be at least 5″ to 6″ thick if placed against the wall or 2″ to 3″ thick with at least a 4″ air space. There should not be any opening at the bottom insulation termination to allow a chimney effect to take place. The complete hopper should be insulated.

2. I like the use of a windbreak in addition to proper insulation. Elimination of the wind effect will add to the available heat around the hopper components.

3. Heaters should be installed at least at the bottom apex of each hopper — from the flange to 6′ high for the inlet, and at least the bottom 3′ for the outlet hoppers. Power density can range from 50 to 80 watts per sq ft for this area. All four walls should be heated. Tubular elements have the advantage of easy removal, but the strip or blanket type are also utilized. The object is to derate the surface heat density of the element so reliability is increased. This will require more heater components per area. If the strip type of heater is used, attempt to provide easy access for check-out and replacement. Thermostat control is desirable (it is usually set for a 250 to 300°F range at the outer wall surface of the actual hopper), but the heater design should be capable of continuous operation without failure due to high temperature. One danger to guard against is the build-up of dust around the tubular elements which can cause failure through excessive heating. Tracing with steam heat has been used successfully in some applications, but reliability factors are present. With the electric heater applications, I prefer to see some visual means to monitor heater operation, such as electric current meters on specific groups of heater elements. This is especially important with the large installations.

4. Integrity of the insulation cover and heater circuits should be confirmed periodically, and especially prior to every start-up of the process.

Design Factors and Internal Obstructions

Repetitive hopper problems can be based on design deficiencies, some easily corrected; others must be counteracted by greater emphasis on the heating and maintenance techniques. Success is the name of the game, and if no difficulties are experienced with installations that have components similar to those I relate, there is not need for concern. If you are having troubles, then modifications may be in order. Briefly:

1. Accepted design of a 55° or greater slope for the hopper wall is usually satisfactory. I have observed slopes, often on one or two walls of a pyramid hopper, that gave trouble because the slope was about 45°. Extra care must be given these slopes, possibly with a low intensity wall vibrator near the upper half of the hopper. In extreme cases, two hoppers are made out of the original large zone.

2. Insufficient opening at the outlet flange of the hopper has given troubles. I prefer a minimum 12″ diameter or 12″ square opening. In extreme cases, the openings are enlarged to minimize the bridging and packing problems at this location. Utilizing small width screw conveyors can produce bridging problems at the top entry of the screw housing.

3. The corners of the hopper can aggravate the retention and subsequent caking of wall build-ups. I prefer the gradual radius instead of the 90° corner in critical applications.

4. Any pipe or obstruction in the bottom two (2) feet of the pyramid hopper can reinforce the start of a dust build-up. The use of aeration stones in the lower portion of the hopper can add to dust build-up troubles rather than help.

5. One practice that has caused as much trouble as helped is the use of a steel grid in the lower portion of the hopper to prevent clinkers or bottle weights from blocking the apex opening. Some of these grids had openings too small or were placed too close to the hopper wall. No grid is recommended, but if it is still desired, the openings should be at least 4″ square with the outer frame placed at 4″ from the walls. Position of grid should be at least 4′ above the flange, but in no case should be above the access door.

6. Most hoppers contain center baffles that tends to provide two dust storage halves in the direction of gas flow. The purpose of the baffle is to minimize gas sneakage through the hopper. There is some doubt for this application, as I shall discuss in the gas distribution chapter. However, its use can provide difficulties with the free flowing of material, especially if this baffle extends to within 12″ to 18″ of the flange opening. This location effectively narrows the passage for the movement of material. This situation is especially important in the batch removal of material from hoppers where a packing condition can exist between baffle and wall. If the material on one side is free to flow, the other side of the baffle can continue to build-up with dust when proper level detectors do not exist. As a minimum, I suggest a 3′ distance at the apex, or at least 1′ above any grid. There is no magic dimension to consider except what is required to relieve hopper problems.

Air Inleakage into Hoppers

Entry of outside air into any part of the hopper system is considered poor practice. Aside from the effect on performance, excessive air can cool down the inside wall surfaces, or condense moisture in some of the high water-vapor installations. Unfortunately, most of the screw conveyor installations coupled with process precipitators are conducive to this condition. Here are some places to look for trouble:

1. With the high negative pressure systems, be alert to the sound of air inleakage anywhere on the hopper system. Special areas include flanges, slide gates, and inspection flaps.

2. Systems without positive seals at the transition between the hopper and conveyor should have some means to prevent air flow between the storage bin and conveyor. This can be obtained with a dust plug or a positive device at the bin entry itself. It will depend on the length of the conveyer system and the number of inspection ports whether a tight seal can be maintained. The degree of suction or negative pressure at the hopper will usually determine the adverse effects of this condition. Training of operators to the importance of air in-leakage cannot be stressed too highly.

3. Modern hopper systems are utilizing aeration stones to a greater degree, and I have some concerns over their use. The injection of any air into the hoppers can cause difficulties at

different times of operation, especially at start-ups. Maintenance of the seal between the stone and its support base is important.

4. Integrity of the positive shut-off valves and double-flap valves must be continuously checked for air-inleakage. Alignment of the valve gate to its seat, dust build-ups, and the erosion of the seat surface can produce a slow growth of air-inleakage. It is well to inspect and correct any defects during annual outages.

5. Air-inleakage at the access doors is a common problem. After an inspection, the re-seating and positive sealing of the door should have a high priority.

Hopper Wall Vibration

The use of vibrators or impact blows on the walls of hoppers can either help or hurt a specific situation as to the removal of material build-ups. I prefer some means to manually apply some shock or vibration to the walls. But, the automatic mode of vibration is satisfactory only for hoppers that are in a continuous mode of evacuation, or when there are safeguards to keep from operating the vibrators with dust build-ups of a couple of feet deep in the hoppers. This is especially true with fine-sized material, coupled with reduced openings at the apex of the hoppers. Some key points:

1. Two smaller vibrators are preferable on opposite walls of the hopper, than the use of one large vibrator. Intensity and duration of vibration should be initially set at minimal levels. The vibrators should be located slightly above the half-way mark of the hopper, as measured on the wall surface. Experience will further determine the best intensity and frequency operating levels of the vibrator.

2. If an automatic vibrator system is used, it should be keyed to a break in the vacuum of a batch system, or to the time in the process cycle when the hoppers have been cleared. As pointed out above, the use of an automatic system can pack an existing loose build-up into one that is difficult to dislodge. Once the dust build-up packs into a dense volume, manual means are usually required to get it free again.

3. Use of strike pads on all four sides of the hopper are usually recommended. These strike pads can protrude through the insulation to receive a manual blow with a small hammer or

sledge. See Figure 5—8. The operator will soon learn to tell by sound whether the hopper is free of material in the lower half. After a full hopper build-up when wall deposits still exist, the pads can be lightly struck before the use of the vibrators to help keep from overloading the apex zone with dust. The pads should be placed approximately 5' to 6' feet above the apex flange and safe access to the pads should exist for personnel.

Clean-out Ports

Most installations that experience hopper difficulties soon add a number of clean-out ports around the apex of the hopper to help facilitate the removal of material. There are several methods used, including 2" to 4" diameter pipes placed at various angles to the hopper walls. See Figure 5—9. I usually suggest that the clean-out pipes be placed downward to eliminate any chance for someone to extend a long pipe toward the high voltage frame. There is little chance of danger in clearing up dust build-ups from the hopper without deenergizing the associated field. If the build-up had already reached the high voltage frame, the

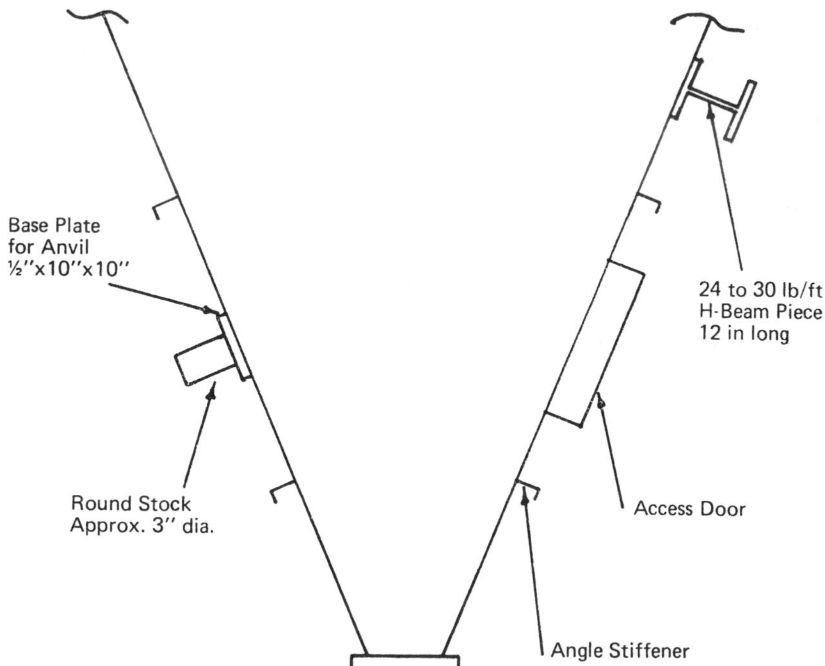

Figure 5—8 Two Methods that can be Used for Strike Pads for Manual Blow to Walls of Hopper.

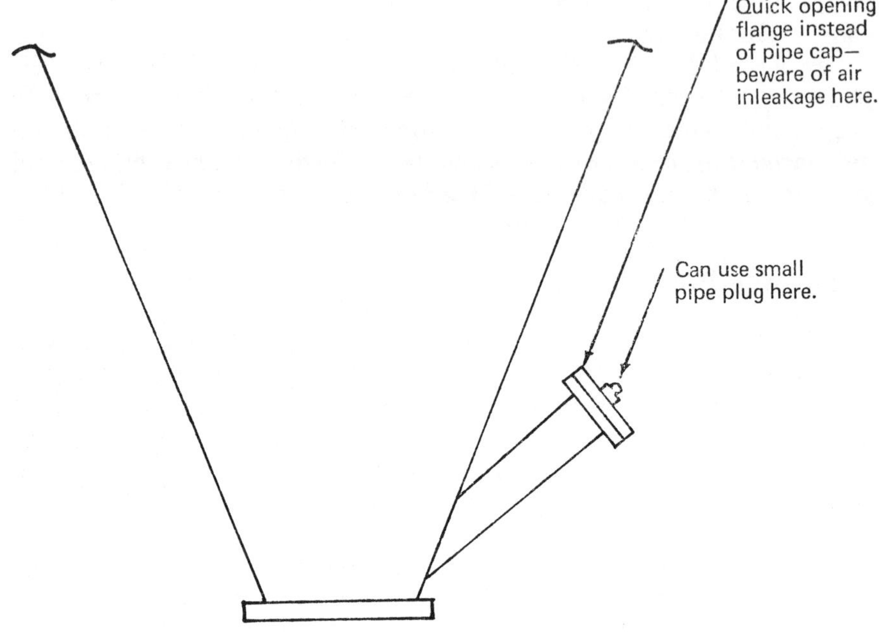

Figure 5–9 Suggestions for Clean-out Pipe
at Apex of Hopper.

power supply should be shut down and locked out until the hopper
is judged free of material. Deenergization will usually occur with
the under-voltage relay in the circuit.

The deenergization of the field, prior to the grounding condition, can add to the dust load in the hopper. I do suggest
shutting down the rapper system for a short period while the
hopper is being worked on to free the material. A steel pipe or
air lance entering the clean-out port should have a flexible ground
strap attached to eliminate any fears for the probing of the
lower part of the hopper while the field remains energized.

Another suggestion with the use of large clean-out ports is
to apply a quick-acting flange as a pipe cover in case a large
amount of hot material is dislodged, and forces itself out of the
opening. Rather than use the full port opening, it is sometimes
advantageous to provide a small adapter plug in the large cap or
flange cover to check for dust or static pressure conditions.

Other methods of freeing up hard dust build-up at the apex
sometimes utilize compressed air connections above the hopper
flange. The compressed air should be as dry as possible.

Level Detectors

While it is always better to place time and money in the prevention of hopper difficulties, the detection of build-ups by some method is desirable in most applications. These devices can utilize gamma radiation, sound, capacitance, pressure differential, temperature, or even paddle-wheel methodology for the detection of excessive build-up. Any method that does not require components within the hopper appears the most desirable. The radiation or sound reflection technique offers the best reliable methods to date. Several comments:

1. Use detectors as maintenance tools rather than to identify full hoppers. For example, if an automatic batch cycle allows a maximum 3′ of build-up in the inlet hoppers, I normally suggest locating the detectors no more than 5′ above the apex flange. The object is to alert the operator before a major hopper fill-up exists, yet minimize frequent detector alarms. A rule of thumb would allow the lapsed time from normal dust height to alarm level to be equal to the same length of time it takes the dust to rise from the apex to the normal height. Remember that the pyramid design allows for a greater volume of material to accumulate in each foot of hopper height. One problem arises in the uneven build-up that occurs by the slope and corner effect of the hoppers. The detector should recognize this effect and be placed a little lower at the center of the wall or a little higher near the corners.

The use of low placed detector levels will normally activate all the inlet hoppers, and even the outlet hoppers, during an extensive shut-down of the main dust evacuation system. There is nothing wrong with this situation, since the operators will be alerted as the system starts up to the clearing action of the hoppers and can therefore isolate the hoppers that are still in trouble.

2. In a continuously evacuated system where no appreciable build-up exists, the detector may be placed only 2′ to 3′ above the apex. If one considers that the weight of material will compress and increase troubles at the apex, the sooner an abnormal height is detected, the better. The amount of build-up that can exist before the problem of removal becomes critical is based on the particle size distribution of the dust and the maintained temperature at the apex. The outlet hoppers may be more critical than the inlet in many installations.

3. Where a center baffle exists in the hopper, it is generally wise to place a level detector on each side of the baffle unless a radiation type is used.

4. For very deep hoppers, where the design of the internal structure can receive damage from excessive material build-up, deenergization of the field could be initiated with a 4' long steel rod hanging down from the high voltage frame. These rods, generally ¾" to 1" in diameter, are placed so the end of the rod is about 2' from the corner of the hopper walls, and at least two rods per field are used. An undervoltage relay must be used with this technique. I prefer the rod not be tightly secured at the frame, but to be able to slide through a simple guide.

Evacuation System Effects

The best features of hopper design cannot overcome any weak spots in the evacuation system used to transport material from the flange to the final destination at the plant. This includes valves, vacuum or pressure sources, and all other moving parts of the system. Troubles can occur on either side of the hopper flange, but the net effect on the precipitator is the same. The recent trend to pressurized systems has increased the number of critical components, but this system appears successful if careful training and maintenance techniques are stressed. Over the years, a number of problems have been observed, and the main ones are discussed briefly for your information.

1. The high pneumatic vacuum system has been used for years, and has been quite successful. Problems have surfaced with the adjustment of the bellows at the termination of the piping system. Sometimes the pull of the vacuum at the apex throat will drain a "rat hole" or small opening through the center of the hopper build-up. This will result in a foundation for more dust to accumulate until an eventual shorting of the high voltage frame can occur at the corners of the hoppers. The "rat hole" condition mostly occurs from too much time between the clean-out cycles, especially in cold weather and inadequate heat at the hopper apex. One problem with the batch removal cycle is that heat dissipation can occur through the hopper flange to the steel piping of the evacuation system. This cools down the outer zone of the material which can cause condensation to form hardened material. If the complete hopper area is not enclosed for critical installations, I usually recommend an insulation enclosure or barrier, that may even contain heat, for the immediate area around the hopper flange.

2. Regardless of the system, the use of very dry compressed air for the gate valve operators cannot be overstressed.

3. A periodic inspection and adjustment for the seating of the gate valves is usually required. For example, fly ash can produce metal erosion of the gate and seat surfaces.

4. Minimize the amount of gate and seat erosion potential by keeping the time delay for holding the gate open as short as possible after the final break in vacuum.

5. Use the outages as much as possible to replace piping in the main transport system. Long repair periods during operating periods sets up the potential for hopper difficulties from excessive build-up of dust.

6. The original design should strive for parallel vacuum or pressure sources, as well as transport systems. The ability to transport dust to alternate depositories is important. The cost of an extra blower in place as a spare may be cheap insurance against costly outages.

7. The vent system of the pressurized system has provided some troubles. The problem with the vent pipe return to the precipitator hopper is the possibility of piping blockage during dust build-ups. When a vent from the pressure tank is blocked, the opening of the top valve during the cycle will probably result in a massive reentrainment of material from the hopper. The effect on precipitator performance by the vent blow-off is usually quite small compared to conditions of blocked vents. Insulation, heat, and design of the piping all have a part in successful venting. I prefer a vent return to a common header, which then connects to the inlet of the precipitator. The header should be well insulated and utilize heated intake air. If an exhaust fan is required, a replacement should be made on a periodic basis since blade erosion is normally great unless low speeds are used. If an orifice is used in the vent line, make sure it can be easily checked and cleaned. Clean-out ports and provisions to periodically blow out the vent lines are desirable.

If possible, use one line for both the pressurization and venting of the tank, which can minimize piping blockages even though two valve operators are still required. Use of elbows of not more than 45° will minimize blockage in the piping. Another important item in successful venting is the use of sufficient time delay between the venting and pressure cycle.

8. Air intake for the compressors or blowers should be obtained from a clean area. Too often, the air from the hopper area will include reentrained ground level dust.

9. Another problem with some pressurized systems has occurred when the bottom orifice is not set large enough to pass the lumps or small slabs of hard material which often arise during start-ups. The transport system can usually handle this type of material. At least a 4 sq in opening may be required in order to keep out of trouble, but some experimentation may be in order for each specific system.

10. Screw conveyors have been relatively reliable for the process industries. I prefer to see a low rpm operation with some means of speed control. Difficulties with bearings may require access doors, but positive closure should be designed. Quick acting flaps are nice, but become important contributors to air in-leakage problems. Insulation and even surface heat may be required in systems handling high moisture-laden material to help minimize caking from condensation. Dust caking at the bearing housing may accelerate wear by forcing material into the bearing itself. If the conveyor is used to dispense with floor sweepings, the opening should use a relatively fine mesh screen to prevent large foreign matter from fouling the screw. Probably the metal parts left over from outage work have stopped more conveyors on start-ups than any other problem.

Rapper Maintenance

Maintenance procedures for the care of rappers and vibrators can be very important for all precipitators. Reliability of the rapper system starts with proper design by the manufacturer, but plant personnel can often correct repetitive trouble spots. Remember that while the rapper system may not be the critical portion of many collectors, performance will suffer if rappers cannot be kept in operation on a routine basis.

The early generation of rapper devices provided a variety of difficulties. But the small-sized precipitators that existed then had relatively few rapper components compared to modern designs. For that reason, the modern rapper system should command an early priority for the discovery of weak spots. Proper identification of the individual rapper, and an easy procedure for check-out, is important. This task is more difficult with 200 plus rappers, and the time it takes to locate a defective rapper can nullify an effective program. A key point is not to rap excessively when it is not required for performance. I have seen very few jobs where excessive rapping allowed satisfactory performance to be attained. However, the degree of rapping should be sufficient to keep collector surfaces within satisfactory build-ups, usually less than ½″ thick during operation. During the operating period from the first start-up to the initial internal inspection, I normally like to see slightly higher rapping intensities than what may be finally required. Some of the critical areas include:

1. Compressed air used for pneumatic rappers should be dirt and moisture free. The normal moisture trap used in piping is usually not sufficient. High efficiency driers including an expansion chamber are good.

2. A clean ambient environment for the rapper control cabinets is especially important for components such as cam timers and armature relays. Static controls are used extensively now and can be sensitive to high ambient temperatures.

3. A periodic check and adjustment of the anvil gaps of electric vibrators are recommended on a yearly basis. Check the manufacturer recommendations.

4. Check for the binding of rapper shafts that extend through the precipitator shell. This can occur with a shift of the internal structure allowing the shaft to come in contact with the shell. This condition is sometimes apparent by observing the plumbness of the rapper. Another type of binding between the shaft and the precipitator shell occurred with the horizontal rapper application at the bottom structure of the collector plates. If insufficient clearance exists between the bottom of shaft and shell, the downward growth of the electrode system can allow the shaft to come to rest on the lower edge of the hole. At least 1½″ clearance is required on most jobs, and that opening must be sealed with a packing gland.

5. The packing gland, or boot enclosure, that provides a gas seal between the rapper shaft and the shell does tend to give problems at these locations. Penetrations of dust into these junctions can produce a hard mass, which will transmit part of the shock or vibration to the steel shell. Corrosion can initiate substantial damage to the shaft at this point. Periodic inspection, repacking, and boot replacement will minimize this problem. On pressurized installations, it may require some attempt with a purge air system to keep these glands free from hard encrustations.

6. Another loss of rapping energy could come from a poor connection between the rapper shaft and the frame being struck. There were some installations that provided for the rapper shaft

to fit into a pipe socket without any solid connection. In other cases, the integrity of the weld soon broke under severe rapping intensity. These internal junctions should be inspected periodically to detect whether the connection is intact. A loose junction will sometimes allow dust to build-up under the shaft and further soften the rapper blow. Usually, gusset plates are required between the shaft and frame to ensure a reliable connection.

7. Additional loss of intensity may occur at the clamps securing the insulator of the high voltage rapper system. Sometimes the bolts are double-nutted and tacked during construction, and looseness later develops during operation.

8. A breakage of bolts has occurred on many occasions because of excessive rapping intensity, usually coupled with oversized bolt holes. Make sure the bolts are sized correctly for a matched set of holes of a flange connection. Certainly, alloy steel bolts will help.

9. Failures of the springs or coils in the electro-magnetic type of impact rapper has caused troubles in the past. Some designs have utilized springs to return the steel plunger to its ready position within the coil enclosure. High rapping intensities can cause spring fatigue failures or the jamming of the spring. This problem has been minimized in recent years with better spring technology, but some problems still arise. If you experience this type of trouble, do not live with it, but work with the manufacturer or spring designer to overcome the failure rate.

Failures in coils may occur at the wire leads or in the inner coil insulation where there is a breakdown because moisture has entered. Again, do not live with this type of repetitive problem, but attempt to obtain an improved design if high intensity rapping is required. However, I have often found rapping to be higher than required, and lower intensities should be attempted first to evaluate their effects on the failure rate.

10. The plane of impact can be important to the extent that flaring damage is experienced at the end of the plunger or hammer, and strike pad. If the two parts are not aligned correctly, then some peening or deformation of metal will take place. This condition can eventually cause failure by setting up wear in other parts of the mechanism.

11. One problem that has surfaced, on occasion, with the vertical electro-magnetic impact rapper is that induced magnetism can cause the plunger to remain attached to the strike rod. Some change in the steel composition may be required. The surface

condition between plunger and strike rod can be a factor. I would also check that the length of the plunger within the coil is satisfactory. In fact, the variation of rapping intensities may lie in minor differences of the height of the plunger within the coil housing.

12. The use of rotating hammers within the precipitator proper of the rigid frame designs make periodic inspections of the mechanisms and bearings of this type of rapper system mandatory. Recent improvements have minimized some of the wear problems experienced in the past, but the service in gas and dust conditions make any rotating part subject to undue wear. A key concern would be the alignment of the drive shaft. The bias of the impact marks on the strike pad or uneven wear on the hammers will usually indicate a problem with excessive bearing wear or shaft alignment.

Maintenance of Insulators

The surfaces of precipitator insulators that are exposed to flue gas conditions provide potential electrical leakage paths to ground. Insulators can consist of tile, alumina, or ceramic material coated with a highly glossed porcelain finish. These insulators can take many shapes and designs but the object is to basically supply a seal or support function. Recent designs of the tub shape for the support of the high voltage frame has found wide acceptance. These tubs of about 20" to 24" high, about 20" in diameter, and having a wall thickness of about 1", provide both the gas seal and frame support.

Other insulators are used for the isloation of the high voltage frame rapper from ground and can vary from shafts of alumina or porcelain with 2 to 3" diameters and about 30" long, to various other synthetic insulating materials. Insulators used for anti-sway devices of the high voltage frames at the hopper location vary from 30" long alumina shafts to shafts of teflon, nylon, and micarta formed in either round or flat bar shapes. Some of the applications have been successful, some not. Much depends on the type of material and process characteristics. For example, a batch process with high moisture gas that occasionally drops into dew-point conditions, will make any insulator surface questionable for that environment. If a proper heat or purge system exists, then practically all types of insulators are usually satisfactory as long as the mechanical strength or length of leakage surface is sufficient.

I am often amazed at the reliability and performance of most insulators used in the precipitator application. These insulators often get covered with material and withstand various shocks without failure. They will take a tremendous amount of compression force, but can easily fail with severe tension or shear force. But the most critical factor for insulator application is the maintenance of the electrical integrity of its surface.

The danger of any electrical leakage to ground is that the localized stress caused by the temperature gradient could lead to a cracking of the insulator surface. A longitudinal crack can then provide a low resistance path to ground to further carbonize into an electrical short of the field. A key point I would like to make is that if you have experienced no difficulties with insulators on your system, do not made modifications unless a drastic change in the process is contemplated. However, if electrical tracking of the insulators is common, and replacements have occurred, then one or more of the following comments should be of some help.

1. Insulation of the insulator compartment is important, especially with the dog-house or coffin-box designs. Thicknesses of 2-3" insulation is usually sufficient.

2. I do not have strong preference for either individual heaters or heated air purge as a means to keep insulator surfaces above dew-point conditions, but one or the other should be used. The individual heater element is more difficult to monitor, but the purge piping is susceptable to an uneven air distribution or failure of the total heater system at critical times.

3. Heaters can consist of wall strip, tubular space element, or a tubular element placed around the base periphery of the insulator. Usually 900 to 1000 watts per insulator is sufficient. The object is to provide as uniform a heat environment around the insulator as possible. Any space heater element should not be close to the insulator because of either a localized heating effect or a potential flash-over to ground. Usually a 5" spacing can be considered a minimal distance. It is wise to place at least one or two thermostats for the cycling of the complete system, and generally about 300°F is satisfactory as a nominal control level.

4. The use of heaters or heated purge air must be co-ordinated with the cycling of the process. With a fast batch operation as in the 20 minute BOF steel cycle, the heating

system is best left on continuously. Under steady state operations, the thermostat control could shut-down the heaters after a specified internal compartment temperature is reached — usually at about 230°F. This is true for the negative draft units. For the pressurized collectors, the purge air is continually in operation, and the heating elements can either cycle or not, based on the process. If the flue gas temperature normally operates at the 220 to 250°F range, it may be wise to keep the heaters on continuously at the 280°F level.

5. Another factor in the heating arrangement of the insulator compartment is how the intake air is handled, and how the floor of the comparement is insulated. The negative draft unit will normally have vent elbows into the compartment. About 100 cfm of air is required for each insulator if an air sweep is included in the design. Usually the cover of the tub will contain small holes, or pipe fittings, that allow the air to purge the inner zone of the insulator. If this is done, the air can be directed downward in a swirl pattern across the surface of the insulator. It is well to keep the quantity of air at minimal levels and preheated if possible. On some installations, no air movement within the tub is needed if the heating arrangement is adequate. Be aware of the opening at the high voltage frame rapper shaft, as it normally passes down through the center of the tub insulator main support pipe. A large opening at this location could starve the air available for the sweep of the insulator surface.

The use of insulation material on the floor of the insulator compartment has several ramifications. In the small dog house, for example, floor insulation could be deleted. But in the larger insulator penthouses, floor insulation is generally recommended, as long as sufficient heat is present at the insulators. The immediate floor area around the insulator itself can be exposed to allow the internal heat of the precipitator to rise in that localized zone. The main purpose for insulating the floor of the penthouse type insulator compartment is to allow maintenance personnel access for check-out, or other minor work, while the heat of the precipitator is bottled-up. It also minimizes floor corrosion.

6. Probably a number of insulators have been damaged by their proximity to the compartment air intake. Make sure that the intake air does not impinge directly on the insulator surface. However, this is not as critical with the heated purge system. It is well to place a deflector plate or shield between the incoming air and the insulator.

7. There are still a number of tile bushings in existence on older installations that provide a gas seal around the support pipe, while porcelain insulators support the high voltage frame. These porous refractory tile bushings are quite susceptible to moisture and dust uptake and must be carefully heat-protected. It is especially important not to provide a tight seal between the pipe and the tile cover. Usually a ¼" opening around the pipe with sufficient air intake will keep trouble to a minimum. It would be wise to replace this type of insulator with the modern tub design.

8. The inner surface of the tub design insulator is exposed to gas conditions. The outer surface can also get coated if the insulator compartment becomes pressurized for any number of reasons. During operating periods of low negative draft, the chimney effect can allow dust particles to rise into the compartment. This can especially occur in large compartments with a large number of insulators. There is no need of concern as long as heat is present in the compartment. However, it is good practice to clean all the surfaces during outages. Cleaning can be accomplished with plain soap and water. Difficult cleaning may require a fine steel wool. Various solvents have been used, but are generally not recommended.

One point against leaving the insulators coated with material during an extended outage is that some material is hygroscopic, and it will form a sticky deposit if the heat is not left on during the complete outage.

9. The heater electrical circuits are usually separate from the high voltage power supplies. Since these heaters are normally on a 480 volt supply, make sure these circuits are locked out before any cleaning activity.

During the cleaning period, the leads to the heater should be inspected for decay of wire insulation. It is especially important that no damage occur to the elements or connections at the base of the tub insulators.

10. Some insulator failures have occurred from either a shift in the structure impressing tensile stresses on the surface, undue force on one portion of the insulator, or metal expansion. Failures for these reasons have been minor compared to cracks caused by the surface leakage problem.

11. Replacement of an insulator is a critical procedure. Extreme care must be provided to prevent misalignment of

the high voltage frame, and this is especially important with a two-point suspension system. An ⅛″ deviation at the top support pivot can produce a spacing problem at the bottom of the precipitator.

12. Post insulators used for the support of high voltage leads, either in the penthouse type compartment or in the pipe guard enclosures for the connections from the T-R set, are usually reliable and trouble free. However, they should be cleaned periodically. Ventilation should exist for the guard enclosures to minimize any sweating effect, leading to an accumulation of moisture. These vents usually are ½″ or ¾″ elbows placed downward so rain cannot enter.

13. Feed-through insulators, which are generally of the tapered porcelain design, have sometimes failed by either metal expansion or electrical leakage problems at the center flange. This is normally rare in recent years. If this type of insulator can be eliminated, do so by all means.

14. The insulators used in the hoppers for the stability of the high voltage frame make for a difficult application. I usually like at least a 24″ leakage path of the insulator surface at this location. Steady state operation and fly ash applications are generally satisfactory, and failures are minimal. The most trouble occurs during start-ups with premature electrical tracking through conductive deposits. Cleaning and inspection of the anti-sway devices during shutdowns is highly recommended, but access to these insulators can be difficult in most precipitator designs.

There is no easy answer to this special insulator situation, but if at all possible, I suggest eliminating this potential cause of trouble in the design stage. Four point suspension of the high voltage frame and certain stiff electrode designs can prevent frame oscillation on the weighted wire installations. Some installations may not require the anti-sway insulators based on stability of housing and frame. If insulator trouble is repetitive, I suggest operating without the insulator tie on one or two fields to determine the effect of this change.

Several methods utilizing knife-edge insulators or bumper insulators have been used successfully on some jobs. Maintenance of heat and elimination of in-leakage air in hoppers may help out. On critical jobs, heated purge air at the insulator can be attempted.

15. There have been a number of attempts to coat insulator surfaces with silicone type preparations to help counter moisture

effects. Some success had been experienced, but there is some question on the overall advantage of this technique to keep out of trouble in the precipitator.

Start-Up and Shut-Down Procedures

Start-up and shut-down periods for precipitators are the times when trouble will often occur; this holds true for all types of collectors and equipment. It is a time of fast transient changes in process characteristics, gas temperature, and moisture conditions. It is a time when flue gas will come in contact with cold metal surfaces, and condensation problems are prevalent. There is no one simple method to keep out of trouble for all installations. Each application must be evaluated, and the best procedure might only occur after several trials. There are a number of critical areas and safeguards that are common to most installations, and I will discuss these key points. Make no mistake; effective start-up and shut-down procedures can often determine whether the high performance of the precipitator will occur routinely in the periods between outages. The frequency of shut-downs will determine the scope of the potential problem.

The Start-Up

1. I highly recommend that for the initial start-up of any new industrial installation, the precipitator not be energized until 230 to 250°F is reached at the inlet of the collector. This first start-up is special in that much can go wrong in the process, and the confusion that may occur during this period can result in a number of severe problems in the collector. Don't take a chance on the first start-up. Gas and dust conditions should be monitored for potential combustible mixtures. It is desirable to make a comprehensive internal inspection following the first shut-down, using earlier suggestions. The next start-up could include some modifications in procedure, based on the observations of the initial operating period.

2. Insulator surfaces should be protected by the early start-up of the heater, or purge system, at least several hours before light-off of the process. This is also true of hopper and evacuation system heaters. The key is sufficient temperature, and during ultra cold weather periods, the heater operating period may have to be increased. Do not rely on just throwing breakers in, but

physically check that these systems are functioning properly if monitors are not available. After several hours in even severe weather, some elevation in temperature is normally detected on metal surfaces near the heater locations.

3. The hopper evacuation system should be started before light-off of the process, especially if the fans have been operating for any extended period. I strongly recommend a visual check of each hopper some time prior to start-up, and after all the internal work or inspections have been completed. Any lumps of material should be physically removed.

4. During the start-up period, and even long past the point of energization for some installations, I often suggest collector surface rappers not be operated. This situation has more merit when hard encrustrations were left on the collector plates in a patchy type pattern.The temperature transient, after a cold start, will normally pop some of those build-ups off the metal surface. Rapper operation can substantially increase the amount of this hard slab dislodgement during this early period to help foul-up the evacuation process, especially with a critical installation. Even though some of the hard lumps will breakup, the character of this material retards easy flow or movement through the hopper or evacuation system.

When the rappers are initiated after the precipitator temperatures are near normal, a minimal amount of new material is also dislodged along with some larger lumps. It is usually found that the new fine-sized particles act as a medium to carry the lumps along through the system. The initial inspection will usually point out the need for this rapper caution. Of course, the shut-down procedure I will discuss later can often determine the extent of build-up left on the plates, but some applications such as found in high resistivity or fine-particulate jobs make removal of all the material quite difficult. Each installation must be judged separately, but the rapper start-up about 15 to 20 minutes after the flue gas material starts to enter the precipitator is a good rule of thumb. If no hard encrustations exist, then the collector plate rappers can be initiated prior to light-off. The high voltage frame rappers should be started prior to light-off in any case.

5. Efforts to decrease the amount and duration of carbonaceous material that occurs in any fossil-fuel light-off and warm-up period can be very important to the precipitator well-being. This fine soot can provide an initial coating of insulator surfaces and electrodes that defies easy removal. Whether this coating is

critical depends on many other factors, but the elimination of, or at least minimizing of this carbon residue should be a prime objective in the burner and air mixing design and operation.

If the start-up period drags out with an excessive production of carbon particulate caused by incomplete combustion, the greatest danger lies in the hopper area. Energization, collection, and rapping down of substantial amounts of carbonaceous material during the early period could cause conditions for spontaneous combustion in the hoppers. This is especially true if air in-leakage or even the presence of aeration stones exists. This is another reason for holding off the rapper system on some jobs until a certain amount of fly ash is also collected. This can help dilute the effects of nearly pure carbon deposits in the hopper. If combustion took place, it might be observed by the formation of porous clinkers not unlike large pieces of coke.

6. Knowledge of all aspects of the process as it relates to the potential start-up problems, including excessive stack opacity, cannot be overstressed. If a gradual procedure for bringing the process on stream can minimize stack emissions, then premature energization of the precipitator can be avoided. The use of required excess air instead of excessive air flow can help. Avoidance of sharp increases of air flow can minimize stack puffs. Deployment of proper oil igniters, and providing sufficient turbulence at the burners, can help keep the production of carbonaceous soot to a minimum. Provide the maximum heating of the flue gas during the early periods of dust generation. This type of philosophy will pay off for both the protection of process equipment as well as the precipitator itself.

7. Knowing just when to energize the precipitator during the start-up of the process is often a worrysome time, and I have observed a variety of procedures. Some installations have all fields energized, at full power, to light-off. Others do not energize as a rule until gas temperatures reach 180 to 220°F at the precipitator. Each process can be different depending on the gas and operating characteristics. I prefer to only energize sufficient precipitator surface during the start-up cycle to keep the stack opacity within code requirements, until at least stable operation is achieved. This usually occurs within $\frac{1}{4}$ to $\frac{1}{2}$ of the process rating of the system, after which the remainder of the precipitator is energized. Trial observations will dictate when and how much should be energized relative to the generation of effluent material. In most fly ash applications, partial energization occurs with the start-up of the first mill.

8. Regardless of when the precipitator is energized at start-up, the power supplies should be started one at a time on a low manual setting and watched for at least a half-minute before moving to the next control panel. With proper heater protection or effective clean-down of insulator surfaces during outages, this procedure may be less critical. Electrical leakage to ground over the insulator surfaces will normally respond within seconds by the sporatic varying activity of the meter needles showing voltage dips toward zero and with the upward ammeter swings. As soon as this is observed, shut the control down and go to the next power supply. It is possible by a series of steps of energizing and deenergizing within short time periods of one second or less, that the insulator can be dried out by the leakage current. This routine is not generally recommended, and should not be necessary with good insulator protection. I normally attempt a second energization of the field that showed insulator tracking. If the same characteristics occur, it is well to wait until more heat is present in the precipitator before another attempt at energization. The response on the meters is quite different than what is observed with a spark-over between electrodes, so it is usually apparent that insulator leakage is the culprit. I suggest leaving the power supplies pegged at about 250 volts until about one-half the rated process operation is attained, then swing all controls on automatic or normal power settings.

9. Which fields to energize at the start-up will partially depend on internal inspections, type of material handled, primarily related to particle size, and the installation size itself. I prefer to energize the fields least apt to cause difficulties with electrode build-up during normal operation. While this may mean the outlet fields in some cases, I normally suggest energizing the center fields during the critical start-up period. In a three field precipitator, the inlet field would probably be the next to energize, if fine-particulate has been found to normally coat the outlet field electrodes. In a four field precipitator, the center two fields are usually the first choice, and this should be sufficient to keep the stack acceptable.

The Shut-Down

The shut-down procedures of the precipitator system has much to do with the success of keeping deposits on the collector surface to an acceptable or workable thickness. Hot and newly deposited material will be much easier to remove than material that has cold hardened during the outage. Remember that heavy deposits left to harden can reduce the rapper effectiveness during the next operating period. However, the normal deposit cohesiveness is enhanced by the electric field and too often the precipitator is left energized throughout the temperature decay cycle. In other words, a strong electric field will battle the rapping forces needed to dislodge the material. The removal of these electrical forces as soon as possible after dust generation ceases in the process is very desirable. Unfortunately, deenergization coupled with rapping at relatively high air velocities through the precipitator will result in unacceptable stack appearances. Some compromise in an effective shut-down procedure should be reached, especially when shut-downs occur less frequently than once per month.

Here are some comments that can be of help:

1. Maintain the hopper evacuation system at full operating capacity at least until the final cold shut-down. I generally like to see the hoppers in operation for about 2 to 3 hours after the fans shut off, or at least go to minimal ventilation draft. The rappers should be shut down about an hour before the hopper system is shutdown. This last phase of shut-down will generally occur after precipitator temperatures are below 120°F.

2. Now, if the maximum material is to be removed from the precipitator surfaces, then some coordination must occur between the amount of air flow through the system and deenergization of the precipitator. The object is to minimize stack discharge, and yet clean the plates while the material is still hot.

3. One method entails reduction of the air flow through the system by a factor of 2 or 3 as soon as dust generation ceases. Then reduce all the power in the precipitator to minimum settings just above the undervoltage trip-out levels or approximately 180 volts on the nominal 480 volt supply. Keep rappers on at normal settings. Adjust scope of these conditions to keep stack below 20% capacity. After about 10 minutes or at least two rapper cycles, start deenergization of power supplies on a frequency to keep stack appearance satisfactory. During the first trial of this procedure, radio communication between a stack reader and the control room will help determine the optimum shut-down cycle. After all the fields have been shutdown — reenergize at least the last half of the precipitator and return fans to the normal cooling procedure. The rappers will have stayed in operation during this whole cycle.

4. Another method I prefer includes bottling up the whole system for about an hour before the cooling begins, and then

cleaning down the precipitator on an accelerated basis. On large systems this will add very little time to the overall cooling cycle, yet is simpler and more effective. Usually fan dampers will have to be closed under nearly a full fan throttling. Whether the fans are completely shutdown depends on the size of the installation. Begin an accelerated precipitator deenergization with one or two sets at a time. Increase rapper frequency and intensity through this period if this can be easily accomplished. There should be some minimal draft during this period so a slight seepage would be noticed from the stack. At the end of the hour, reenergize the latter half of the precipitator and begin normal cool down of the system.

5. There can be variations in the above methods that should be discussed between operator and maintenance personnel for the best shut-down implementation at each plant. The object is to obtain the maximum precipitator cleaning before temperatures fall below 200 to 250°F in the collector.

6. During any exaggerated shut-down, additional care or monitoring must be shown to the hopper evacuation process. Depending on the severity of material build-up, the hoppers can receive large amounts of accumulation in short periods. Although this material will be hot and easily removed, take no chances on a stoppage at the hopper apex or any other weak spot in the system. Consider a manual vibration or rapping cycle of the hoppers at the conclusion of the special shut-down to clean down any wall deposits.

7. In some installations where severe shut-down procedures cannot be implemented, make some attempt for improved cleaning of the collector surfaces during the normal cool down cycle. Increase rapping intensity and frequency if this is easy to do. High intensity rapping for short periods, approximately one-half hour, should not physically damage any of the structure components. Reduction of power settings can be implemented in a manner to keep the opacity of the stack discharge within acceptable limits.

Outage Clean-Down of Electrodes

If good shut-down procedures are followed on most installations, the degree of build-up on electrode surfaces will usually require no further cleaning during the shut-down. That does not mean that $1/4''$ to $3/8''$ mounds of deposit will not exist, but the build-up will be spotty with most of the surface holding less than

$\frac{1}{8}''$ to $\frac{1}{4}''$ thick mounds. There are exceptions, especially caused with high resistivity materials or other operating characteristics.

Whether any manual cleaning is implemented during the outage depends on several factors. If it is an annual outage with certain planned work on the electrode system, then a water wash of the unit might be considered. I usually recommend against this type of cleaning unless it is necessary to perform major work on the system. Depending on the time of year and thoroughness of the washing, some rusting and corrosion pockets can be accelerated.

If the precipitator maintains high performance levels with varying degrees of build-up, I would not be concerned about leaving the build-up undisturbed. But there are critical jobs where excessive build-ups can be the performance difference during the next operating period. This will generally occur with wall thicknesses of $\frac{1}{2}''$ to $\frac{3}{4}''$ of material that has not yet hardened but was not removed during the shut-down cycle. Inadequate rapping or even non-functioning rappers can be the cause. Some fine-sized material forms a strong adhesion bond with the collecting surface. This type of build-up can be often dislodged with a sharp manual rap on the top or side termination structure of each collector plate. Whether the internal hammer or top impact blow is used in the normal rapping system, the attenuation of the impact through the system can leave thick isolated build-ups. The manual rap with a 2 lb hammer in these localized zones will usually provide an effective cleaning. Again, the rap should always be on a reinforcing support or rib of the collector plate. It is desirable to keep a strong draft in the collector and to work upstream of the manual operation. The hoppers will usually require some manual cleaning at the conclusion of this program.

If $\frac{1}{2}''$ to $\frac{3}{4}''$ of relatively uniform dust build-ups are observed on the collector surface during the initial inspection, then an increase in rapping intensity or frequency is indicated. A clean-down should be implemented before the next change in rapping.

Any discharge electrode cleaning during an outage can usually be accomplished by a slight tap on each wire by a wooden stick. As noted earlier, the beady and ragged coating on the electrode is usually of not much concern. Mounds of material may exist in low velocity areas in the shape of large beads as strung on a string. These can be dislodged, but no gain in performance will be noticeable. Some attempts at cleaning the round electrode have occurred in the past by running a weighted device up and down the wire. Again, the result of these efforts is generally nil in terms of performance.

Electrical System Maintenance

Whether or not the precipitator performs at maximum capability depends to a great degree on the care of the electrical components of the installation. This part of the system can be highly reliable, but minor defects and problems have occurred over the years to provide headaches for the maintenance department. Usually most of these problems will occur within the first couple of months of operation. If the high-voltage components make it through the first year, there is a strong probability that the T-R set will last for years without difficulty. Fortunately, the high voltage transformer has been a highly reliable piece of equipment over the years, and the silicon rectifier has also provided reliable service in recent years.

The move to more complex controls had opened up some difficulties with various circuits and component parts. While the new static electronic component parts can be highly reliable, they can also be adversely effected by the excessive heat and dust found in some environments. The key goal of any installation is to keep all the parts of the electric control system working effectively through process changes and other transient occurrences on the system. It is one thing for the control to deenergize the field for an internal fault, but false trips because of inadequate protection or ballasting of the circuit should not be abided. The following comments can be of help in the overall electrical maintenance program.

Circuit Protection

1. For at least the first 3 or 4 months of operation of a new installation, attempt to have each trip-out of a power supply recorded relative to the time of day, field identification, and pertinent process data at the time of the trip. The key trip-out factors might involve a slag-fall in the boiler, or a severe change in gas conditions.

2. Some common trip-out causes include blown fuses and activation of the heater-trip element in the distribution or main breakers. Hot ambient air temperature may have downgraded their trip rating substantially. In many cases, the sizing of these

protective devices was set too close to the transformer rating. Fuse and breaker designs are generally for short-circuit protection, and I generally like to see a minimum 150% of circuit current rating even with normal time delay provisions. On occasion, the bakelite enclosure of the breaker heater element can be removed for temporary relief. If fuse replacements are warranted, I prefer keeping the same rating with more time delay, if this is possible. In all cases, check with the manufacturer immediately if back-up protection devices trip during the early operating period.

3. The first line of defense for circuit protection should utilize a time-delay relay in the primary or secondary circuits which are usually set for 110 to 115% of current rating at a 4 to 6 second trip-out period. This should be sufficient to eliminate many of the false trips that occur in practice, because many overload relays are usually adjusted for 100% of winding rating under cool conditions. As a key to how much protection is warranted, be more concerned with the temperature of the high-voltage transformer oil. For example, it may be required to keep the overloads near the winding current rating if the transformer is in a hot location, and oil temperatures are near their stated rise limits. When oil temperatures run cool, the 115% setting, as well as the long time delay, is usually satisfactory.

4. Undervoltage protection is very desirable, but can cause false trips if set too high on certain installations. Where high resistivity conditions can occur, primary voltages of 170 to 180 are often observed. I usually suggest a setting about 50 volts below the low range of normal voltages. Time delays of 5 to 7 seconds are sufficient.

5. Fuse protection for the silicon diodes of the primary circuit voltage control have failed routinely on some installations. This should not be, since they are for short-circuit protection and should be sized adequately. After clarifying the fuse size with the manufacturer, make sure the high R-F or current peaks are not occurring during spark-over. The most severe disturbances might occur with large power supplies, either operating at low levels relative to the rating, or connected to large-sized fields. R-F choke coils can be added, and proper ballast reactors might be indicated. Be on the look-out for the arc-over condition caused by troubles with internal components.

6. Surge protection in the secondary current meter circuit has prematurely failed in the past, but replacement with an improved type element will generally solve this problem.

Initial Check-Out with Serviceman

A couple of days spent with the manufacturer's serviceman can be worthwhile if comprehensive trouble-shooting training is accomplished on the controls. I do not mean classroom discussion, but rather, on the job observation of changes and their effects on all the critical phases of the circuit. Maximum use of oscilloscope and voltmeter are recommended. Several points should be stressed:

1. Oscilloscope tracings should be obtained at different operating levels of the control circuit. Voltage and current tracing should be obtained at the extreme settings of the automatic feedback circuits. All scope tracings, coupled with meter readings and control setting markings, should be part of an on-site trouble-shooting manual. Tracings under spark conditions should be obtained to ascertain responses with the different control settings. If photographs can not be made, then sketches of the wave-forms can be satisfactory. Scope measurement points should be clearly recorded or marked.

2. The maintenance manual often includes a number of check-points, where measured voltages will determine proper operation of the various circuits and components. These points should be measured during the initial check-out period, until the plant electricians are familiar and comfortable with the trouble-shooting techniques. If the manual does not provide this section, then utilize the time with the serviceman to develop this information.

3. All critical segments of the high-voltage circuit should be measured for integrity to ground before energization, and all readings made part of the initial check-out manual. I prefer a 500 volt megger on all high voltage frames, which will include the insulator surfaces. Isolate the circuits to the smallest denominator or bus section. Operate the megger for 30 seconds and record the readings at the beginning and end of this period. The reading should be slightly higher at the end of this time measurement. The high voltage T-R set should be checked out under manufacturer recommendations. If major differences exist between comparable equipment during this initial check, then further investigation is warranted. For example, if 5 out of 6 bus sections indicate 100 megohms to ground, and one shows 20, the reasons for the low reading should be ascertained. It may only be some surface contamination present, or a minor hair-line crack not detected visually during the initial inspection.

Control Settings

Usually all control circuit limits of voltage and current are originally set at the ratings of the windings. After a short period of operation, the current may be reduced to a limit about 50% over normal levels. This is usually done for steady state processes where gas conditions are quite stable. For example, if secondary currents during operation remain in a band of 200 to 400 ma on a 1000 ma power supply, I usually suggest setting the limit to 600 ma. The main benefits may accrue with this procedure if some failure occurs in the automatic control circuitry. The voltage limit is usually kept at the rating of the winding.

Some method should be implemented for firmly securing all control potentiometer settings, or keep a record of the settings at each cabinet. Once the controls are set up properly, there is generally no need to readjust on a continuous basis, unless some drift or instability exists in the circuitry. One problem that occurs on some installations, is that easy access for changes in control settings exists for personnel, yet there is no way to ascertain the original positions of these control parts.

Air-Load Electrical Readings

The technique of achieving corona power characteristics during outages is a valuable maintenance tool. A full set of electrical data should be obtained before the initial start-up of the precipitator, with all components in relatively dust-free conditions. There have been a number of misconceptions about this electrical test, and the following comments should help define the limits and procedures for maintenance personnel.

1. Record all pertinent high-voltage transformer information, regarding voltage and current ratings, and the primary tap setting if a multi-tap winding exists. Clearly identify the field or bus section.

2. Obtain the readings under air conditions that can be duplicated on subsequent air-load tests, specifically, internal precipitator temperatures in the 70 to 90°F range, and usually with air flow at either natural draft or low fan levels. Record these conditions.

3. Sub-divide the precipitator into the separate bus sections for the initial air-load. I also prefer this technique for future tests, but since time may be a factor later, it is wise to obtain

readings for the smallest sub-division as well as the normal field hook-up during the clean initial period. The small bus energization gives the greatest voltage level available to help pick out any close clearance. But the air-load test can only partially simulate normal gas conditions. It is possible to show full power without spark-over during the cold test, and yet have lower power controlled by spark-over during operation. Air-load will only pick out glaring internal problems, and many close clearances or subtle defects can escape detection. One main purpose is to assure the absence of a very close or direct ground after any construction or repair activity, and prior to light-off of the process.

4.   Under clean surface conditions, full current rating should be obtained on each of the individual bus sections without any electrical break-down. There might be an occasional spark, possibly due to a rust flake. Subsequent air-load tests should be similar as long as the measurements are obtained at the same low temperature, even though dust build-ups on electrodes can distort the voltage-current readings somewhat,

5.   There are occasions when it is difficult to obtain air-load readings during outages, based on a number of factors. I have seen rust flaking after a wash-down foul up the measurements. If the readings are attempted too soon after shut-down, it is possible to have spark-over because of elevated temperatures, especially toward the upper one-third of the precipitator. On installations without proper insulator heating, air-load measurements can be quite difficult to obtain.

6.   Air-load readings should be obtained with the power input controlled manually in steps, from the start point of corona to the full transformer rating. The pattern of voltage-current obtained by this method can be plotted into a V-I characteristic curve, often used for the analysis of changes in the precipitator. More about this later. Usually the initiation of current will occur within a range of 80 to 130 volts on a T-R set with a 400 volt primary and a nominal 53,000 volts secondary. Without a DC ammeter in the ground return circuit, this onset of corona may be difficult to detect. With only a primary ammeter, the first movement of the ammeter upward can be considered the onset point. Usually the primary ammeter scale will make it difficult to read AC amps up to a value of 10 for most power supplies. A low-amp tong-meter can be used to provide valid readings in this range. When the DC ammeter is used, I usually take readings at 10 to 25 ma steps depending on the size of the T-R set up until 100 ma, and then at 50 to 100 ma intervals. For example:

36 KVA T-R set, 400 volt/53,000, 7.6 mh reactor, inlet field, approximately 80°F, no fan, no heat on clean insulators, fly ash build-up on electrodes, 32 passages, 6′ x 28′ high field, 9″ spacing, 0.109″ dia. wire, surface area 10,750 sq. ft., 7,168 linear ft. per field. Silicon Rectifier controlled.

Primary Volts	Primary Amps	Sec. KV	Sec. ma
—	—	17.0	1
60	—	18.5	10
95	1	21.0	20
110	2	23.0	30
129	8	25.5	40
142	10	27.0	50
185	19.5	30.5	100
212	27.5	33.0	150
240	36	35.5	200
265	43	38.0	250
290	50	40.0	300

The plot of this air-load is shown in the dashed curve of Figure 5—10. The solid curve of this figure is a set of readings

Figure 5—10 Voltage—Current (V—I) Characteristic Curves for ESP.

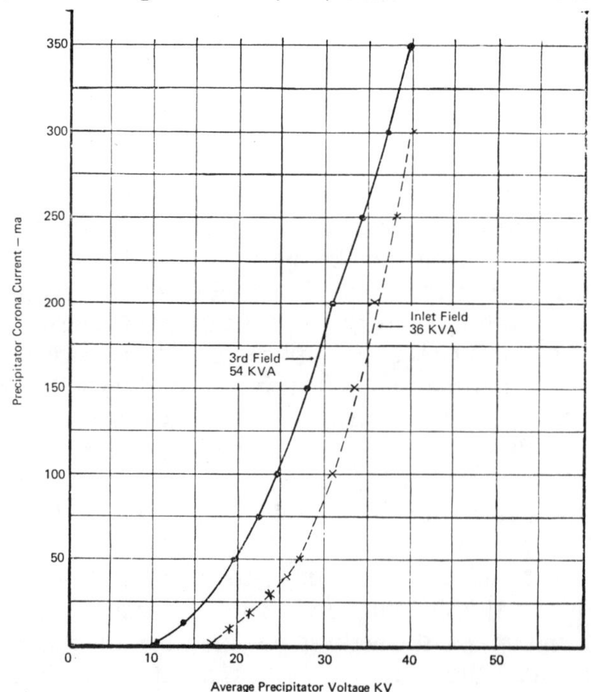

on the 3rd field of this precipitator with the same field dimensions, except powered by a 54 KVA T-R set with a 460 volt primary winding and 5.1 mh reactor. Note the difference of starting voltages. I would be suspicious of the approximately 10 KV value on the 3rd field which could indicate wire damage. I generally look for the onset of corona to occur at 18 to 22 KV for the 0.109″ dia. wire. The reduced slope of the inlet field curve up to 50 ma is probably an effect of discharge electrode build-up, which tends to be less a factor as the voltage increases.

Now the readings obtained on this cyclone fly ash application during normal operation produced the following results as compared to air-load:

	Primary Volts	Primary Amps	Secondary KV	Sec. ma	Spark/Min
Inlet	340	44	45	260	100
3rd Field	260	112	42	710	none

In this case, the normal operation of the 3rd field bore a close resemblance to its air-load values, while an elevated voltage and reduced current existed for the inlet field. In other applications, there will be little comparison between the air-load and flue gas operation readings.

Dust Control in Cabinets

Whether or not problems develop within the control cabinet from dust exposure depends to a large measure on the design of the components, and the size and character of the material. Certainly a highly conductive coal dust can be worse than say, cement dust in its possible effect on the printed circuit boards. Usually the controls will take a substantial deposition of dust without any operational effect. However, it is wise to maintain as clean an environment as possible.

The normal ventilation of the cabinet with a fan at the bottom, and with air discharge at the top, can be a disadvantage. Even with a filter mat for this type of fan, the fine-particulate will get through. Openings at the bottom of the cabinet such as cable trays or large conduit can provide a chimney effect for the material to flow from the lower floors.

In a very severe environment, consideration should be given to a complete seal of the cabinet, as long as major heat producing components, such as reactors, can be removed. Of course, the cabinets would have to be located in a relatively cool area. Another approach to a dusty atmosphere would be to pressurize all the cabinets, with a duct system drawing the filtered air from some clean area.

Clean-out of the cabinet should be accomplished during an outage period with a low-energy vacuum cleaner. It is preferable to use the pressure side of a vacuum cleaner to blow dust out of critical areas. Low pressure compressed air of 10-15 psig levels can be a secondary choice, if the air is completely cleared of moisture and foreign matter.

Temperature of Cabinets

As mentioned before, high internal temperatures of the cabinets can produce early failure of some sensitive electronic components. I prefer to limit the internal temperature to about 110°F as measured at the top of the cabinet. While design criteria of diode chips or other critical components can often operate in the 130 to 140°F range without failure, it is well to provide a substantial margin of safety. Be concerned with any high temperature source located in the vicinity of the sensitive controls. Sometimes the heat sink effect of common walls can produce zones of hot temperature.

High Voltage T-R Set

In many cases, the T-R set is exposed to severe ambient temperatures, and the bushing enclosure should be protected from moisture and subsequent rusting. A small strip heater of several hundred watts can be applied to keep the enclosure dry during extended outages, or during winter conditions. The enclosure should be vented to the atmosphere with about 3/4" elbow.

Be on the alert for slight oil seepage from the bushing flanges that may occur from excessive temperature conditions in the tank. It is well to keep a periodic record of transformer oil temperatures, especially if the tanks are in enclosed areas and are operating at full load currents. In some cases, the current may have to be decreased slightly to keep within the 45°C rise above ambient, normally stipulated on the nameplate. Another approach used with radiator fins is to utilize a cooling fan in lieu of the power reduction. Be sure to reset the temperature gauge after a temperature excursion. If oil does get on the bushing surface, some evidence of electrical tracking might be observed. Clean all bushing surfaces during these inspections.

Another inspection that should be made periodically within the enclosure is the integrity of all connections of the high voltage bushings so as to eliminate any spit sparking at these junctions. If any access is made to these enclosures while the rest of the precipitator is energized, make sure a grounding clamp is placed on the H.V. leads of the feed-in bushing before any work is performed. If the deenergized T-R is downstream of an energized field, a sizable static charge can occur without proper grounding. All of these safeguards should be discussed with the manufacturer's serviceman.

An oil sample is usually obtained from each T-R tank for dielectric break-down tests, before the sets are placed in service. I suggest a repeat of these samples after about six (6) months operation. The same sampling technique should be used each time under the recommendation of the manufacturer. After these tests, I usually see no need for further oil tests for 2 to 3 year periods, as long as difficulties with the individual T-R does not surface. Again, the history kept by the electrical department can determine a change in direction. For example, if heavy arc conditions or very high oil temperatures have occurred a number of times on a specific T-R set, it might warrant a precautionary dielectric test. I would rather see some safeguards in the system to keep these types of disturbances to a minimum. A dielectric test would be suggested if any evidence of carbon or other foreign substance is noticed in the oil.

A spare transformer should be checked for oil break-down when it is received, and then prior to use. Often months or years may pass before this unit is installed. Even T-R sets in place, but inactive for months, should probably be tested. Batch operation with frequent start-ups and shut-downs should be considered for testing more frequently than steady state installations. It does not take much moisture to lower the dielectric qualities of the oil, and frequent excursions between high and low temperature can cause some breathing to take place in the tank. Some safeguards are necessary, including being careful not to open up an extremely hot tank containing the synthetic fluid, and being careful to prevent oil contamination if opening up the cover plate for a silicon rectifier check or replacement. Withdrawal of correct samples for either mineral oil or the synthetic fluids can be different, so check with the particular manufacturer of the T-R set.

Adding oil to the T-R tank is a rare occurrence, unless a leak developed. Of concern today is the synthetic insulating oils that have been used for years for its non-flammability characteristic. These synthetic fluids can have an irritating effect on the skin, and the new T-R tanks are now generally supplied with mineral oil. In some installations, both types of insulating fluids will still exist because of a recent retrofit. The type of fluid is shown on the T-R name plate, but a stencil or sign should be prominently placed on each tank if the two types are mingled.

H.V. Connection System

The transfer of power from the T-R set to the precipitator has either used a high voltage cable, or a pipe conductor enclosed by a light gauge steel enclosure. Early precipitator applications, that had mechanical rectifiers housed in a room some distance away from the precipitator itself, invariably used an oil-filled paper impregnated, lead-sheathed high voltage cable. These cables, about 2½″ in diameter, required spring loaded oil reservoirs, usually at the precipitator end, so as to maintain a proper oil supply. A number of these cables are still in existence, but replacement is practically impossible with this type of cable construction. Problems occurred at severe bends, oil leakage at the reservoirs, compound seepage at the bottom termination, and excessive voltage stress near terminations. Overall, these cables performed reliably, but when failures began to occur they often came in groups. Building an effective stress-cone at the cable terminations was always difficult, and since the electrical characteristics of the cable was unusual, it often showed a semi-ground when tested under no load conditions.

In recent years, the dry-coaxial high voltage cable has been used as replacement for the oil-filled application. While troubles have reduced somewhat with this cable, insulation breakdowns at severe bends, at clamping locations, and at the stress cone areas still exist.

The construction of the transformer and rectifier within a combined housing, with either both components immersed in the same insulating fluid or kept separate, has allowed physical placement of this T-R set on the top or at the side of the precipitator. Therefore, a direct stiff-type of pipe connection has replaced the cable, primarily in the last 25 years. Little difficulties have surfaced with this type of high-voltage transfer system if minimal care is given. Specifically, keep the outside surface of the pipe guard enclosure well painted in corrosive atmospheres. The minimal wall thickness of this pipe guard is also susceptible to thinning from within by rusting action, especially if air ventilation is not provided by small openings, usually supplied by the manufacturer. A shedding of rust flakes or peeling away of a rust skin can cause localized breakdown in the guard enclosure. Internal moist conditions can also cause the decay of the short braided connections or support insulators often found in these enclosures. Look for holes in the side or top of the pipe guard enclosure that might develop from any construction activity in the area. The proper gasket reseal of any covers that are opened for inspection is important.

High Voltage Disconnect Switches

A number of installations will have blade high-voltage disconnect switches, other than the type found internally with certain transformers. These switches can be used to make connections between half and full-wave operation or just disconnect a field from a common T-R supply. This switch, operated in air, often is spring loaded, or has some other device to provide a positive action when withdrawing the blade from the pan discs or clip. Nevertheless, this type of switch should be normally considered a no-load device and operated with the T-R set deenergized. However, it has been common practice to open and close one of these switches or circuits when two or more parallel paths are connected to the same power supply. Regardless of the speed of switch action, some arcing will take place between blade and clip before firm contact has been made. Whether this will cause difficulties in operation depends to a large degree on the design of the switch. More often, the tension of the pan discs or clip will determine if spit sparking takes place during operation to further erode away the metal surface. Dissimilar metal for this application can hasten damage. Spit sparking at the switch is sometimes heard. Another key point is to make sure the switch is aligned properly, and that the mechanical stop for the switch handle allows a full blade contact with the clip receptor. Some safeguard is needed to keep from opening all the switches connected to the same T-R set while the power supply is energized.

Mechanical-Rectifier System

There are still enough mechanical-rectifiers in existence to make a discussion of this device more important than just for historical interest alone. As with most of the practical aspects of this book, I am assuming that personnel exposed to the mechanical-rectifier is familiar with design and terminology obtained from the manufacturer's literature. There are some aspects of this equipment that is both interesting and perplexing. Power conversion is much poorer in the M-R supply than in the modern T-R sets, for the high voltage transfer of current must be completed through the gaps between the tips of the rotor and the stationary shoes of the rectifier. For the full-wave rectifier, four (4) gaps are required for this transfer, while seven (7) gaps are part of the half-wave circuit. The theory of rectification is basically the same for the mechanical rectifier as the modern diodes, but the voltage drop across the gaps might range from 20 to 30 KV versus losses from 1 to 5 KV for the modern rectifiers Consequently, the transformer secondaries for the M-R sets will operate at 25 to 30 KV higher levels than those found at present.

Most troubles found in the M-R sets are mostly concerned with tip erosion, phase adjustment, electrical tracking of the rotor and stator support arms, and the synchronous motor commutator and brush wear. The bakelite arms must be wiped down periodically with a clean cloth. Another problem has occurred with the wound-wire resistors usually placed in each leg of the M-R circuit to dampen the peak voltages. Vibration and excessive heat is the reason for most of these failures of these resistors. It is well to replace all the resistor rods in a rack at the same time. Some trouble has also occurred with the R-F choke coils placed in the output legs of the rectifier.

There should be about 1/8" clearance between rotor tip and stator shoe for best operation. As the tips erode, additional voltage drop is taken across the rectifier. Additional wear occurs on the stator shoe until a groove becomes quite visible which further helps reduce the pricipitator voltage. I have observed one precipitator where the gap spacing produced practically open circuit conditions.

The key point on mechanical rectifiers is to prevent positive polarity from being placed on the high voltage frame. This can occur by reversing two opposite leads on the commutator brushes, by relocating the jumper wires between rectifier tips, by rotating the rotor hub about 90° with respect to the shaft, or by reversing the primary voltage leads going to the transformer. In some installations, polarity control relays, or controls, exist to show if the right polarity exists. Positive polarity will usually show a slightly lower voltage and higher current characteristic than would exist for negative polarity on the discharge electrodes. A back-up check for this condition consisted of a stiff wire being placed on the upper stator shoe and pointed toward the steel case of the motor. If the gap between the point of the wire and the motor case was about $2\frac{1}{2}''$, positive polarity would usually cause a flash-over while the negative polarity would not. If the flash-over did not occur with either change, then the gap was reduced until this method gave a valid indication. These wires were usually left in place as permanent safeguards against polarity reversals during maintenance activity.

Chapter 6
Trouble Detection and Solving

Introduction

In previous chapters I have shown a number of problems that can occur in precipitation, for which a number of corrective ideas or solutions were briefly given at that time. This chapter intends to go into the symptoms of major troubles and their possible solutions, so as to bring many loose ends together. It is not an easy task. A number of problem areas produce similar symptoms. But a systematic trouble-shooting approach can help sort out the false from the real indicators.

I do not intend to go into detailed maintenance procedures that are already more attuned to good plant practice. For example, normal maintenance repairs to electrical control circuits and fuse replacement need not be covered. I will look at the most common problems observed over the years in attempts to evaluate and correct equipment difficulties. These difficulties are the types that often confuse plant personnel, and allow repetitive failures to occur. In many cases, I will again refer to information discussed in previous sections, in efforts to define and point out the right direction to take.

The following information is only intended to maintain the precipitator at whatever performance level it is capable of, with all fields in service. Later chapters will discuss methods to upgrade the collection efficiences of existing precipitators.

Important Trouble-Shooting Approaches

Because of the high-voltage danger, familiarity with all the safety aspects of the system cannot be over-stressed. Even portions of equipment inside each control cabinet will be at a 480 volt potential, so care must be taken in any measurement procedure. The manufacturer's manual should be well studied for guidelines in the handling of certain control difficulties.

If any troubles occur initially with control circuit components, fuses, or any other low voltage trouble source, correct these problems post-haste, since the high voltage portion of the precipitator tends to supply enough potential difficulties of its own.

Problems that crop up randomly in both the low voltage and high voltage systems present a chaotic situation. The usual ratio of low voltage problems to internal troubles, with reliable control circuitry, should be about one in thirty. So whenever the voltmeter goes to zero, the odds greatly favor troubles within the precipitator.

The effective use of the control panel meters is an important diagnostic tool, so that strong familiarity with the normal readings obtained under all the transient conditions as well as process variations, cannot be over-stressed. Know thy precipitator as yourself and everyone benefits.

Record-Keeping

I mentioned the importance of record-keeping in Chapter 3. The object is not to get overloaded with paper work, nor to retain records that have minimal long term benefits because they were not keyed to the process. The common practice of having a shift worker take a set of control panel readings has merit. Information of major electrical disturbances, and certainly trip-outs of the supplies, should be submitted to the electrical department, or person in charge of the precipitator, at the start of the daylight turn. If electrical personnel exist around the clock, then trouble-shooting procedures should be begun soon after the discovery of a failure. Between readings, most modern installations will have a comprehensive alarm system that will denote difficulties. When alarms sound, the time should be recorded, and the operator should attempt to reenergize the precipitator field. Any pertinent information should be noted on a trouble slip submitted to the responsible person.

While the shift readings can serve a purpose, I prefer one person to be responsible for one set of comprehensive readings each day. These should include the electrical data of the precipitator and pertinent operating data such as process loading, gas flow rate, gas temperature at precipitator, fuel use, and ambient conditions. A stack opacity measurement either by chart or a direct visual check should be keyed to the readings. The readings should be obtained at near the same time of day and should normally be accomplished within an hour. In large installations, the operating data should be checked at the start and conclusion of the electrical readings. If the batch operation is of short duration and involves a fast rate of change in electrical characteristics, it may also be wise to record readings during the non-operating part of the cycle.

The electrical readings should be recorded in specific order from the inlet to outlet fields, regardless of how the control panels are physically located in the control room. The estimate of spark-over levels should be at least noted on the comprehensive set of readings, based on the earlier suggestions of Chapter 4.

The person in charge of the precipitators can best ascertain subtle changes in the readings by plotting the individual voltage and current readings of each field on a running daily record. A plot of the KVA or KW will be of less importance in trouble evaluations. Figure 6—1 shows one method that can be used. I prefer the primary voltage and secondary current readings, since only two plots need be maintained. Always note any spark-over at the time of readings by a small "s" above the recording of the voltmeter reading.

Evaluation of Meter Readings

By now, you should well understand that the control panel readings are a reflection of everything that is occurring in the precipitator. The magnitude, as well as the trends of the readings, will generally tell a story of normal precipitator performance as well as abnormal conditions. A sound knowledge of the effective

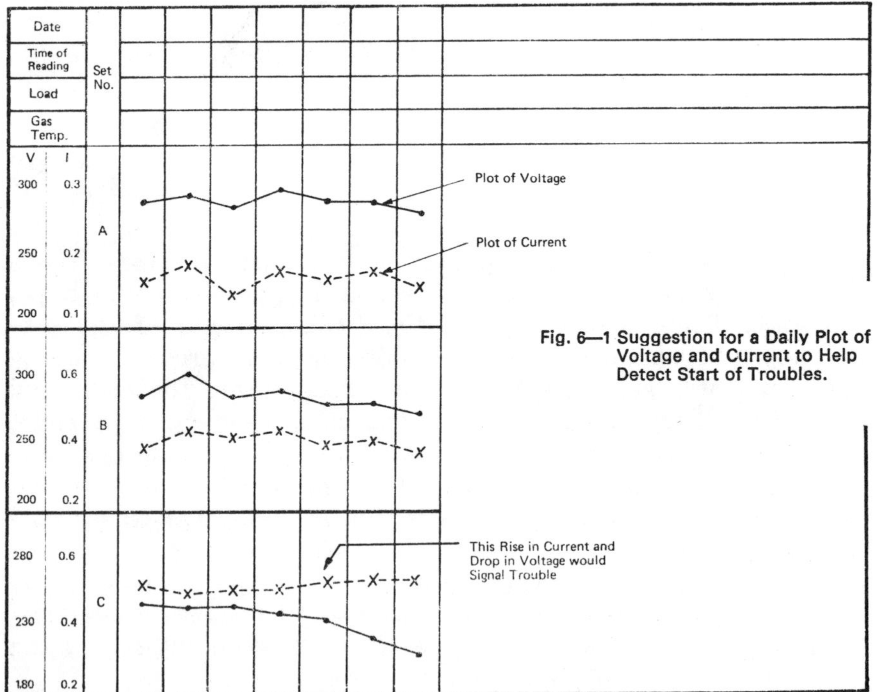

Fig. 6—1 Suggestion for a Daily Plot of Voltage and Current to Help Detect Start of Troubles.

voltage-current characteristics can allow one to judge emissions on reproducible process operations almost as well as actual stack tests.

The primary winding voltage and current readings should provide valid reflections on what is occurring in the secondary circuit of the high-voltage transformer. However, the presence of secondary as well as primary meters does provide added monitor capability. In all difficulties within the precipitator, the two voltmeters and the two ammeters will work in unison for specific characteristics. That is, when the primary voltage is low, the secondary voltage should also be low, while the ammeters could both be showing relatively high values. There are some departures from valid meter readings, especially in the AC primary voltmeter, when a waveform is radically different from a 60 cycle form factor. I would not be overly concerned with these variations, since relative readings and comparisons between fields are the tools primarily used in the evaluation techniques.

Probably 80 to 90% of the problems that occur in precipitators will tend to reduce voltages and raise currents at the same time. I am not talking about effects of resistivity, but rather the multitude of difficulties associated with electrode failures, insulator leakage, or hopper build-ups. While most of these problems can be effectively designed out of precipitators, there is no such thing as a 100% trouble-free unit. Unfortunately, there are still many precipitators not fully protected against the above group of triple trouble sources.

Locating Trouble

As with most other equipment, the process of elimination is frequently used as an effective trouble-shooting procedure. The first thing to determine is if the trouble exists in the precipitator, or in the T-R set or associated controls. By opening the circuit between the rectifier output and the support insulator, the T-R set can be separately energized without the connected load. While this is a questionable practice with some of the early static rectifiers of the 1950's, modern silicon units have the necessary protection to allow an open circuit check. This should be done on manual operation at minimum setting and involve a brief energization. I would obtain approval from the manufacturer, in writing, for this procedure. This technique is easily accomplished with high-voltage isolation switches. However, disconnecting leads in the high voltage circuit might be required with some designs, in order to check different parts of the circuit or precipitator. These isolation procedures are relatively safe, after everyone involved has been well indoctrinated in all the safety aspects of working with the high voltage system.

The dog-house type of compartment usually allows the disconnect to be made at the support insulator. In the case of high voltage cables, this isolation point is important. However, with the infrequent troubles associated with the H.V. pipe and guard system, the disconnect made at the transformer is generally sufficient to pin-point the difficulties between the T-R set and the precipitator. The disconnect at the transformer can either be by use of the internal tank switch, if one exists, or by removing one lead at a time from the H.V. bushings at the top of the tank. In the case of the dual bushing T-R set without the internal switch, make sure a full-wave jumper is placed between the bushings before the procedure of energizing one bus at a time is employed. Where one (1) T-R- set is energizing two (2) bus sections F-W from a single bushing, it might be required to manually disconnect the leads from a common junction point. Figures 6—2 and 6—3 also show how disconnect switches are sometimes arranged for H-W and F-W hook-ups. The internal transformer disconnect switches will not allow half-wave operation on one bushing alone, but that does not mean an open circuit is not possible with a broken connection beyond the switch.

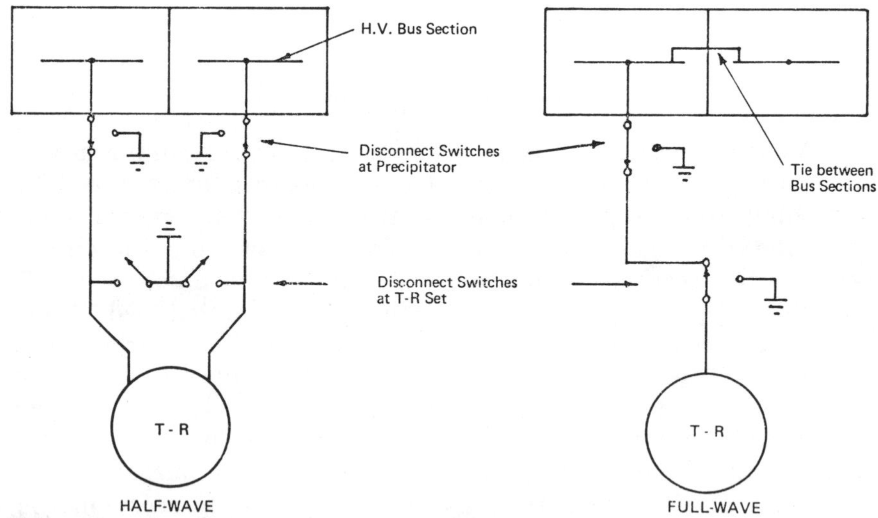

Figure 6—2 Precipitator Arrangemnts Showing Half-Wave and Full-Wave Energization of Two Bus Sections.

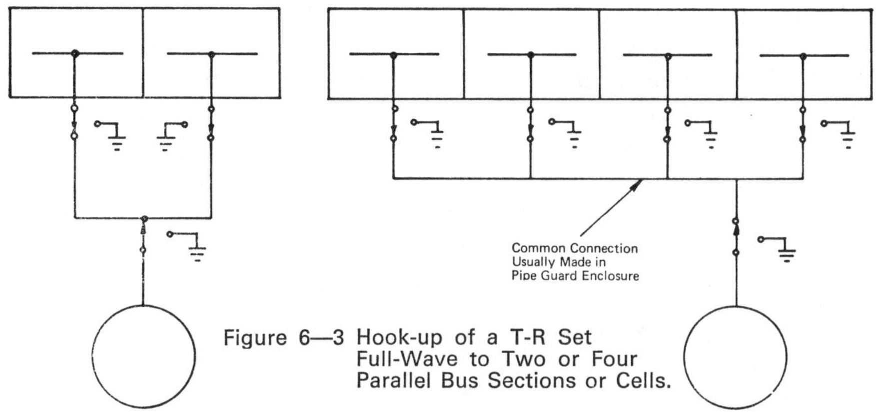

Figure 6—3 Hook-up of a T-R Set
Full-Wave to Two or Four
Parallel Bus Sections or Cells.

Normal Readings

As noted in Chapter 4 on Resistivity, patterns of the electrical readings from inlet to outlet should form some reasonable trend of voltage and current between fields. This will be true for all types of installations or process characteristics. Also remember that in large installations, wide differences in gas temperature could produce observations of widely varying voltage-current readings. However, it is the intent of this chapter to concern itself with transient troubles that occur after the start-up of the precipitator.

There are a number of internal difficulties that will produce an erroneous electrical pattern caused by either the misalignment or warpage of collector surfaces. I believe the information and discussions of Chapter 5 should identify these problems during inspections. However, at any given time, a distortion of readings should always alert one to further investigation. As an example of this point, consider a large fly ash precipitator observed to have the following readings in adjacent chambers during operation of about two-thirds boiler load.

Conditions: About 2% sulfur coal, gas temperatures in the 275-285°F range, almost absence of spark-over.

	Chamber 1			Chamber 2		
	Pri. Volt	Pri. Amp	Sec. Amp DC	Pri. Volt	Pri. Amp	Sec. Amp DC
Inlet	259	128	.73	262	135	.78
Center	298	228	1.37	228	220	1.37
Outlet	260	232	1.37	252	220	1.41

Note the low voltage in the center field of Chamber 2. After the voltmeter was checked to be correct, I awaited the first outage. Subsequent inspection disclosed warped collector plates in a number of areas, but the center of Chamber 2 was considered the worse. We could not do much in one day except remove 63 wires in the center section, located at the closest clearances to collector, or 18% of the total discharge electrodes for that field.

A repeat set of electrical readings were obtained after this temporary fix at a higher steam load and gas temperature level:

	Chamber 1			Chamber 2		
	Pri. Volt	Pri. Amp	Sec. Amp DC	Pri. Volt	Pri. Amp	Sec. Amp DC
Inlet	295	149	0.90	310	148	.87
Center	322	225	1.40	330	216	1.43
Outlet	280	230	1.38	305	225	1.46

The voltage level of the center field of Chamber 2 showed a substantial gain in comparison to the other fields and relative to the earlier readings. I would not consider the pattern shown typical for all fly ash installations with three fields, but those patterns can be considered normal for the internal alignment that existed at the time on that specific collector.

Correction of the internal difficulties or gas steam differences that produce disjointed readings are discussed in other chapters. For the present, our goal is to discuss the problems that can occur at any time during operation which will show a transient change on the meter readings if it is not a trip-out of the power supply.

Trouble-Shooting Symptoms

The following symptoms, potential causes, and possible solutions are considered the most common found in precipitators. The control panel meters will reflect the subtle and sharp changes that might occur internally. The problems considered here will be confined to only a small part of the precipitator rather than caused by general gas or dust conditions.

1. Symptom

The slow rise of current and a simultaneous fall in voltage over an extended time period of hours to days. No spark-over.

Possible Cause

If this is observed on one of many fields, electrical leakage over an insulator is possible. However, if the condition occurs at least several days after the precipitator has been in service, I would check out the hopper system immediately after discovering the trend. Collector surface build-up could be suspect if electrical characteristics stabilized after a few hours.

Discussion and Solutions

This is one of the tough symptoms to handle. If the hopper is judged clear, then it becomes a waiting game to see when and where the trends finally settle. At times, this type of abnormal condition may stabilize at a 10 to 15% reduction in voltage, and hold until the next outage. But if the rate of voltage decay is relatively fast, it usually indicates a carbonization of the electrical path that leads to a lower and lower resistance value. The voltage generally decays to a 55 to 75 volt range as shown by the primary voltmeter. The time span might be 2 to 36 hours between the start of tracking to the condition of low resistance.

Correction during an outage is the prime way to correct an insulator problem. Whatever the cause, early detection and deenergization, in the case of the fast decay case, might save further difficulties. For example, if the leakage to ground is through a dust build-up in the hopper, early deenergization might prevent a clinker formation. Therefore, there is a good chance the field can be returned to service if the hopper is drained of dust.

2. Symptom

Gradual decrease in voltage over long periods with control at the same spark-rate. Current stays relatively the same.

Possible Cause

Localized metal erosion of the discharge electrode is a strong possibility.

Discussion and Solutions

A careful internal inspection of the discharge electrodes should be made during the next outage and the defective electrodes removed.

3. Symptom

Pulsating needles on panel meters at about a one (1) second period caused by spark-over. Should see about 20 to 50 volt dips on the meter.

Possible Cause

Slack in the discharge electrode.

Discussion and Solutions

This condition can occur with the weighted-wire design when the bottle weight hangs up in the guide assembly. One way for this to happen is if the expansion of the electrode allows the weight to bottom out, or allows a bottle weight protrusion or pin to come in contact with the guide ring. On some designs, hard dust deposits on the guide frame can hang up a weight. Another reason might be the use of an insufficient bottle weight for the length and diameter of the electrode. Weights less than 20 lb for a 30′ long and nominal 0.1″ diameter wire can be suspect. I usually recommend a 25 lb weight for this size electrode.

Actually, the pulsations come from an electrical break-down that tends to drive the electrode toward the opposite collector, which then produces another spark-over. This condition can proliferate until the wire gets close enough to the collector, in severe cases, to maintain an arc. Wire damage usually occurs, and failures in the middle zones of the electrode will be observed. Excessive long periods of rapper operation could initiate sufficient movement in the wire to strike a spark, even within a normally spark free precipitator. A very fast response in the automatic control can minimize this problem. One method to determine if a slack wire is the cause is to reduce the power input or deenergize the field in order to break the pattern, and see if the condition vanishes.

4. Symptom

Pulsating needles on control meters on about a four (4) second period. Should observe severe needle movement with primary voltage dips of 150 to 250 volts on the 480 volt system.

Possible Cause

Oscillation of the high voltage frame, with the bottom of the frame moving toward close proximity to the collector frame.

Discussion and Solutions

While this condition is a rare occurrence, it can exist on high narrow frames with two point suspension, and without anti-sway safeguards. On some jobs where anti-sway insulators frequently fail, the removal of this apparatus can set up the possibility for this action. In some cases, broken anti-sway insulators can be the cause. Much depends on whether a tremor or shock occurs in the steel work to initiate the movement of the frame. I have also observed a relatively stable frame start oscillating after about 15 seconds of heavy vibration on a hopper wall.

What is bad about this condition is that severe metal erosion will generally occur on the electrode surfaces at the points of the close proximity of the frame swing. Stabilization by anti-sway insulators is the prime solution. In batch operations, and other difficult applications for these hopper insulators, the solution might involve some complex steps. Additional support insulators of the high voltage frame might be considered, as well as bottom frame insulators that can be kept operable by use of heated purge air. Another technique that was attempted years ago was the ability to remove and replace defective insulators through the hopper wall. This would be used in an extreme application. Long and sharp-edged alumina bar insulators should hold up in many difficult process environments.

5. Symptom

The instantaneous move of the control meters to zero voltage and high currents, as limited by the control, after the precipitator had been in stable operation.

Possible Cause

In practically 98% of these cases, the difficulty will lie with a metal to metal contact of the collector surface by the discharge electrode.

Discussion and Solutions

While this symptom can happen instantaneously, it often may be the aftermath of a swinging wire or a necked-down electrode. The broken end of the discharge electrode will generally bond or weld to the grounded collector at the point of contact. If the wire breaks at the top or at the middle point, the shorting out of two fields in series is possible. While good design will minimize these conditions, occasional electrode losses must be faced.

In a few cases, the broken electrode can be extracted without the need of a complete outage. This usually depends on the design of the access doors, and the terminations of the discharge electrodes. However, in most cases, an outage in the process will be required to clear the short. This is the advantage of a highly sectionalized precipitator, providing the electrode break does not take out more than one field. Many years ago, I cleared a couple shorts of this type by burning off the electrode with welding machines, although I do not recommend this technique at present.

6. Symptom

Heavy bursts of spark-over, lasting 5-15 seconds on cycles of 3 to 30 minutes.

Possible Cause

Build-ups of dust on ledges, or in hoppers, that reduce clearances for periodic electrical flash-over to ground.

Discussion and Solutions

While not common, this spark-over condition does occur with obstructions in hoppers, or a flat edge on the high voltage frame that allows dust to build-up into a pyramid mound. This situation may occur near the upper point of a hopper build-up, where the dust almost comes in contact with the high voltage frame and where flash-over blows away some of the dust. This symptom also looks like a localized break-down between electrodes caused by the action of a vibrator on the motion of a wire. It is possible to isolate the cause of periodic bursts by rapper operation. Much depends on how many fields are involved in helping to identify the cause. If this situation occurs right after a full hopper, then the possibility of a hard dust build-up lodged in one of the corners must be suspect.

7. Symptom

Steady rise in voltage and decay in current over a fairly long time period.

Possible Cause

Build-up of material on the electrodes must be the prime suspect.

Discussion and Solutions

A rise in resistivity will also raise the indicated voltage, and cut-back corona current, prior to the start of spark-over. This condition occurs pretty rapidly for existing build-ups on the collector surface. A steady build-up of dust on the collector surface, caused by a rapper failure, will tend to show the pattern indicated, as long as the resistivity range is between the conductive and spark-over bands. In highly conductive cases, a heavy build-up on the collector surface can instead reduce the voltage and raise the current level.

I would be more concerned with the build-up of material on the discharge electrode tending to decrease corona formation, and thus producing the abnormal pattern. A loss of vibration in the high voltage frame may have occurred. Isolation of the cause can be evaluated by observing the effects from the increase of rapping intensity for each electrode system separately. It may take the deenergization of the field to clean down the collector surfaces effectively. A thorough internal inspection should be made at the first opportunity.

8. Symptom

Full voltage and no current.

Possible Cause

An open circuit in the secondary high voltage system.

Discussion and Solutions

Although this symptom occurs much less frequently than the low voltage-high current condition, this situation may occur occasionally with the precipitator designs that contain isolation switches, small braid connectors, resistor racks, or cables in the high voltage system. An open in the actual rectifier or transformer secondary is a rare occurrence. Periodic inspections of the high voltage transfer system is part of a preventative program.

Use of a megger or ohmmeter will usually isolate the break in the circuit.

9. Symptom

Low voltage and low current characteristics without spark-over.

Possible Cause

In most cases, this condition is caused by control circuit problems.

Discussion and Solutions

The most frequent problem is a feed-back circuit failure in the automatic mode. Reverting to pure manual operation will usually eliminate the symptom.

10. Symptom

Indication on the meters of relatively low voltage, high primary current and low secondary current without spark-over.

Possible Cause

A chance of single phasing, or difficulty in a secondary winding or rectifier must be considered.

Discussion and Solutions

With the reverse thyrister rectifier type of low voltage control, the failure of one diode could cause this condition. In cases of the dual-bushing T-R sets, an open in one leg of the secondary is possible. This is a case where preliminary readings of resistance in the control circuitry might be useful in later isolating the trouble.

11. Symptom

No voltage and no current indication.

Possible Cause

Control circuit difficulties would be the problem in almost every case.

Discussion and Solutions

Interlocks, fuse, breaker, and open circuits in the control wires are the most common cause. Usual maintenance techniques will correct this condition.

Control Circuit Parameters

I have already brought up that the fast response and complexities of the precipitator controls may or may not be important, depending on the design of internal precipitator components and process characteristics. Reliability is the important ingredient. But be alert to certain control meter swings that tend to falsely indicate heavy arcing in the precipitator. This occurs with either excessive amplification of the input signal or inadequate feedback response. This could show 50 to 100 volt dips in the primary voltmeter and will usually involve a slow return to the base level. Some knowledge of how the normal secondary voltage and current waveform looks like in the precipitator, by use of an oscilloscope, can be important in evaluating troubles. The precipitator field with its capacitance and resistance characteristics does react like a filter to smooth out the pulsating D.C. voltage waveform. Figure 3—1 of Chapter 3 shows a pictorial representation of the precipitator voltage relative to the 60 cycle imput pulses. The peak value of this waveform can be calculated by the formula shown on Page 35. Note that the minimum secondary voltage in a half-wave installation will decay to a lower value than for the full-wave condition for any given average voltage level. While the voltage gradient field will rise and fall, the secondary current, as viewed by the scope would show isolated conduction pulses as shown in Figure 4—4 of Chapter 4. These pulses will usually vary from about 15% to 90% of the full current conduction time available in the cycle, depending on the operating level of the power supply. Since the load is capacitance oriented, the peak of the current will lead the voltage peak by some phase angle.

Figure 6—4 shows an actual oscillogram of a full-wave secondary voltage at a 33 KV average voltage. Figure 6—5 shows this same field at a 39 KV average at a later time. The effect of a spark-over on the voltage waveform is shown in Figure 6—6, with the collapse of the field toward the zero reference and subsequent recovery. An oscillogram of the secondary current is shown in Figure 6—7 for a field operating at about 85% of the power supply rating.

Insufficient impedance or spark quenching ability in the circuit could cause excessive voltage downtime in the field. Figure 6—8 shows the result of this condition under a poor control response, as shown by a trace on the secondary current circuit. Note that the dual spark peak, just prior to the cut-off of power, was over twice that of the normal current waveform peak. Figure 6—9 shows how the waveform would look with a control circuit able to extinguish a full cycle of power input at the onset of a heavy spark or arc. Note that the first pulse upon reenergization is higher than the next 6 to 8 pulses, due to the recharging current of the field. Also note that the ramp characteristics take about 0.3 seconds to return to the spark control point. Another automatic control response without the ability to punch holes in the waveform might look like the waveform of Figure 6—10. Note that the bursts of sparks tend to step the power level down until the feed-back eventually allows a return

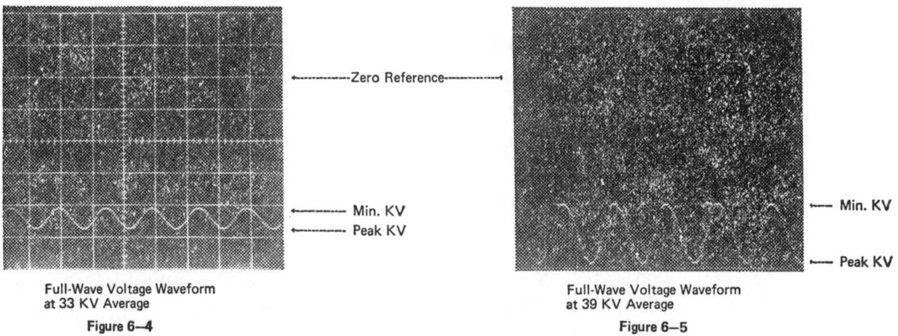

Full-Wave Voltage Waveform
at 33 KV Average

Figure 6—4

Full-Wave Voltage Waveform
at 39 KV Average

Figure 6—5

Oscillograms of Precipitator Voltage on Full-Wave at Two Voltage Levels on a Fly Ash Installation.

into the electrical break-down range. This type of control gives a slight pulsation to the meter needles, but is considered a normal characteristic. About 50 seconds of a circuit response to a rapid change in process conditions is shown in Figure 6—11. The step action of this automatic control is a little severe after the spark-over ceases, but the decay of power and its eventual recovery is clearly demonstrated.

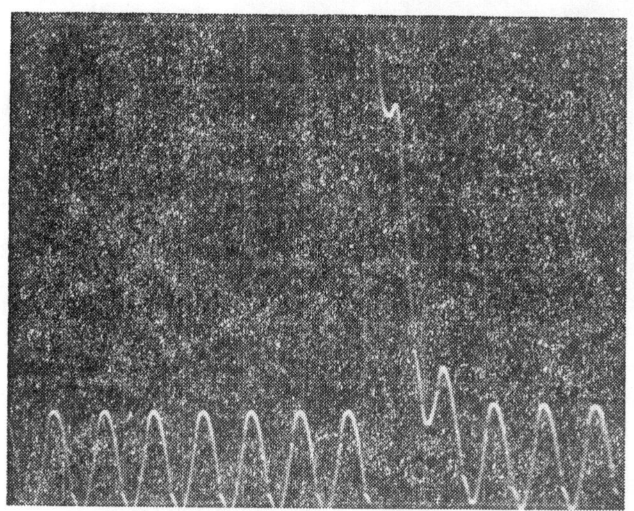

One Break-Down on Full-Wave.

Figure 6—6 Oscillogram of Voltage Collapse
During Spark-Over.

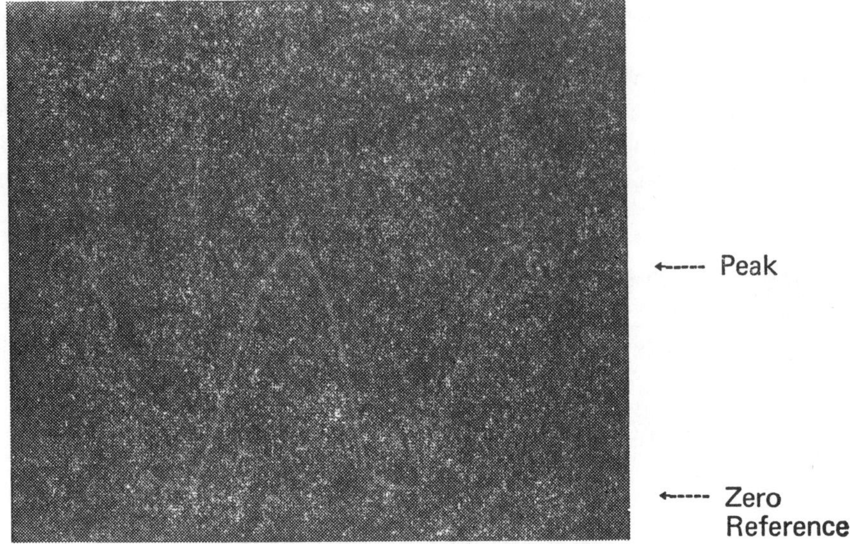

Current at 85% of T-R Rating — F-W

Figure 6—7 Waveform of Precipitator Corona Current.

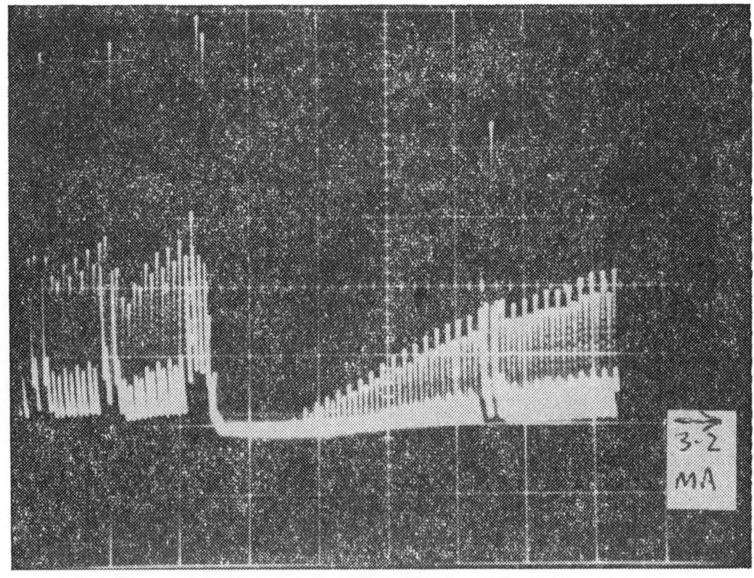

**Figure 6—8 Oscillogram of a Precipitator Corona
Current Waveform with Poor Response
of the Automatic Control.**

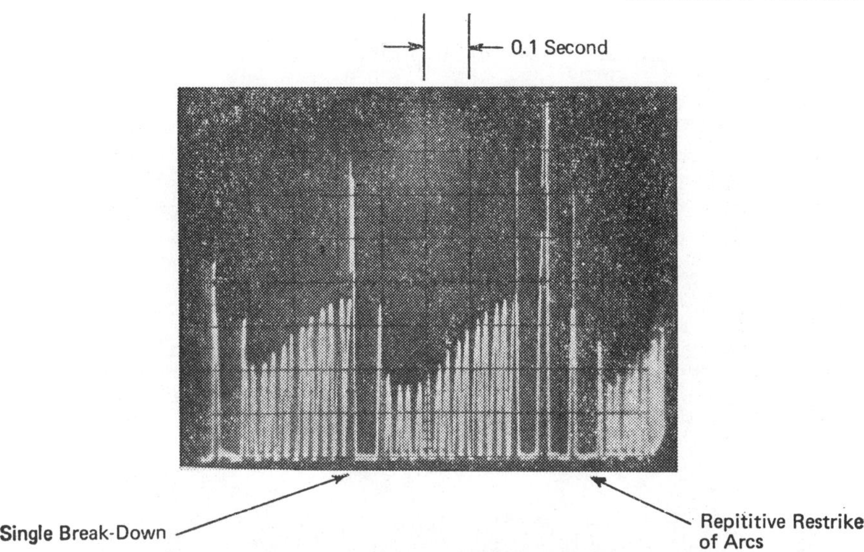

Figure 6–9 Oscillogram of an Automatic Control Able to Extinguish
a Full Cycle of Power Input at Onset of Arc.

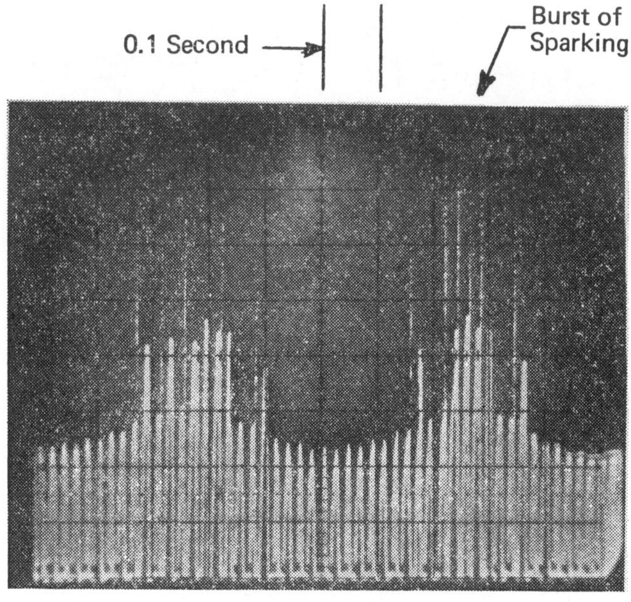

Figure 6–10 Oscillogram of an Automatic Control Response
That Might Be Considered Sluggish but Allows
a Break in Sparking.

149

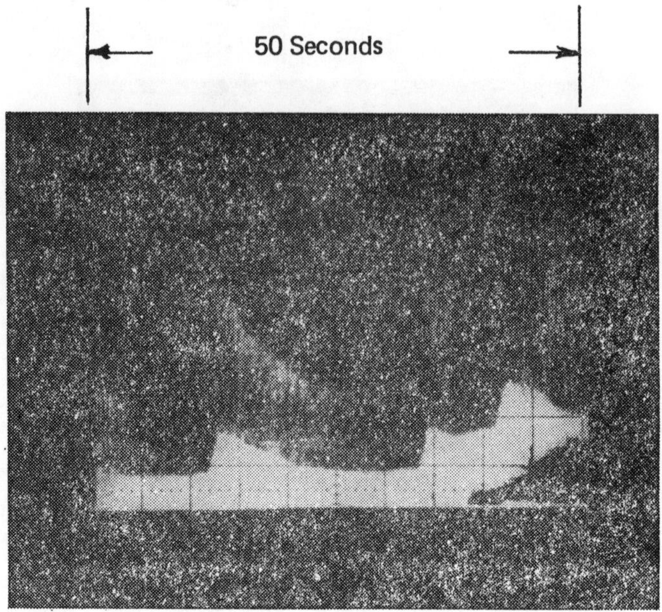

Precipitator Current Waveform

Figure 6—11   Response of an Automatic Control During Rapidly Changing Process Conditions.

Chapter 7
Corrosion Factors

Introduction

Long life of the metal surfaces in the precipitator is related to proper design, in conjunction with sound operating and maintenance practice. Aside from the potential costs of component replacement from corrosion, the cost to the reliable performance of the collector may very well be tied up in electrode failures and damage that either results in field outages or reduced power input.

Corrosion usually includes a loss of metal from either large metal areas or localized zones. Three major causes must be continually guarded against:

1. Temperature excursions that result in acid dew-point or water vapor condensation.

2. Air inleakage.

3. Insufficient application of heat or insulation at exposed wall surfaces.

By just a routine knowledge and interest for the prevention of corrosion, this area of potential difficulty can be kept manageable. I have observed minimal damage in most of the fly ash and cement precipitators, where steady state operation is the norm. Batch operations, especially in conjunction with high moisture contents in the flue gas, are difficult applications.

Evaluation

Careful inspections of the internal components can pin-point the onset of corrosion pockets. By utilizing some of the following information, you can avoid allowing the early evidence of corrosion to grow into costly proportions. Inspections after drastic changes occur in the process and raw materials are highly recommended. If detection of pitting or metal reduction is observed, I suggest utilizing a corrosion expert to help analyze the situation. Losses of discharge electrodes in wide spread areas of the precipitator, or at junction points between wire and shroud, should raise a corrosion alert.

Samples of material from electrode surfaces can be sent out to laboratories for a chemical analysis. But remember that practically all materials handled by the precipitator contain sufficient chemical constituents available for corrosion. The way in which the internal atmosphere or air inleakage is controlled is usually much more of a corrosion factor in the basic industries.

Types of Corrosion

Briefly, the various types of corrosion action would include the following:

1. *Uniform Attack* — Large metal surfaces in contact with acid or caustic conditions. High temperature oxidation frequently takes this form too.

2. *Pitting* — Small areas under intense attack, involving the local breakdown of protective coatings or oxide layers. Such areas can act as starting points for stress corrosion.

3. *Concentration Cell* — Typically occurs in crevices.

4. *Intergranular* — Selective attack along grain boundaries; may be associated with cooling from welding activity.

5. *Stress Corrosion* — High tensile stress, leading to premature metal breakage.

6. *Fatigue* — Cyclic loads leading to metal failure and breakage.

7. *Selective Attack* — One element of alloy is usually involved.

8. *High Temperature* — Particularly where non-protective oxide layers may be present due to spalling, sulfidation, carburization or decarburization factors.

9. *Impingement Attack* — Scoring by flow media exposing bare metal, eroding metal or breaking through scale layers.

10. *Fretting* — At tight metal fits usually subject to high frequency vibrations, possibly galvanic and/or concentration cell actions.

Typical examples of various corrosion types of electrostatic precipitators may be illustrated as follows:

1. *Uniform Attack* — The thinning of collecting plate surfaces, and hopper, shell, and roof plating.

2. *Pitting* — Usually occurs near air inleakage areas, and usually noticed on the discharge electrodes.

3. *Concentration Cells* — Joints and corners of the collector surfaces.

4. *Intergranular* — At welds and collecting plate supports.

5. *Stress Corrosion* — Discharge electrodes at tight bends, especially with certain stainless steels and the presence of chloride.

6. *Fatigue* — Discharge electrodes at the terminations, or in rapper anvil weld areas, or any type of collector surface hanger.

7. *Selective Attack* — Discharge electrodes, for example, where copper coated wires are used with NH_3 conditioning.

8. *High Temperature* — Not too common below 700°F, but can occur on roof plates.

9. *Impingement* — Discharge electrode erosion due to abrasive dusts.

10. *Fretting* — At discharge electrode terminations and fittings.

11. *Dissimilar Metals* — Primarily where copper high voltage connections are made to the frames.

Materials of Construction

Practically all materials of construction in the basic dry-type duct precipitators utilize a low carbon or mild steel grade. Cost and ease of fabrication are key considerations. Deviations have occurred in the design of discharge electrodes, where materials from the high carbon spring steel to the 400 series stainless steels are used for a variety of reasons. Operations such as acid mist precipitators use special lead covered electrodes for corrosion protection. The best materials of construction should be checked out with corrosion experts before any change or retrofit is contemplated. Based on field observations over the years, the following information should provide some directions.

1. If a stainless steel grade application is indicated, stay away from the lower grades such as 304. Stainless grade 316 has been used with success, but I have seen evidence that the 400 series may be more desirable to help minimize damage from chlorides.

2. The use of a high grade carbon steel for hopper surfaces, such as cor-ten, may be desirable in applications where condensation effects are expected.

3. Refrain from using a different metal for the shroud than exists for the discharge electrode. This has occurred in the past, and spit sparking has caused electrode damage.

4. Be on the alert to the type of steel of the bolts used in securing the collector surfaces.

Critical Areas of Corrosion Attack in Precipitators

A large number of physical components of the precipitator become exposed to the potential attack of corrosive atmospheres. The following discussion identifies critical zones of precipitator that are the most vulnerable to metal damage — typically, (1) outer shell walls, (2) roof plate, (3) collector plate surfaces, (4) high voltage system, (5) hoppers, (6) access doors, (7) expansion joints, and (8) test ports.

Outer Shell Walls

The shell walls of the precipitator, generally 3/16" to 1/4" thick, provide pockets of potential corrosion at the support steel and baffles located in the space between the outer collector and the wall. This wall surface is the interface between the inner gas conditions and the outside air environment, and thus is most susceptible to condensation problems. Ledges where dust build-up can occur offer the best corrosion pockets.

Roof Plate

Of special concern for possible corrosion attack is the roof header plate that provides the gas seal at the top of the precipitator and is the support for many insulator and rapper devices. Several factors make this plate critical:

1. The upper support structure provides numerous pockets and ledges for material deposits.

2. During normal operation and during shutdowns, air intake (possibly moist) is often provided in the upper compartments or penthouse.

3. The internal height of the precipitator produces a slight chimney effect during process start-ups, which allows the warmer, moist gas to rise to the cold roof plate. Repeated operations aggravate the corrosion potential at this location.

4. Integrity of the roof deck insulation is difficult to maintain.

Collector Plate Surfaces

The large area of thin metal causes concern for the magnitude of possible corrosion of the collection plate surfaces. There are critical corrosion zones for the collector surfaces which are dependent upon the physical location within the precipitator as well as the specific location of the plate. Briefly:

1. The plates adjacent to the outer shell walls are subject to the maximum temperature variations and possible dewpoint conditions.

2. The lower portions of the collector plates, especially toward the gas outlet of the precipitator, become sensitive to temperature variations.

3. The top portions of the collector plates can be affected by the flow of air through support insulators.

4. The type of fabrication used to reinforce or stiffen the collector plate can provide crevices for corrosive activity.

5. Misalignment or bowing of the collector plate can cause localized metal erosion from excessive electrical breakdowns between the electrode systems.

High Voltage System

Corrosion attack in the high voltage system can take several forms:

1. Stress corrosion of the high voltage electrodes can occur at excessive bends, or can be aggravated by stresses formed during fabrication.

2. The use of dissimilar metals at the terminations of wire electrodes can set up pitting and excessive galvanic activity.

3. Metal fatigue provides a potential area of electrode failure, because of excessive vibration or whipping of the wire,

often related to the design of the equipment. A critical point on the wire is about 10″ to 12″ below the upper shroud.

4. Metal erosion can occur from electrical activity, especially in localized areas.

Hopper Areas

The surfaces of hopper walls will usually be exposed to the lowest gas temperatures in the precipitator. Modern practice is to provide sufficient heat insulation, as well as external heaters, to help maintain the inside wall temperatures above the dew point levels of the gas. Smooth surfaces (including rounded corners), and the severe slope of the hopper walls, generally provide minimal pockets for localized corrosion attack. There are several points of interest:

1. Potential air in-leakage, through the valving at the discharge of the hoppers, can cause difficulties.

2. It is extremely difficult to maintain the integrity of the heat insulation system at the apex of the hopper, especially if the initial design of the hopper appointments are less than optimum. If small sections of insulation are removed for maintenance purposes, subsequent wall cooling can provide a potential for corrosion.

3. Exposure of the walls to any wind-chill factor can cause material encrustations to form on the interior surfaces. This condition can initiate corrosion attack at the interface of the build-up and wall surface.

Access Doors

Access doors, either on hoppers, or for entry to the precipitator proper, are potential corrosion sites. Specifically:

1. Seals at the door plate edges can deteriorate with time and allow cold air in-leakage to occur.

2. The design of the door frame usually allows a build-up of material to occur and then react with the cooler surfaces where moisture-dilute acid condensation may take place.

3. It is difficult to provide and maintain a proper insulation seal at access doors, especially at the hopper locations.

Summary of Physical Design Factors

Prevention of corrosion attack of the steel surfaces in dry-type precipitators assumes greater importance and emphasis as the size of precipitators increases to meet modern collection efficiency requirements. Usually economic factors must be weighed in the application of special construction materials because of the size of the precipitator involved. Often less expensive materials are acceptable if external heat input is used correctly. But a thorough knowledge of how the process variables effect potential corrosion conditions is most important in the design of the precipitator components.

Case Histories and Solutions

It has been seen from earlier discussions that the existence of corrosion is closely related to both the precipitator design and the complex characteristics of the flue gas or particulate matter to be collected.

The basic application of dry-type precipitators for the utility, cement, and steel industries encompasses a wide range of process variables that often produces synergistic effects for the production of corrosion conditions. Table 7—1 shows some typical operating parameters. The following examples cover some common field experiences and solutions for these industries.

Utility Field

The large sizes of precipitators used to capture fly ash stress the need and difficulty in the control of corrosion. Steady state operating conditions are characteristic of the larger boilers, but the periodic start-ups and shut-downs of the equipment promote corrosion.

The geographical location of the boiler can be important in determining the degree of insulation protection. As boiler efficiencies have increased in recent years, the reduction of flue gas temperatures at the inlet of the precipitators has placed a greater burden on protection from acid condensation. Flue gas streams can have between 10 and 45 ppm of SO_3, depending on the sulfur content of the fuel and other system factors that promote conversion of SO_2 to SO_3. There have been studies that have identified the acid dewpoint, related to concentration of H_2SO_4 and flue gas temperatures, but the subject is complex.[9, 10] Table 7—2 shows how I have looked at this relationship.

TABLE 7–1

SUMMARY OF TYPICAL INDUSTRIAL PRECIPITATOR OPERATING CONDITIONS PERTINENT TO CORROSION PROBLEMS

ITEM	UTILITY	CEMENT	IRON & STEEL
1. Gas Temperature, $^\circ$F	240-750	300-700	200-600
2. Gas Flow Rate, 1000 acfm	150-5000	100-500	100-500
3. Gas Compositions (major)			
Moisture (H_2O), % vol.	5-12	5-35	5-30
Oxygen, % vol.	4-10	4-12	5-19
Carbon Dioxide, % vol.	9-15	10-20	3-22
SO_2, % vol.	0.05-0.35	0.01-0.05	0.01-0.05
SO_3, ppm vol.	3-50	neg	neg-10
4. Fuels Fired	coal oil nat. gas	oil nat. gas	coke-oven gas nat. gas oil coke
5. Available Corrosive agents—principal	Dilute H_2SO_4 sulfates moisture oxygen SO_3 alkali (some coals)	alkali moisture oxygen	oxygen moisture SO_2 fluorine
6. Cyclic operation; periods of low load or shutdown with some cold start ups	yes, common in smaller plants	infrequent	yes, variable process conditions

Table 7—2

Sulfur Content in Coal %	Gas Temperature Range at Precipitator Inlet °F	Corrosion Potential
0.5—1.0	220—290	Minimal trouble
1.0—2.5	220—290	Look for trouble
over 2.5	220—290	Troubles probable
1.0—2.5	290—400	Minimal trouble
over 2.5	290—400	Look for trouble

Specific areas of concern and possible solutions include:

1. Older fly ash precipitators commonly used bolted roof plates to provide access for the replacement of internal components. This arrangement provided ample opportunity for uniform metal attack, which usually results in openings for the in-leakage of air and water. The usual solution is to replace the plated area with a solid steel cover, and apply at least 4″ of insulation block.

2. Exessive air intake into the insulator compartments can cause the presence of moisture and oxidation of metal surfaces at these locations. This condition can be minimized, by the reduction of the intake opening to allow approximately 100 acfm to sweep the bushing surface. In many cases, appropriate application of electric heater elements will minimize difficulty in this area. While I like insulation on the hot roof of a penthouse, some corrosion potential is transferred from the hot roof itself to metal surfaces of the penthouse.

3. The normal steady state fly ash precipitator being operated at 290°F and above should experience minimal corrosion damage on collector plate surfaces, except for possible crevice attack over an extended period of time. A periodic program for the removal of ash deposit from the collector surface by use of a water wash can cause excessive rusting and metal loss from the light gauge sheeting.

Collector surface damage has been detected at the bottom of the plates when excessive air in-leakage occurred through the hopper system, especially on high sulfur applications.

Localized areas of metal erosion have been observed on collector surfaces where reduced electrical clearances have caused excessive arc-over. This erosion can lead to holes of substantial size that further aggravate the spark-over condition.

4. Failures of high voltage discharge electrodes from corrosion attack in fly ash precipitators have been far less frequent than when caused by other reasons. Stress corrosion failures in stainless steel wires from the the attack of chlorides have been observed. The failure usually occurs at a bend in the fabrication of the wire. Poor quality alloy steel used in electrodes can produce fatigue cracks, and subsequent failure at locations of stress due to wire whip or defects in the design of the terminations. A tight electrode termination may be a disadvantage in this case. Wire electrode erosion has occurred at locations opposite the upper and lower terminations of the collector plates in

fly ash precipitators, exhibiting high spark-over characteristics. Often pitting and thinning of the electrode occurs prior to failure. Proper application of a steel shroud or ferrule on the wire at these critical locations will eliminate the problem.

The material specification for the shroud should be similar to the steel material used in the electrode to minimize electrolysis between the two components. Especially important is the use of a proper bond between the electrode and shroud to minimize spit sparking at this junction.

Cement Field

High moisture contents, high gas temperatures, and alkali and chloride constituents produce potential conditions for corrosion activity in cement precipitators. The areas most effected simulate the utility application, with more emphasis on the outer walls and hoppers. Briefly:

1. Roof plates and outside walls provide critical zones for condensation of moisture from the gas stream if proper insulation is not applied. Mild carbon steel plates 3/16″ thick have corroded to thin sheeting in a period of several years, under certain conditions. Changes in the raw material mix can enable corrosion producing constituents to attack cement precipitators at an accelerated rate. When the source of carbonate comes from sea shells or coral, evidence of greater corrosion is usually observed. Periodic soundings of the thickness of metal surfaces is highly recommended.

2. Maintenance of insulation and seals on the hoppers is considered critical to help minimize the corrosive attack on the walls. As gas temperatures decrease at the exits of wet process kilns because of additional heat exchange surface, application of external heat on the walls of the hopper will help minimize corrosion conditions.

3. Stress corrosion attack of certain electrode systems have occurred, related in part to the chloride constituents of the gas stream. Other observations of electrode systems have disclosed a general corrosive activity on the edges of square wires over a period of years.

Iron and Steel Applications

The precipitator for iron and steel applications is exposed to many gas and process variables conducive to the formation of corrosion products. High moisture contents of the gas stream coupled with varying temperature conditions provide fertile ground for corrosion. The use of iron ore or waste materials provides an ample source of chemical constituents that can form acidic conditions with excursions below dewpoint levels. Some of the problems that can be realized in this industry:

1. The basic oxygen process for the production of steel is both an interesting and challenging application for the dry-type precipitator. During a relatively short heat period, usually 18 to 23 minutes, a vessel containing 150 to 250 tons of blast furnace hot metal and scrap is quickly refined to steel with release of large quantities of heat and gases. The heat produced during the decarburization portion of the oxygen blow may require 500 to 800 gpm of water in order to contain the flue gas temperatures in the range of 500 to 550°F at the inlet of the precipitator. This temperature lasts for only minutes in the precipitator before a decline in temperature occurs that may eventually reach levels of 120 to 150°F during the off period of the cycle. It is the cycling of high temperature-high moisture giving way to low precipitator temperatures that provides the best conditions for corrosion attack.

Where the outer walls and roof plate become most susceptable to corrosion in the steady state processes, it is the internal collector surfaces that become vulnerable in the BOF because of the light metal gauge used for the plates. The lower portion of the collector surface is most sensitive to corrosion activity especially toward the outlet sections of the precipitator. These areas generally attain the lower temperature patterns during the period immediately prior to the next heat. Close control of air movement through the precipitator between the heats will minimize dewpoint conditions. There is an advantage for two vessels to alternately blow into the same precipitator system. Of importance is the tight control of air in-leakage in through the hopper and screw conveyor system. The maintenance of heat input on the outer wall surfaces of the hoppers will help reduce dewpoint conditions in the lower portions of the precipitator. It is desirable to use corrosion resistant steel in these critical areas.

2. Similar corrosion problems exist in the dry-type sinter plant precipitators. Although the sinter plant is more of a steady state operation compared to the basic oxygen process, gas temperaure excursions frequently occur. The passage of air through the cooling burden of a stopped sinter strand can event-

ually take a precipitator dust build-up into dewpoint conditions after about 20 to 30 minutes of down time. Moisture and gas constituents of the sinter operation provide susceptable corrosion areas along walls and isolated segments of collector surfaces.

Air in-leakage through hopper valves are a classic problem with sinter plant precipitators, primarily because of high negative static pressures required in the gas system.

3. The hot scarfer precipitator faces continuous excursions of moisture, iron oxide fume, and powder flux from the steel surface conditioning operation, and mild steel electrodes will corrode rapidly. The precipitator operates in a saturated gas mode in many cases, and high nickel alloy and stainless steels are required. The wet precipitator has been considered for this application in recent years.

4. The pipe designs used in the blast furnace and even detarring applications are candidates for electrode and shell corrosion, primarily by a sulfur product attack during dewpoint conditions. Insulation and heat for critical areas during the periodic dips in operation is highly recommended.

Although corrosive conditions and problems in varying degrees are facts of life in the practical application of many electrostatic precipitators, good preventive measures can assure that this equipment will have a long life in industrial service. Corrosion damage on duct work and breechings as well as in other process equipment can also effect the performance of the precipitator by the increase in dilution air.

Chapter 8
Methods To Improve Performance

Introduction

The previous chapters have discussed a number of inputs to help you understand the precipitation process better. Knowing ways to effectively inspect and trouble-shoot is very important in the overall picture, but not knowing what to do with a precipitator that does not perform up to expected collection goals has concerned users of the equipment. Everyone loses if a rational step-by-step attack on the problem areas does not begin soon after they surface.

Solutions do not come easily, for there is no black magic to make problems vanish. I have been primarily involved with problem jobs over the years and have observed the good and the bad of the many approaches used to bring existing precipitators into code compliance or to meet guarantee specifications. The fact that I have been called in possibly two to three years after problems occur has allowed me to analyze why a particular installation has worked itself into a hole as far as reasonable solutions are concerned.

Practically all modern precipitators on established processes have sufficient surface area to do the job. This is because most designs are based on empirical data that includes some hidden margin potential. It is when one or more of the fundamental factors in precipitation has been distorted that problems arise. Identify these problems, face up to them, and success stories happen.

Part of the problem has been misconceptions and a lack of knowledge to keep pace with the exponential growth of activity in this field. I often ask a client how many contacts were made before I arrived, and then I bet that the number of different ways submitted to attack the problem matched the number of contacts. I usually win the bet. It is little wonder that nine years after the Clean Air Act, we are still fighting the same performance problems.

There is no need to discuss the basic reasons why so much confusion has existed; the precipitation process is more grey than black and white. The act of crawling through dark and dusty precipitators is one of the least desirable activities for anyone. Unfortunately, this is often how inputs to a performance problem can be detected, and some personnel must be willing and able to look beyond the easily reached zones. At the least, personnel need to stick with complex field situations long enough to implement a solution. Since I have been personally involved over the last ten years in saving clients about $200 million in new precipitators which did not have to be installed, I believe some of the techniques and approaches I used in the field have indicated some measure of success. There were never any easy solutions; most of the major corrective work took months of field activity.

This chapter intends to summarize some philosophy, some corrective directions to take, and how to put together some of the information previously covered. The prime areas discussed will be further detailed in the chapters to follow. I cannot guarantee that if all the suggestions are closely adhered to, that precipitator success is assured. Based on my experiences, the precipitator is very much like the stock market — out to fool you at every turn. But I can state, that if the fundamental factors involved in precipitation are placed in proper perspective, there will be more successes than failures.

The Time Factor

I call the period from that first knowledge of performance difficulty to when a program of corrective steps are implemented, the "Critical Time Factor." Some installations pass through this critical period with possibly the manufacturer and user still at odds as to who's at fault. And the next thing that occurs is a decision made for a retrofit precipitator, because there is no time left to investigate and correct the original problem. The sooner the purchaser and manufacturer get together in a joint program to study, and yes, even share some of the costs of a corrective program, the better the final economic results, as well as the reduction of area pollution.

I am not only discussing design problems of the precipitator components, but the grey areas of low electrical power input, or possibly operating in the wrong gas temperature range for the existing dust characteristics. Work towards achieving full power input, and most of the other problems will diminish in size.

It is usually a lengthy period from the time of specification to the actual start-up, and changes can occur in the input of raw materials. The object is to get the precipitator working under existing operating conditions and to find what has to be modified to reach those performance goals. I believe it is prudent for all parties to work together to evaluate and draw up options for corrections. In some cases, relatively minor changes can cure the problem when there is time to proceed slowly.

Knowledge Implementation

One of the basic factors in correcting precipitator problems is to recognize the complexity of the situation and not jump to conclusions. When things look too clear to you, then that is the time to beware.

Unfortunately, this point is lost to many people who have been faced with precipitator decisions, and classic mistakes have resulted. A healthy respect for the precipitator adversary is the first step to achieve gains in performance. Moreover, complete all the steps in a well thought out program because synergistic factors will generally compound the benefits. Make modifications on the smallest division of the precipitator that can be effectively evaluated. Do not base decisions on any one evaluation, but make sure the data is reproducible after at least two different investigative periods. Remember that every process or equipment change in an evaluation program could produce a counter-effect over an extended time period.

Elimination of Component Failures

Immediately investigate and implement a program for the elimination of precipitator component failures. It usually never gets any better, and efforts to gain maximum performance benefits will become nullified by frequent outages of the electrical fields. Many of the comments of the earlier chapters will help, but discussions between user and manufacturer must address any weak spots that surface during the early months of operation. In some cases it may only take a change in operating procedures. As stated before, component failures in precipitation need not be a normal happening.

Determination of Performance Parameters

When the desired performance is not attained, a series of investigative steps should be initiated to help determine key parameters of the precipitator. Briefly:

1. The composition and characteristics of the flue gas stream is the easiest starting point. Excessive air in-leakage can be determined, and usually, this causes as many performance problems as any other single factor. I am usually concerned with dilution air that begins to occur after the last piece of process equipment. A gas temperature and velocity profile at the inlet face of the precipitator can also provide important input to this initial analysis.

2. Determine what the maximum precipitator performance is under clean electrode conditions and full power input. This may even require a water wash-down of the collector, if heavy collector deposits exist.

3. If high resistivity conditions exist, the test might have to be obtained within an hour after the process levels off. Even then, spark-over will have reduced the power level in the first fields.

4. Determine the effect of rapper reentrainment on the precipitator losses. This can be done with a short outlet test of about 10 to 15 minutes duration, or the period of at least one (1) complete rapper cycle. Immediately after the normal rapping test, repeat with the collector rappers shut-down. The object is to pin-down gas distribution problems, or excessive losses caused by other problems. As mentioned before, the loss due to rapping should be relatively low compared to the total losses.

5. Clarify if a basic resistivity problem exists, or whether localized electrical break-downs are controlling the level of power input. Use some of the techniques described in Chapter 4. Remember that spark-over conditions caused by resistivity will show some uniformity, while internal problems or temperature differences tend to show erratic electrical characteristics between fields.

6. Determine the gross characteristics of the material entering the precipitator. I would first look for quantity and critical substances such as combustible. The size and broad chemical analysis of the sample is of secondary importance for this initial analysis.

7. Several other diagnostic tests would include the comparison of collection efficiency under different levels of gas flow rate or power input. The performance of the collector should follow expected patterns, providing that variations in the process conditions do not adversely effect other test factors. For example,

if the reduction of gas flow rate in a fly ash boiler results in a significant drop of flue gas temperature, then the comparison results could be suspect. The object of these comparison tests is to show up a gas sneakage problem or other fundamental design factors.

Modifications of the Process

As soon as it is apparent that the precipitator will not reach desired performance levels because of a resistivity problem, an analysis of the process should be made. The object is to determine what minor, or even major, modifications may be implemented to overcome the problem. Of course, the potential corrective steps must be placed in some economic order considering both capital and operating costs. The results of this study can be compared to the alternate corrective methods such as the injection of gas conditioning agents, or even additional collector surface in extreme cases.

The main directions that offer the most potential are minor changes in the raw materials, major reductions of the material in the gas stream, improvements in the moisture levels of the gas stream, and modification of the gas temperature level at the precipitator inlet. Each of these directions will include a wide range of options, and several of the corrective changes might be considered at the same time.

I especially lean toward the option that tends to improve production goals or process characteristics. Blending of coal supplies or reduction of ash contents are two changes that can achieve these improvements. The ability to reduce flue gas temperature by extracting heat from the gas stream has much merit.

Proper Evaluation Tools

All the good intentions in the world cannot be implemented without getting a minimum number of facts on which to base corrective judgments. Large expenditures for precipitator corrections that dwarf the allotment for investigations, can be as bad a move as the sums of money spent for major studies when relatively simple field corrective steps are indicated. Early directions to take should be based on a short preliminary survey. Use of proper evaluation tools and adherence to some of the earlier recommendations for the determination of performance parameters, will help during this survey.

Evaluation methods require a minimum set of apparatus to help evaluate short-term performance characteristics of the precipitator. This will generally require in-stack test equipment, utilizing small filter mats instead of the thimble at the outlet of the precipitator. The opacity meter will also help isolate trouble areas, if these detectors are available for separate precipitators instead of the common stack. I know observation windows will be useful in these evaluation techniques, especially for dust loss and spark-over patterns. See details of this method in Chapter 10. A number of different techniques and evaluation tools are discussed in the later chapter.

I suggest that only a minimum of test data be obtained to pin-point the directions to take. Too much data can sometimes confuse the decision-making process. It is especially important to measure and record all pertinent process data and precipitator readings for at least a half hour preceding the actual test period.

Observe and plot the electrical patterns of the precipitator power input before a costly test program is initiated. Remember that too much time consumed in the mechanics of a comprehensive test program will leave little time for daily experimentation. It is much better to run shorter evaluation tests and then have analysis time available to determine the best set of conditions for the next day.

An important evaluation tool is the comprehensive internal inspection, discussed in Chapter 5. Patterns of dust build-up in flues and plenum chambers, as well as a close inspection of the fields that were in electrical difficulty before the shut-down, can offer suggestions for possible corrective action.

Gas Distribution Effects

Even though I place the effect of gas distribution on performance as a distant second compared to proper power input, I believe that most precipitators have potential gains in this area. Usually a baffle installation can be relatively simple, does not require continuous maintenance, and can even reduce pressure drop in the system if vanes are placed in an elbow to correct a mal-distribution.

As I shall cover in Chapter 11, the original model study results may not match the actual patterns found in the field, for a variety of reasons. The important point is that attaining optimum gas distribution where it did not exist before can simulate the same effect as additional collector surface. Use of several of the measurement techniques covered in Chapter 10 will help define the scope of the problem.

Part of the distribution problem may show up as gas sneakage through the hopper system. Therefore, the direction of the gas vectors at the inlet face of the precipitator is as important as the magnitude. Aside from the hopper sneakage problems, a poor layout of the outlet flue can cause cross-currents of flow within the precipitator. See Chapter 11.

Improved Power Input by Gas Conditioning

It seems redundant to say that the name of the game is high power inputs to the precipitator — but it bears repeating. Unless the collector is designed for low power input in terms of size, most precipitators will not perform as expected, unless proper voltage and current levels are attained. A rough rule of thumb is that the emission loss will decrease by ½ for a doubling of effective power.

Unless these desired power input levels can be obtained through changes in raw materials or gas composition, then a change in the electrical characteristics of the dust by artifical means can be an effective and economical approach. The art and science of gas conditioning has been available for many years, but the recent emphasis on low sulfur coal has brought this technique into ever increasing use. Even a conditioning agent to counteract highly conductive ash has been used in recent years. Chapter 12 will cover many aspects of this interesting phase of precipitation. While I believe the first attempts to alter resistivity characteristics should come through process changes, the artificial method has its place for specific installations, where economical process modification is not possible. Before conditioning means are employed, the physical precipitator system should be brought into the best possible shape, especially if frequent component outages occur.

Gas conditioning may be utilized to a greater degree in the future, if these systems are accepted as a substitute for surface area in new installations. However, I also forsee more emphasis in the design stage of new processes incorporating provisions to effectively alter gas conditions. See Chapter 14 on optimum designs for some recommendations in this area.

Chapter 9
Process Factors

Introduction

The importance of the whole process in the performance of the precipitator has been pointed out in previous sections. This chapter details how various parts of the process effect electrical operations as well as other aspects of precipitation. I believe that by discussing the process as it functions in the basic industries, I can show how the same analysis techniques can be used for other industries as well. I will discuss pertinent gas and dust factors and the effects of certain process equipment on the characteristics of the effluent. However, I will not delve into greater process depth and calculations, since that information is beyond the scope of this book and is best left for further study.

It is impossible to separate the precipitator from the process. When I first got started tagging along with some of the older precipitator servicemen, I observed the emphasis they placed on the process. They took pride in pointing out operating characteristics and parameters, and they spent a lot of time studying the process. Those early lessons were invaluable for me. Looking back over the years, I have been involved with some major precipitator success stories that were brought about primarily by modifications to the process.

However, I don't believe that a complete knowledge of the process is necessary. Rather, it is generally sufficient to have a working understanding of the dust generating action, flue gas compositions, and effects of certain equipment on the precipitator. For example, I have found that industrial management would prefer considering process and equipment modifications to substantially reduce stack emissions rather than consider additional collector surface. In many cases, the modifications to the process will result in slightly improved production or even gains in the quality of product. The object is to gain the confidence of the plant management by knowing what you are talking about. If you make a foolish request or use poor judgement, the door to further modifications may be closed. Another point I would stress is that the differences between plants will make a single solution most unusual for all similar precipitator problems. Be alert to attacking flue gas characteristics by many different avenues and options.

Even though a great amount of information has been published on precipitators, very little has addressed itself to the process. Control of the process to prevent temperature excursions can be a much better approach than adding power supplies or surface area. I know some of the following information will add to the understanding of the process.

The Fly Ash Application

The most common use of precipitators is in the collection of fly ash from fossil fuel boilers. Yet this use is the most difficult application for the electrical collector. Combustion itself involves a complex set of variables, and if the infinite number of fuel compositions is considered, the problem becomes even greater. I hope to sort out some of these variables to help increase the working knowledge for the fly ash application.

There are generally five main categories of fossil fuel precipitator applications, the pulverized-coal fired unit being the most common. Table 9—1 shows some typical effluent characteristics of each type. Briefly:

1. Oil — While not generally required for conditions of low ash residual fuel, there are a small number of precipitator units being used for the high ash oils. Concentrations of ash in the flue gas are in the low range and power input problems are generally minimal. The use of fine-sized particulate matter

TABLE 9—1
RANGE OF EFFLUENT CHARACTERISTICS FROM
FOSSIL-FUEL COMBUSTION BOILERS

TYPICAL RANGES EXITING PROCESS BEFORE COLLECTORS

TYPE BOILER OR FUEL USED	EXCESS AIR % by Weight	EXIT FLUE GAS TEMP. °F	GAS FLOW RATE acfm x 10^3	DUST CONC. gr/acf	PARTICLE SIZE % by Weight Less than 10 microns
Fuel Oil	20-40	300-500	100-800	0.005-0.1	95-100
Waste Material					
Wood	40-60	250-450	30-100	0.1-0.4	20-40
Municipal Refuse	40-80	300-600	40-300	0.1-0.8	20-40
Stokers					
Spreader	50-130	300-600	80-300	0.2-1.5	20-30
Underfeed	60-130	350-700	50-200	0.1-0.5	30-40
Cyclone	20-50	250-500	100-600	0.3-1.5	40-80
Pulverized Coal-fired	30-80	250-450	100-4000	1.0-4.5	30-60

requires close supervision of the combustion techniques to minimize the carbonaceous soot type material that can cause maintenance problems.

2. Waste material — There have been a number of older precipitator installations in this field, and in the future, there may be a rapid increase in them, especially in municipal waste burning plants. The fly ash coming off the combustion of waste material can vary from the very small size particulate to large low-gravity flakes of unburned char. The process is often quite variable, which means that uniformity of power input also changes rapidly.

3. Stoker—The application of precipitators to the coal spreader stoker is not new, but in recent years, this use has increased. Ash concentration to the precipitator is relatively low, but higher percentages of unburned carbon grit is possible. Mechanical collectors are frequently used as a pre-cleaner. Since intimate mixing of air and fuel is difficult, the presence of carbonaceous fine-particulate matter is also possible. Power input is still a function of coal quality, but fine carbonaceous material coupled with higher levels of SO_3 can produce problems with low resistivity.

4. Cyclone — I consider this type of boiler to be located somewhere between the stoker and the pulverized coal-fired designs. The coal is swirled into the furnace through cyclonic burners. The size of coal particles and specified low fusion temperatures of the closely controlled fuel presents a much different dust effluent than the pulverized coal application. The ratio of top ash to bottom ash is generally the reverse of pulverized coal-fired boilers, in that about 20% of the ash produced is carried through the system. The ash is generally finer in size, but it could contain an abundance of carbonaceous material depending on the fuel-air mixing and burner upkeep. Low excess air usually exists in the boiler, and that can further intensify the combustible problem. I have observed several installations with 2 to 4% sulfur coals where ample power input to the precipitator was available. Where sulfur content is 1.5% or below, the electrical characteristics may be less than optimum. I will describe an example of a cyclone boiler precipitator in the Case History Chapter.

5. Pulverized coal-fired — This is the most common application with the greatest challenges for the electrostatic precipitator. In fact, most of my experience has been with the bituminous coal grades east of the Mississippi. I have observed every fly ash

characteristic from the ultra-high resistivity to the most conductive situation, and it has not been difficult for me to see what this variable range has done to the basic performance guarantees of the manufacturers. I will present this process in relatively detailed form because of its importance for the precipitator application.

The Pulverized Coal-Fired Boiler

Whether it is the early pulverized coal boilers of the 1920's, or the recent large critical pressure and temperature units that are being considered, the basic combustion concepts remain the same. The object is to finely pulverize the coal, and burn it in suspension with an intimate mixing of sufficient air in the furnace zone. The combustion chamber looks like a large cube with tubes filled with water; these tubes basically comprise the boiler wall construction. Sufficient heat input by the spontaneous combustion creates a fire zone that reaches a gas temperature range of 2200 to 2600°F. There is a partial transfer of heat into the water walls; the high temperature gases then pass through a superheat steam tube section and possibly a reheat section in some of the larger boilers. This process converts the flashing of heated water into either saturated or superheated steam, depending on the final temperature and pressure. Further extraction of heat from the flue gas usually occurs in an economizer tube section and air preheater, where the incoming combustion air is heated before it enters the furnace zone. All the phases of the combustion process effect the performance of the precipitator, and so I will consider each of them.

The Fuel

Reproducible coal characteristics is an elusive condition. Over the course of a day, week, month, year, or decade, the boiler system can see a wide spectrum of ash, sulfur, and other important constituents based on a change in suppliers, variations in the same mine, and the coal storage and mixing procedures. This variability makes for a critical precipitator application, and the care and thought given to this end of the process can work wonders, not only for the precipitator, but also for the process.

Coal Variability

The coal purchaser must be made cognizant of the tight specifications required for successful precipitation. In recent years this task has become more difficult since larger boilers

Table 9—2

Representative Proximate Analysis of Eastern Bituminous Coals (as received)

AREA	Moist. %	F.C. %	Vol. %	Ash %	Sul. %	BTU/lb
Western Pennsylvania	6.64	—	—	19.30	2.03	11,164
Western Pennsylvania (washed)	5.88	—	—	7.75	1.05	13,117
Western Pennsylvania (after mills)	0.58	52.89	23.67	22.86	2.84	11,579
Western Pennsylvania (after mills)	1.45	—	—	19.40	1.97	12,013
Middle Pennsylvania	10.09	56.10	20.46	14.30	1.38	11,153
Western Pennsylvania (after mills)	1.42	47.62	34.22	16.74	2.56	11,819
Mid-West	4.91	48.55	33.60	12.94	1.29	11,804
Northern W. Va.	6.31	45.57	37.12	11.00	3.79	12,436
Middle W. Va. (after mills)	1.77	—	—	9.00	0.97	13,586
Mid-West	12.87	44.89	30.16	12.08	2.59	10,709

require a steady and dependable supply of fuel from many mines. Until the 1940's, a major portion of the utility coal supply came from deep mine sources, and the quality and band of coal characteristics generally included lower ash and higher sulfur contents. After this early period, the move to extensive strip mining tended to produce a much more variable fuel with higher ashes, and often included a lower range of sulfur leached out in some cases by the exposure to surface water. Table 9—2 shows some typical proximate analysis of coals often found in the eastern bituminous mine fields.

It is both interesting and frustrating to know that the analysis of coal from the same deep mine source can vary in composition over the lifetime of the mine. I have observed trends of 15 to 30% variations in both ash and sulfur values over a 5 to 7 year span. In some cases, the ash and sulfur will trend in the same direction while they can be divergent in others. Even though core sampling is usually taken over the expected area to be developed, this information can only tell part of the story. Certainly, the number of simultaneous shafts being operated in different parts of the mine complex can minimize variations at the boiler, but economics will generally dictate this approach.

Coal Sampling

The task of arriving at a valid representation of coal entering the plant or the boiler is most difficult. During any evaluation period on the precipitator, the match-up of actual coal consumed in the boiler to the observed electrical characteristics assumes an important part of any analysis program. If feeder discharge coal is obtained, it is better to take frequent small samples on about a 15 to 20 minute cycle from each feeder during at least an hour's period. The composite coal sample should be riffled into an acceptable lab quantity. Do not rely on main conveyor belt samples, even though they can represent valid daily averages. Obtaining a good sample at the discharge of the mills can also be difficult, unless the sampling technique takes into account the pipe contour to help overcome any bias caused by large particle segregation. I will say more about this in the pulverizer section.

Coal Quality

Proximate analysis on an "as received basis" has been a common method of reporting coal quality over the years. These figures include the surface water content received, the percentage of ash, fixed carbon, sulfur and BTU per lb, and often, volatile matter is also listed. The proximate analysis is generally sufficient for our purposes, but the ultimate analysis is required for further coal quality comparisons and in the heat balance calculations used for boiler efficiency.

The BTU content, based primarily on the values of the fixed carbon, volatile matter, ash, and sulfur fractions, will basically determine the quantity of coal required to attain the desired steam generation in a particular boiler. A general rule is that the higher the ash content, the lower the BTU content of the coal. This relationship presents a double bogey, since the need to burn more coal with high ash contents again increases the net concentrations of material handled by the collector system. Major efforts in recent years toward more sophisticated coal cleaning plants, primarily to reduce sulfur fractions, have resulted in the major reductions of ash contents.

Coal Washing

The file is still open on the effects on precipitator performance by the use of extensive coal cleaning techniques. Major reductions in the sulfur content is often tempered by greater percentage reductions in the ash content of washed coals; some trade-off can be expected. The percentage of sulfur tied up in either the organic or pyritic form of the coal will determine the magnitude of reduction during the high degree of gravity washing, and unfortunately, the pyritic portion normally constitutes the largest part of the sulfur reduction. Some changes in ash chemistry is bound to occur in severe washing, but little has been observed in low level washing. I would be mostly concerned in the reduction of the alkaline portion, including sodium. Of concern is what will be the bottom line in stack emissions, and I would prefer a decrease in precipitator efficiency if it could handle a 7 to 9% ash and still meet code requirements. Some reduction in power input can be expected with effective coal washing.

Coal Blending

Since very few large installations will receive a uniform coal quality, some thought and care in providing a steady mix to the boiler can be most important to the overall system. This may require special receiver methods, to possibly blend truck shipments with railroad car unloading. In many cases, more effective blending can be accomplished at the bunkers. If there are several bunkers feeding one boiler, an attempt to feed different coal into each of the bunkers can be utilized. If coal also comes in by barge, a method to shift barges to blend the coal from different mine sources can be considered. There are a number of different blending schemes, but they will usually be difficult and costly. The goal is to mix coals of widely different characteristics, and the precipitator will benefit from as tight a uniform blend as possible.

Bunker and Feeder Design

Since I consider uniformity important in all aspects of coal handling, the design of the bunkers, feeders, and transition piping to the mills can effect segregation of the fines and coarse segments. The ability to handle incoming wet coal is important. The object is to minimize mill upsets, or starvation, because of hang-up or coal blockage in the lower apex of the bunker or feeder.

Pulverizers

The coal pulverizers, or mills as they are often called, convert the raw coal into a powdered form that can be carried by air suspension to the burners. Several types of mill designs exist, but the object is to provide an optimum range of particle size with less than 0.5% by weight larger than 290 microns, and about 70 to 80% less than 74 microns in diameter corresponding to the 200 mesh screen. While it would be desirable to have all the pulverized coal passing the 50 mesh and about 85% passing the 200 mesh, this condition is usually obtained at higher mill maintenance costs.

Finely sized pulverized coal will usually result in a reduction of combustible, or unburned carbon, in the fly ash. Finer grinding may be required depending on the burner design or the availability of excess air. I tend to want a finer grinding with a conductive ash condition and less fine grinding with a high resistivity ash in the precipitator.

High density segments of the coal, as well as foreign material, will reject from certain mills more than others. Since it is possible to lose a percentage of the pyritic sulfur by this route, it is well to determine this situation if the precipitator characteristics do not appear to match expectations based on the as-received sulfur values.

Sampling for a valid representation of pulverized coal at the discharge of the mill is critical. Utilize recommended traverse procedures of the pipe based on centers of equal areas. If the coal finess results appear erratic, it is wise to determine the isokinetic relationship of the sampling probe and compare it to the actual air velocity within the pipe. The coal analysis of the size fractions are usually different. Also, if one mill is compared against another, make sure that the mills are operating at comparative load levels.

Pertinent mill data is usually important in order to compare test results or other evaluations. Feeder speeds and mill pressures can help ascertain changes in the coal during an evaluation period.

Burner

If high combustible contents exist, or evidence of lower levels of SO_3 formation is present, the burners can be the culprit. The primary air, transporting the pulverized coal from mill to burner, usually constitutes about 10% of the required combustion air. The secondary air is usually swirled into the combustion zone by means of louvers placed around the burner. The design of the burner, how far it is inserted within the burner box, and its relationship with the secondary air can be important in determining the intimate mix of coal and air. Balance of wind box secondary air between burners has caused some problems, depending on the duct configuration of the windbox system.

The level of boiler operation can often determine the optimum settings of the burner apparatus. On smaller boilers with front wall burners, it is possible to impinge coal directly on the back wall, and therefore double or triple combustible contents. Visual observations of the flame patterns can indicate a poor combustion condition. With intimate mixing and the correct air-fuel ratio, the combustion zone is bright without evidence of dark zones and indicates high temperature conditions.

Furnace Characteristics

A wet bottom boiler is considered better than the dry bottom for effluent characteristics. The effect is subtle and relates to slagging conditions, some moisture evaporation, and possible reduction of the top ash portion to the collector.

The large furnace zones (existing above the 2 million lb per hour steam generation) aggravate the start-up problems of the system. It is much more difficult to raise the required heat for optimum combustion within these giant boxes, so extra care in start-up and burner adjustment is needed.

Soot Blowing

Soot-blowing, or the cleaning of ash deposits from boiler tube surfaces, can be considered a maintenance tool. Depending on the size of the boiler relative to the ash characteristics, soot-blowing can be done once a shift or almost on a continuous basis. Wall blowing on some of the larger units has relatively minor effects on the precipitator. If the retract blowers in the upper tube sections are used fairly regularly, most precipitators will handle the extra dust loading with minor effect on performance. However, when the retract blowers are used after long down-periods, minor disruptions may be observed in the precipitator.

Much depends on whether compressed air or steam is used as the blowing medium. Steam usage will probably have the greatest effect on highly conductive ash, where the extra moisture can help reduce what little holding ability exists for the ash on the collector surface. In some cases, steam usage can improve the power characteristics of a high resistive ash. But whether a disturbance occurs on many of the smaller boilers depends on how severe the boiler conditions are modified during the soot-blowing period. Some experimentation is suggested to minimize the quantity of excess air usually employed during these soot-blowing periods.

Air heater soot-blowing usually is most severe in the regenerative rotating air heaters. While the time impact on the precipitator may be short, some consideration could be given to more frequent cleaning, if the normal once per shift operation is too severe. In some cases, the soot-blowing can be accomplished on the air side.

Air Heaters

I usually consider the air heaters as part of the precipitator because of the importance they play in the system. Blockage, air in-leakage, and gas temperature and velocity distributions are the main factors of the heat exchangers that effect the precipitator. Two main types of air heaters are generally employed to heat the incoming combustion air.

Tubular Air Heater

A typical tubular air heater, as shown in Figure 9—1, has the incoming air striking what is called the cold-end tubing of the heat exchanger. The hot gases pass through the tubes on

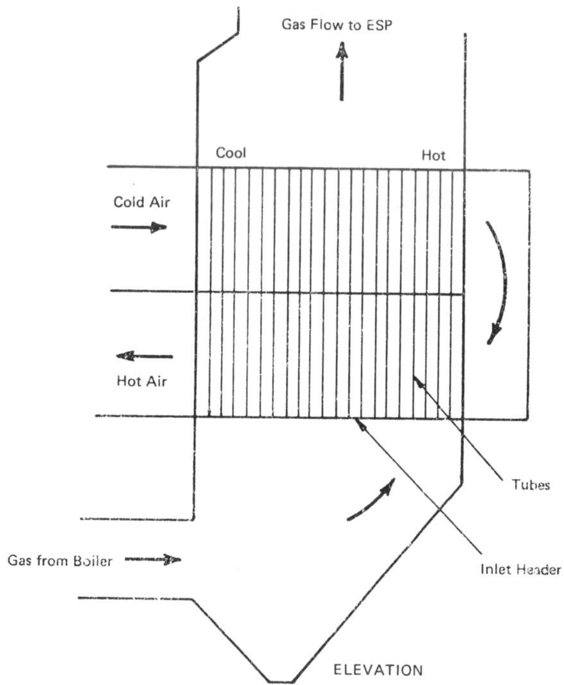

Figure 9—1 Typical Layout of Tubular Air Heater Showing Areas of Cool and Hot Flue Gases.

their way to the precipitator. It can be seen that a temperature differential of the gases will exist at the discharge of the heater, and this differential can be maintained all the way to the collector, depending on the turning vane and flue design.

Air inleakage on a new tubular air heater should not be observed, but erosion and corrosion take their toll in time, and leakage rates of 5 to 15% by volume can occur. Most of these leakage openings occur at the inlet header plate and the first few inches of the tubes, and since the air side is under positive pressure, the air flows into the hot gas stream.

Another problem area that can effect precipitator performance is the tube blockage that generally occurs in the cold-end sections. Condensation occurs within the tube until it becomes completely blocked by hardened fly ash. The magnitude of blockage can effect the gas distribution into close-coupled precipitators. Another subtle problem sometimes occurs with a concentration of combustible and heavy particles at the inlet of the tubular air heater, as noted in Figure 9—1. Again, the design of the flue and vaning at the discharge of the air heater will tend to keep these patterns intact. I believe localized tube erosion results from particle segregation in this type of elbow arrangement.

Rotating Regenerative Air Heater

The use of a rotating regenerative air heater for heating incoming combustion air is quite common. These devices contain a number of finned metal baskets that continuously rotate through the gas and air streams, alternately taking up and giving out heat. Approximate exchange of heat is in the 400°F range. Figure 9—2 shows the basic design of this device and how the temperature bias will exist at the gas discharge. The diameters of the air heater can range up to about 36 feet and can be installed in either a horizontal or vertical position, although the horizontal position is most common.

The temperature spread of 70 to 80°F can provide sensitivity in some large systems, since the high velocity gas flow will minimize the transfer of temperature within the usual short span between the air heater and precipitator. If the average gas temperature noted on the control room chart, or computer, is 290°F, then measurements will generally range from 250 to 330°F at the immediate discharge face of the heaters. How this variation effects precipitator performance will be shown in the Case History Chapter.

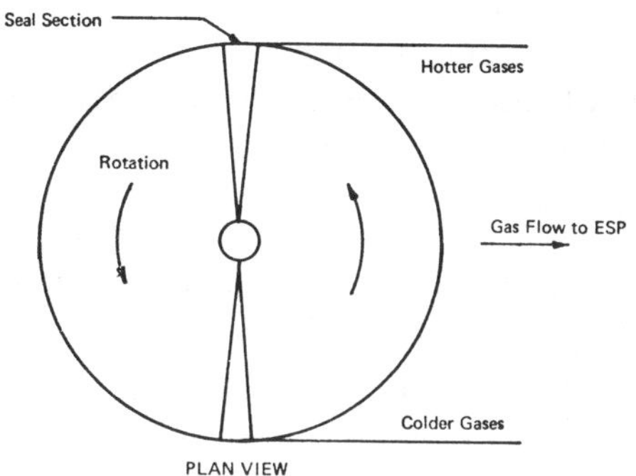

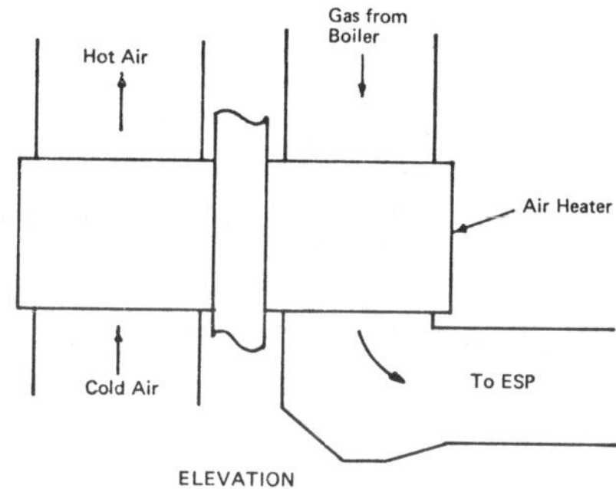

Figure 9—2 Typical Layout of Rotating Regenerative Air Heater
Showing Hot and Cold Zones of Flue Gas.

The design of parallel air heaters which receive and discharge gas in common is a recent concern. This situation has allowed some shifting of gas between air heaters, depending on the pressure drop characteristics of each heater. Even though induced-draft fans exist at the tail-end of the common system, blockage in one heater will tend to push more gas through the clean heater until pressure drops equalize. This condition is best diagnosed by the pressure drops observed across the air-side of the heaters and compared to the F.D. fan operating levels.

Air inleakage between the air and gas sides of the rotating air heater is generally more critical than for the tubular heater. A tight system with proper seal adjustment may have about 10% leakage. Poor seal conditions can show upwards of 20% air inleakage. This higher leakage will usually effect precipitator performance adversely.

Gas Temperature Control

There are a number of methods available to alter the level of gas temperature control to the precipitator. Uniformity of the pattern will usually allow one to operate at lower average temperatures. From a boiler efficiency standpoint, the lower exit gas temperature at the discharge of the air heater is generally desired, but this must be weighed against corrosion and ash blockage problems that tend to increase at these low temperatures. As noted before, the level of SO_3 will determine the acid dew-point temperature, but as sulfur levels of coal decreases, it becomes easier to operate at lower gas temperatures without problems.

Just how much control a specific installation needs depends in large part on the direction the gas temperature change is desired. It is usually easier to raise flue gas temperatures by most of the normal means available to the operator. The most economical method appears to involve the pre-heating of incoming air to the F.D. fans, although every effort to raise flue gas temperatures imposes some economic burden. Some temperature modifications that can be considered before applying any artificial means to change gas characteristics briefly include:

To Moderate Temperature Patterns

1. Bias the F.D. fans to push more air through the air heater where a gas temperature reduction is desired.

2. Speeding up the rotating air heaters will reduce the temperature spread. Doubling the speed of the normal regenerative unit will reduce the temperature spread by about 35 to 45%. Maintenance costs will probably go up slightly.

3. Baffles that cause changes in gas direction and produce turbulence can moderate temperature spreads if sufficient draft is available. About a 1 to 2 inch pressure drop is required by this technique.

4. Gas can be transferred from hot zones to the colder areas by means of small duct or pipe transitions. Whether this correction can be applied effectively depends on the contour and length of flue between the air heater and precipitator plenum. Some effect on gas distribution can be expected, which must be addressed.

To Raise Temperature

1. An increase of excess air through the furnace will usually increase flue gas temperatures. This approach presents other potential problems, such as an increase in combustible particles.

2. Reduction of speed for the rotating regenerative air heaters will raise flue gas temperatures but unfortunately also

tend to increase the spread. Stepless electronic control of the drive motors appear one way to proceed if 20 to 40°F or more temperature is desired.

3.   The best temperature control I know comes from preheating the incoming air at the F.D. fan intake. There is a limit on the heat capacity of the fan components, but closed-loop heat exchangers at this location can overcome severe ambient weather and many of the start-up problems. If hot water or low pressure steam is used for the heat exchanger, then some cross-tie with another boiler on stream, or an auxiliary boiler, must be available. The ability to control at 130 to 140°F air temperatures, under all kinds of cold weather, is desirable.

4.   Most air heaters will have by-pass air ducts where part of the incoming air does not go through the air heater. Designs usually call for a 15 to 25°F rise in flue gas temperatures by this technique. Other techniques are available to shift part of the air or gas flow around the heaters to alter the gas temperature patterns. Some of economizer section gases can be diverted around the air heater to raise the flue gas temperatures at the cold-end discharge of the heaters. The choice of technique used depends on economic considerations of weighing boiler losses against the trade-off in precipitator performance.

5.   The use of steam or hot water coils in the duct work between the F.D. fan and air heater is often found. Unfortunately, coil leaks sometimes make this apparatus difficut to maintain.

6.   Although used as a last resort, closure of some of the tubes of the tubular air heater will allow higher flue gas temperatures. Unfortunately, increased pressure drop and tube erosion may result. Of course, this tends to occur anyway, as the tube blockage increases in time.

To Lower Temperature

1.   Operating at as low an excess air as possible will reduce overall flue gas temperatures; unfortunately, this approach often works against the improvement of electrical characteristics in the precipitator with a reduction of SO_3 formation.

2.   The shut-down, or reduction, of some of the techniques noted in the previous section for raising flue gas temperatures should be explored.

3.   On a major scale, additional boiler tube or economizer surface can be considered. Additional heat can also be extracted

with a low level economizer section placed between the air heater and precipitator. With these methods, boiler efficiency can be improved and a high resistivity condition moderated.

4. While a cool down of the flue gases can be attained with dilution by use of ambient air, this is not considered a good procedure. Some installations have lowered flue gas temperatures by radiant cooling, by increasing the wall surface area exposed to ambient conditions. Use of water sprays is a method not normally used in fly ash applications, but which has merit for critical installations in the 330 to 380°F temperature range. I will say more about this later.

Process Data

The pertinent process information that can be obtained in a pulverized coal-fired station control room include the following:

1. Production rate — steam flow or feed water flow, also gross megawatts if power generation is the end product of the system. This includes in-house power consumption. Air flow rates are usually noted in lb per hour, in the same terms as the steam flow rate, and on the same chart. In recent times, the feed water flow may be the sole indicator.

2. Main steam pressure and temperature, as well as feed water inlet temperature.

3. Any O_2, CO_2, or other monitors of excess air.

4. I.D. and F.D. fan amps and pressures.

5. System reading of drafts from furnace to precipitator outlet.

6. Flue gas temperature before and after air heaters, as well as at the precipitator (if this is available). Usually the air heater exit temperature is judged to be the precipitator inlet temperature, but some care must be exercised in this approach.

7. Fuel consumption rates is sometimes given in percentages, or as lb per hour. Check if any auxiliary fuel such as oil or natural gas is also being used.

8. Pertinent pulverizer data including amps and pressures.

Base Relationships

The amount of coal consumed, or the gas flow rates or particle concentrations to the precipitator can be estimated for most installations. There is no substitute for accurate measurements, but valid estimations are important tools. The following tables show relationships that are based on many field observations on pulverized coal-fired installations. Table 9—3 shows

Table 9—3
Flue Gas Characteristics at Inlet of ESP on Pulverized Coal-Fired Boilers

Steam Flow Rate lb/hr	Gas Flow Rate K acfm	Flue Gas Temp. °F	O_2 % by Vol.	CO_2 %by Vol.	Moisture %by Vol.
400,000	221	430	9.5	11.0	6.6
394,000	254	440	8.0	11.0	10.6
175,000	120	360	8.5	10.3	6.2
4,027,000	2200	302	6.9	11.8	6.4
184,000	113	297	9.0	11.0	7.3
332,000	232	362	7.0	13.5	6.0
220,000	138	308	6.5	11.5	4.9
406,000	268	403	9.0	11.0	7.0
4,505,000	2351	289	5.3	13.5	6.1
288,000	156	330	11.7	7.3	7.1
930,000	585	280	8.0	12.0	5.0
1,253,000	732	290	8.5	11.0	6.4
215,000	124	450	7.5	11.0	4.0

anticipated gas flow rates based on excess air and temperature parameters. Table 9—4 lists a guideline to determine the average ash concentrations which can be expected at the inlet of the precipitator.

Ash Analysis

The fly ash can be characterized by color, size, water soluble fraction, and particle size distribution. Color usually varies from light tan to a dark grey. The tan color, related more to the iron content, usually denotes a satisfactory ash for precipitation. Chemical constituents can be obtained from combusting a small coal sample, and running a pertinent analysis spectrum. Table

185

Table 9—4
Concentrations of Ash at Inlet of ESP on
Typical Pulverized Coal-Fired Boilers

Boiler Steam Flow Rate lb/hr	Gas Flow Rate K acfm	Ash in Coal % by Wt.	Conc. at Inlet gr/acf
184,000	111	12.34	2.19
200,000	123	16.32	3.49
270,000	161	14.30	2.29
1,340,000	811	12.08	1.55
102,000	72	10.03	2.30
185,000	99	6.55	1.80
105,000	66	7.25	1.50
205,000	149	11.63	2.80
80,000	45	5.93	2.00
295,000	157	16.40	1.49
175,000	128	7.75	1.51
184,000	113	12.34	2.19
173,000	110	14.14	3.27
200,000	120	16.32	3.49
406,000	268	10.69	2.00
700,000	354	13.40	2.07
980,000	338	14.10	1.84
4,505,000	2351	11.00	2.53
343,000	176	17.72	3.52

9—5 shows some results of this type analysis on a range of low sulfur eastern bituminous coals. It is preferable to obtain a chemical analysis from ash sampled during a flue gas traverse at the inlet of the precipitator. A difference in the chemical analysis of the ash will probably occur depending on which sample method is used. For example, deriving a mineral ash analysis from one coal sample instead of the inlet to the precipitator sample resulted in a higher Fe_2O_3 content, a higher CaO, a much lower K_2O, and a much lower SO_3 percentage. Any bulk samples obtained from hoppers makes it difficult to obtain a true representation of the chemical make-up of the ash. Table 9—6 shows additional chemical analysis results with remarks on the precipitation characteristics of the ash.

Chemical make-up of the ash becomes more important as the sulfur contents decrease toward 1% by weight in bituminous coal. I do not know the upper limit, but I believe it is somewhere around 1.8 to 2.0% sulfur, depending on boiler conditions. At that point, other constituents begin to assume critical properties. The water soluble sulphate portion of the ash is a valid indicator and should be over 0.7% by weight. The sodium content reported as

Table 9—5
Chemical Analysis of Ash by Combustion of Coal Sample—% by Weight

Coal Location	% Ash	% Sul.	BTU/lb	SiO_2 & Al_2O_3	TiO_2	CaO	K_2O	Na_2O	M_gO	P_2O_5	Fe_2O_3	SO_3
W. Va.	8.63	0.84	13,453	87.2	1.64	0.97	2.0	0.36	0.67	0.53	5.6	0.52
W. Va.	9.49	1.05	13,538	84.1	1.58	0.65	2.3	0.25	0.78	0.58	8.6	0.32
W. Va.	11.00	1.49	13,140	80.6	1.7	1.00	2.8	0.24	0.88	0.36	11.4	0.64

Table 9—6
Analysis of Coal and Inlet Ash Regarding Effect on Performance of Precipitator—
Eastern Bituminous Supplies

Coal Area	Steam Flow K lb/hr	Coal — As Received Moist. %	Ash %	Sul. %	BTU/lb	Analysis — Inlet Ash to ESP — % by Weight SiO_2	Al_2O_3	CaO	Fe_2O_3	M_gO	K_2O	Na_2O	SO_3	Remarks on Precipitator Performance
West. Pa.	4,100	0.6	22.9	2.84	11,579	50.0	20.8	1.5	16.2	1.0	2.7	0.6	1.6	Highly Conductive at 270° F. Good at 320° F.
West. Pa.	4,300	1.3	18.7	2.01	12,368	50.3	26.0	2.6	12.4	1.2	3.0	0.5	0.6	Conductive below 280° F. Good at 310° F.
Mid Pa.	173	10.2	14.1	1.6	11,380	52.3	—	2.8	—	1.9	—	—	0.3	High Spark-over at 320° F.
Mid Pa.	270	10.1	14.3	1.4	11,150	47.5	—	1.4	9.4	1.1	—	—	0.4	High Spark-over at 310° F.
W. Va.	680	7.4	13.4	1.8	13,265	47.7	31.0	2.3	10.5	1.0	2.0	0.2	0.5	High Spark-over 330° F.
W. Va.	700	—	13.4	1.9	11,100	46.5	18.3	1.4	12.6	3.0	0.6	0.5	0.8	Medium Spark-over Levels at 330° F.
N. W. Va.	4,500	6.3	11.0	3.8	12,500	42.8	19.5	6.0	25.8	—	2.2	1.1	—	Highly Conductive at Low Gas Temp. Good at 310° F.
West. Pa.	1,250	5.8	14.9	2.1	12,040	—	29.4	1.0	17.2	1.1	1.8	0.4	1.2	Good Elect. Cond. at 290-300° F.

187

Na$_2$O has assumed some importance, if it exists under about 0.5% by weight in low sulfur applications. Certainly the ratio of basic materials to the alkaline portion must be considered a factor, and the importance of high ferric oxide contents cannot be over-stressed.

Particle Size Analysis

There have been many attemps at measuring the particle size distribution of fly ash obtained at both the inlet and outlet of the precipitator. This is no easy task, but certain key facts tend to help characterize the ash by size and physical appearance. Briefly:

1. There are three (3) basic groups of particles comprising fly ash. One is hollow spheres that include the fine-size fraction, usually in the group that is under 5 microns in diameter. The solid jagged, irregular ash particles that are the basic residue from the coal particles comprise the second group. The combustible, or grit particles usually found in the larger size fractions make the third group. The combustible particle is the product of incomplete combustion not unlike what is formed in coke operations. All of the above can be found in well-operated boilers. Where poor combustion exists, it is possible to form carbonaceous sub-micron particulate matter, normally called soot.

2. By weight, the hollow spheres may only constitute a few percent, while the combustible content could range from 0.5 to 20%, depending on the combustion parameters. Since the combustible particles are quite porous, it is possible to have these particles carried to the outlet of some precipitators in size ranges up to 100 to 200 microns in diameter.

3. The density of the actual fly ash portion of the sample that is controlled by the non-sphere fraction ranges from 2.2 to 2.6 grams/cubic centimeter, while the combustible usually exists in the 1.0 to 1.4 grams/cc density range. However, even though the grit combustible particle has a relatively low density, its large size and surface characteristics give it a substantial role in metal erosion and fan blade wear.

4. The normal fly ash distribution, including combustible fractions, will contain about 2% by weight below 1 micron in diameter, while about 40 to 50% by weight will be less than 10 microns.

5. If the precipitator performance is less than 98% in collection efficiency, the outlet size fractions might show similar distribution patterns to that measured at the inlet. As efficiency increases to the 99.5% range, the precipitator outlet could show a wide distortion in size fractions as compared to the inlet. Specifically, the combustible particles could comprise the bulk of the emission rate depending on precipitator characteristics. The fine-sized segment could also be sizable if low power levels exist in the collector.

Flue Gas Analysis

The gaseous composition of the flue gas from a fly ash boiler varies within a relatively tight band. Major constituents include water vapor, CO_2, O_2, SO_2, and nitrogen in addition to minor trace gases. Table 9—7 shows how the CO_2 content will vary for different levels of excess air for several common fossil fuels. The gas constituents for the coal-fired system will basically depend on the percentages of hydrogen, carbon, and sulfur in the coal, as well as on the excess air employed in the combustion zone and the effects from inleakage air throughout the system. Some key points include:

Table 9—7
Effects of Excess Air on CO_2 in Combustion Products (Dry Flue Gas)

Fuel	% CO_2 cf air per unit fuel	% Excess Air						
		0	10	20	40	60	80	100
Blast Furnace Gas	% CO_2	25.5	24.4	23.4	21.7	20.1	18.8	17.7
	$\frac{\text{cf air}}{\text{cf gas}}$	0.68	0.75	0.81	0.95	1.09	1.22	1.36
Residual Oil	% CO_2	15.6	14.1	12.9	11.0	9.6	8.5	7.6
	$\frac{\text{cf air}}{\text{gal oil}}$	1480	1628	1776	2072	2368	2664	2960
Bituminous Coal	% CO_2	18.4	16.7	15.3	13.0	11.4	10.1	9.1
	$\frac{\text{cf air}}{\text{lb coal}}$	141	155	170	198	226	254	283
Anthracite Coal	% CO_2	19.8	18.0	16.5	14.1	12.4	11.0	10.0
	$\frac{\text{cf air}}{\text{lb coal}}$	130	143	156	182	208	234	259
Coke	% CO_2	20.4	18.5	17.0	14.5	12.7	11.3	10.2
	$\frac{\text{cf air}}{\text{lb coke}}$	132	145	158	184	211	237	264

1. The theoretically dry CO_2 that can be produced from most bituminous coals would range from 17.5 to 19.5% by volume. At the normal operations of 25 to 35% excess air, CO_2 of 13 to 15% should be expected at the outlet of the economizer section. The O_2 value will be the difference between the theoretical CO_2 and the actually measured CO_2 for all practical purposes. Normally, I consider a dry CO_2 of 18.6% as a practical number. Therefore, a measured CO_2 of 13% would indicate that 5.6% of O_2 should be close to expectations for most cases. As air inleakage increases, values of O_2 of 7 to 8% at the precipitator inlet are not uncommon, though undesirable.

2. In pulverized coal-fired boilers, it is rare to measure any CO in the flue gas. For all practical purposes, even though values of 0.1 or 0.2% may be obtained on occasion with sensitive orsat apparatus, the CO content is usually not considered a valuable measurement, unless poor combustion is indicated by other observations.

3. Moisture content of the boiler flue gases ranges in a relatively narrow band which is primarily controlled by the hydrogen content of the coal. Surface and absorbed moisture in the incoming coal will be a secondary input, while the water vapor in the ambient air will add a third minor portion. Most H_2O contents in the flue gases will range within 5.5 to 8.0% by volume at the precipitator inlet, based primarily on the dilution air factor. If natural gas is used as an auxiliary fuel, the moisture content will be higher. For example, 25% of the steam production produced by natural gas burners can add about 1 to $1\frac{1}{2}$% by volume to the moisture contents. About 9 to 10% by volume is an upper moisture range observed for coal-fired boilers with low excess air operation.

4. Release of SO_2 gases does add to some heat generation, and approximately 90 to 97% of the sulfur content of the pulverized coal consumed in the furnace forms part of the waste flue gases. A portion of the SO_2 gases formed in the combustion zone converts to possibly 0.5 to 3% by volume of SO_3 in some equilibrium fashion. The amount converted in the boiler depends primarily on the free oxygen available, the temperature, and the existence of catalytic agents such as iron. That is why high levels of ferric oxide ash, Fe_2O_3, is considered important, and why large portions of pyritic sulfur tied up in iron-bearing substances appears to be significant. While the sulfur combined gases can be expressed in percentages, it is more common to state the SO_2 and SO_3 constiuents in parts per million of the flue gas.

For example, one percent (1%) of SO_2 would be expressed as 10,000 ppm and 0.1% would be 1000 ppm, which is in the general area for a 1.7 to 1.9% sulfur coal.

General Estimations and Calculations

Some base calculations and estimations considered useful in fly ash applications will be given in this section. While I wanted to keep equations and formulas to a minimum in this book, the great interest and activity in fly ash collection does require information to determine important ash and gas formations. I suggest the use of combustion handbooks for further study in these areas. I also suggest that full confidence be gained in any of the measurements or chart readings used in your calculations. There are also a number of published relationships such as the EPA "F" Factor that can be used. I have included the most common relationships that I have used over the years.

Boiler Heat Input

The BTU heat input to the boiler can be estimated by multiplying the coal feed entering the mills by the average proximate analysis of coal sampled.

$$\left(\begin{array}{c} \text{Feeder coal} \\ \text{in} \\ \text{lb/hr} \end{array} - \begin{array}{c} \text{estimated} \\ \text{reject} \\ \text{rate} \end{array} \right) \text{BTU/lb} = \text{BTU/hr input}$$

As noted earlier, if about 9.3 lb/steam per lb of coal is assumed, a rough check would show:

$$\frac{\text{Steam Flow lb/hr}}{\text{9.3 lb steam/lb coal}} \times 12{,}000 \text{ BTU/lb coal} = \begin{array}{c} \text{Estimated} \\ \text{BTU/hr input} \\ \text{(estimate)} \end{array}$$

An approximate heat input can be obtained by using an estimate of the boiler efficiency and the enthalpy values of the main steam and feed water temperature to the economizer. A practical efficiency value that can be used would be 85% for most modern boilers with some adjustment down for exit flue gas temperatures around 310°F or greater.

For example:

Steam flow rate 907,000 lb/hr

Steam pressure 2250 psig

Steam temperature 1040°F

Feed water temperature 455°F

Estimated boiler efficiency 83%

$$\text{Heat Input} \quad \text{BTU/lb} = \frac{\text{Steam Flow Rate} \left(\text{Enthalpy Steam} - \text{Enthalpy Feed Water} \right)}{\text{Boiler Efficiency}}$$

Enthalpy is obtained from standard steam tables.

$$\text{Heat Input} \quad \text{BTU/hr} = \frac{907,000 \ (1479 - 436)}{0.83}$$
$$= 1140 \times 10^6 \ \text{BTU/hr}$$

For the above boiler, the actual heat input obtained by coal analysis and the feeder rate was 1211×10^6 BTU/hr and if the assumed formula was used:

$$\frac{907,000 \times 12,000}{9.3} = 1170 \times 10^6 \ \text{BTU/hr}$$

The above estimates can usually provide satisfactory working numbers for bituminous coal boilers if the coal quantity and coal feed rate are not known.

Gas Flow Rate

Pitot tube measurements are generally used to determine flue gas flow rates. In some cases, the stoichiometric flow can be obtained from heat balance calculations. I have found that the flow rate in acfm for most boilers can be estimated as 55% of the steam flow rate at about 300°F and 6% O_2. Add about 6% to the flow rate for every 1% rise in the O_2 level.

Fly Ash Generation

The quantity of fly ash called top ash that the precipitator receives can vary primarily between 60 to 85% of the ash and combustible generated in the combustion zone. A slagging boiler prone to boiler tube deposition will generally have less top ash than, say, a dry bottom boiler. A spread of 70% top ash to 30% bottom ash can be a reasonable estimate for most modern boilers. The following relationship can be used to convert the ash of coal into the grain concentration normally expressed at the collector. The mill reject quantity is usually neglected.

$$\frac{\text{lb coal consumed per hour}} \times \text{\% ash in coal} \times \text{\% top ash} \times \frac{1}{60} = \text{lb ash/min to precip.}$$

$$\text{lb ash/minute} \times \frac{7000 \ \text{grains}}{\text{lb}} \times \frac{1}{\text{acfm}} = \text{grains/acf}$$

For example:

    Coal Rate 535,000 lb/hr   Avg. Gas Temp. 300°F

    Ash Content 18.7%   Barometric Press. 29.00 "Hg

    acfm 2,370,000   Flue Suction 3" H_2O

$$535{,}000 \times 0.187 \times 0.70 \times \frac{1}{60} = 1167 \text{ lb ash/minute}$$

$$\frac{1167 \times 7000}{2{,}370{,}000} = 3.45 \text{ gr/acf}$$

The concentration of ash is often reported in grains/Std. cu. ft. with standard conditions of 60°F and 29.92 "Hg. This can be done a couple of ways, but consider the gas flow rate converted from actual to standard conditions.

$$2{,}370 \times 10^3 \times \frac{460 + 60}{460 + 300} \times \frac{29.00 - \dfrac{3.0}{13.6}}{29.92}$$
$$= 1{,}560{,}000 \text{ Std. cf/min}$$

Then:

$$\frac{1167 \times 7000}{1{,}560{,}000} = 5.24 \text{ gr/scf wet}$$

The actual measured concentration for this installation was 5.52 gr/scf wet at full boiler load. The concentration may not vary greatly for different boiler operation levels at least down to half load as long as the air-fuel ratio remains similar.

SO$_2$ and SO$_3$ Generation

The importance of the sulfur trioxide to the success of the precipitator has been noted before. Arriving at an estimation is difficult, but it may be as valid as trying to obtain a field measurement. Even though I have no magic numbers here, I feel that these guidelines can minimize the estimation errors. A good sulfur balance may be desirable to determine the true areas of distribution. For example, the sulfur that is tied up in the coal can be partially removed in the mills by rejection, retained in the bottom ash, adsorbed on the fly ash including the porous combustible particles, and exist as free SO_2 and SO_3 in the flue gas. If acid dewpoint conditions are reached in the air heater and precipitator, further conversion to H_2SO_4 and ash adsorption takes place.

From a conservative standpoint, assume that 95% of the sulfur in the coal reaching the burners convert to gaseous SO_2.

Conditions: Coal Feed to Boiler 535,000 lb/hr

Average BTU/lb coal (dry) 12,531

acfm 2,370,000

Ash content 18.7%

Gas density 0.051

Sulfur content (as Rec.) 2.01%
(Dry) 2.04% pulverized sample

Determine total SO_2 production in lb/min

$$535,000 \text{ lb/hr} \times 0.0204 \text{ lb sul/lb coal} \times \frac{1}{60} \text{ min/hr}$$
$$= 182 \text{ lb/min}$$

This number can also be obtained by dividing the total BTU heat input by the BTU/lb coal value times the second two terms.

To Gaseous SO_2

$$182 \text{ lb/min S} \times 0.95 \times \frac{64}{32} \frac{\text{Mole}}{\text{Wt.}} = 346 \text{ lb/min } SO_2$$
$$\text{Ratio}$$

Weight of Total Flue Gas

2,370,000 acfm x .051 lb/cf = 120,870 lb/min

To obtain ppm by weight

$$\frac{346 \times 10^6}{120,870} = 2860$$

ppm by volume

$$2860 \times \frac{28.8}{64} = 1290 \text{ ppm}$$

assume that 1.5% of this value was converted to gaseous SO_3
1290 x .015 = 19 ppm

Excess Air

The percent excess air in the flue gas or boiler is an important calculation based on orsat measurements. One formula used is:

$$\% \text{ excess air} = 100 \times \frac{O_2 - CO/2}{0.264N2 - (O_2 - CO/2)}$$

For most installations the CO content can be considered zero so that only the O_2 value need be known.

Cement Applications

The production of cement clinker has used precipitators for cleaning waste kiln gases and the marriage has been basically good. I was shocked in 1971 when government publications indicated that the electrical collector could not effectively clean particulate matter from the cement kiln waste gases. If I had manufactured precipitators, the cement process would have been the area of major sales efforts. Unfortunately, early precipitators in many cement applications did not perform up to potential, but that was not primarily the fault of the process characteristics. The basic knowledge contained in this book would have solved many of the prevalent problems observed in the field. Retrofits and upgrading activity has corrected many of the early trouble jobs, but even the early designs were capable of collection efficiencies in the high 90's without major modifications. The Case History chapter will contain a couple of examples of this situation.

But while the dust products in the cement kiln waste gases are easily collected, there are still a number of critical areas that must be understood if overall success is to be maintained. The information to follow in this section should help in this regard.

Process Description

The manufacture of portland cement is basically a calcining operation in which various rock and earth materials comprising a kiln feed is mixed and burned slowly in a rotating kiln, driving off CO_2 and leaving a clinker product primarily made up of calcium silicates. The raw feed is crushed in mills and fed to the back or feed-end of the kiln in either a slurry or dry form. If ground wet and processed in slurry tanks controlled at about 34 to 40% water by weight, it is known as a wet process kiln. A dry process kiln will feed the raw materials in dry form with usually less than 1% water by weight. The two methods can produce wide degrees of electrical characteristics for the precipitator.

The clinker usually forms in nodules that roll into a clinker cooler which functions also as a preheater for the combustion air. A single burner at the clinker discharge end uses either natural gas, oil, or pulverized coal as a fuel (although coal has been primarily used in recent year). The ability to use any one of the three fuels during the last five years of tight fuel supplies is important.

Other types of kilns than the rotating method have been used to produce portland cement, but the rotating kiln installed on a slight slope of about $3/8''$ per ft of length has been generally used in modern times.

Raw Materials

Four main compounds are required in the manufacturing of cement. They include:

1. Calcium oxide or lime, CaO

2. Silicon dioxide or silica, SiO_2

3. Aluminum oxide or alumina, Al_2O_3

4. Ferric oxide, Fe_2O_3

Note that these ingredients also comprise a large part of the fly ash from coal burning. In addition to the above, other constiuents include magnesium, potassium, sodium, and sulfur compounds that are generally found in smaller proportions relative to the main group. Moreover, must of the minor constiuents are vaporized and found in the waste gas materials in greater percentages.

A variety of sources are used for the feed stock. In most cases, the cement mills are located close to the source of limestone or cement rock that contains the large input of $CaCO_3$. Calcium carbonate becomes CaO after the CO_2 is driven off in the kiln. Each 100 lb of $CaCO_3$ produces 56 lb of CaO and 44 lb of carbon dioxide gas. All the components needed in the process come from either many diverse sources or just a few, depending on the chemical composition and quality of material. In every case, a blending of the right proportions is continuously being accomplished in the slurry tanks or feed to the mills. Slurry samples, for example, are usually obtained every 3 hours and analyzed for carbonate content. Table 9—8 shows the major sources and the chemistry breakdown of the critical compounds. The quality of some limestone quarries provides a rock that requires minimal additions from other sources. The fly ash from the pulverized coal used as fuel tends to supply some of the SiO_2, Al_2O_3, and Fe_2O_3 ingredients; this must be factored in with the other inputs. There are many unknowns about exactly what occurs in parts of the kiln, but it is an interesting process, and there is still a kind of art to the operating practices.

Table 9—8
Composition of Cement Materials from Various Sources (Typical Values—Range Can Be Appreciable)

Source	% by Weight								Ignition Loss
	SiO_2	Al_2O_3	Fe_2O_3	CaO	SO_3	MgO	Na_2O	K_2O	
Limestone	17	2	1	44	—	2	0.2	0.5	34
Sand	90	3	1	1	—	0.2	0.04	0.4	4
Clay	60	15	4	4	—	1	0.3	2	13
Shale	61	18	7	1	—	2	0.5	2	9
Iron Ore	12	8	64	0.5	—	0.2	0.2	0.2	15
Slag	39	12	1	45	—	2	—	—	1
Marl	6	0.5	3	48	—	0.4	—	—	42
Sea Shells	10	1	0.5	50	0.2	0.6	0.3	0.1	38

Equipment Effects

The design of the equipment used in the cement process is not ultra complicated, but several equipment areas control the amount and quality of material found in the waste gases. Briefly:

1. Extra fine grinding in the mills appear to generate more material in the waste gases.

2. Higher speeds of kiln rotation will tend to generate more material.

3. The type and size of chain section primarily used in the kiln as a heat exchanger and to provide intimate mixing of the wet slurry and waste gases can either raise or reduce the quantity of released dust. Much depends on the density of the chain system. Short sections of dense curtain chains may help reduce dust generation, while the loop systems might allow more to be released. The water content of the material leaving the chain section should have at least 5% H_2O to minimize excessive dusting and chain damage.

4. More draft than required will tend to generate more dust carry-over.

5. Long dry process kilns will usually contain a minimal chain section, but flue gas temperatures are usually controlled at the back-end by water sprays and dilution air on the older units.

6. Recently the dry process rotary kilns have gone to several new types of heat exchangers that preheat the raw feed by intimate mixing with the waste gases.

7. Insufflation, or the method of returning collected material back into the kiln by blowing it into the burner flame (or including it with the pulverized coal) has been done less frequently in recent years. While beneficial in some ways, it has generally caused difficulties in the precipitator by the increase in circulation of dust concentrations including larger proportions of the alkali content.

Effects of Alkalies

A portion of alkalies are contained in several of the raw materials as noted in Table 9—8. Since only a minimum amount can be tolerated in the clinker, usually under 0.6% by weight, the alkalies reported as potassium and sodium oxides must be volatized and removed in the waste gases. A major part of the volatization occurs in the sintering zone above 1600-1700°F. Part of the alkalies condense later in the chain section and sometimes get carried back to repeat the cycle. However, a large part of these condensed fine particles, either in free suspension or attached to larger dust particles, will find their way into the precipitator.

Below 450 to 500°F at the precipitator inlet practically all the alkali has condensed. With sufficient water vapor, I have not observed difficulty with the precipitation of these particles. The quantity of fines may tend to sensitize the inlet field causing some spark-over. Electrode build-up may become a problem and rapping must be effective. The alkali content of the waste material in the precipitator will increase percentage-wise from the inlet to outlet fields. See Table 9—9 for this relationship. The deposition on the collecting surface and discharge electrodes will also show this tendency. Table 9—10 shows the analysis of dust scrappings obtained from electrode surfaces in a wet precipitator.

Hopper dust removal becomes silghtly more difficult with large proportions of alkali material so that the outlet hoppers must be closely watched.

If some condensation takes place within the precipitator proper, the electrode coating problem might become critical. A flue gas temperature reaching 550 to 600°F in the collector can be considered a critical range dependent to a degree on the sulfur content of the effluent material.

Table 9–9

Comparison of Chemical Analysis of Dust Collected in ESP — Inlet to Outlet Fields

Analysis % by Weight	SiO_2	Al_2O_3	Fe_2O_3	CaO	SO_3	MgO	Na_2O	K_2O	Ignition Loss
PLANT 1									
Inlet Half	13.95	2.72	2.10	40.43	5.39	0.85	3.66	7.50	15.40
Outlet half	8.92	2.11	0.94	22.84	15.44	0.94	7.42	16.98	11.29
PLANT 2									
Inlet Field	3.40	1.74	1.34	10.04	38.20	0.39	3.57	39.3	1.53
Center Field	1.60	0.32	0.32	5.02	42.20	0.28	3.95	43.1	3.10
Outlet Field	0.70	0.18	0.12	2.71	45.10	0.01	4.11	45.8	0.88
PLANT 3									
Inlet Field							0.52	3.09	
Center Field							0.54	4.42	
Outlet Field							0.59	5.63	

Circulation of Fines

The recycling of waste materials from the collectors is a critical area for both operating and precipitator problems. This is especially true in the wet process kilns. Aside from the material mixing problems that may occur, the effects on the precipitator can be substantial since the alkalies and other fine particles can promote a circulating load that may be twice that of a non-return operation. This can also produce excessive electrode build-up and cause disturbances in the electrical field.

If only a mechanical cyclone collector catch is returned, the problems are minimized. In some cases, the catch of the inlet field may be returned. It is doubtful that the circulation of the center and outlet field catch should be attempted. A periodic bleed of the system may be necessary to reduce the undesirable portion of the waste material from the process.

Characteristics of Dust

The color of the waste dust found in cement precipitators varies between a beige or light tan to a whitish form compared to the greyish color of the clinker dust. Since this material has been partially calcined, the deposits have a tendency to become

Table 9—10

Analysis of Cement Dust Scrapings from Electrode Surfaces

% by Weight	Cl	SO_3	Na_2O	K_2O
PLANT 4				
Wire Electrodes				
Inlet Field	0.032	18.94	6.67	11.45
Center Field	0.13	33.29	11.46	22.57
Outlet Field	0.11	31.81	11.40	22.90
PLANT 5				
Inlet Wires (bottom)	2.70	18.99	3.60	6.75
Inlet Wires (top)	3.00	18.60	3.80	7.10
Outlet Collector (bottom)	9.60	18.96	10.15	29.40

hard with any condensation of moisture. Density of the collected material is approximately 40 to 60 lb per cf, depending on the make-up and packing of the particles.

Chemistry of Waste Dusts

Table 9—11 shows the span of chemical analyses of waste dusts found in several installations. The analysis of the raw kiln feed is shown for comparision in the same table.

Particle Size

The dust particles leaving the back-end of the kiln usually have a first separation of the large size material in the dust plenum, which serves as a drop-out chamber. The material entering the precipitator can be classified finer in size than fly ash but exhibiting a similar hetrogeneous distribution. In most

Table 9—11

Comparison of Chemical Analysis of Raw Material, Clinker, and Cement Dust Collected by Precipitator

Analysis % by Weight	SiO_2	Al_2O_3	Fe_2O_3	CaO	M_gO	SO_3	Na_2O	K_2O	Loss
CASE 1									
Kiln Feed	21.20	5.12	3.26	65.56	4.40	—	—	—	—
Clinker	21.24	4.95	3.91	64.47	4.20	0.31	0.20	0.39	0.64
ESP Hopper Dust	16.18	5.30	4.72	42.53	2.56	8.88	1.16	5.92	14.50
CASE 2									
Inlet to ESP	5.52	3.21	1.13	21.80	0.46	28.09	1.03	26.00	10.26
CASE 3									
Inlet to ESP	13.76	4.13	2.01	38.38	1.99	7.83	0.78	7.42	22.99
CASE 4									
ESP Hopper Dust	14.49	3.61	1.48	48.22	0.81	3.74	3.18	6.60	16.99

measurements by the BAHCO aerodynamic classifier, I have found about 50 to 60% by weight finer than 10 microns in diameter. These measurements were obtained from traversed samples in breechings utilizing a thimble, and may be slightly biased on the low side. Actually, 55 to 70% by weight, less than 10 microns, might be closer to correct for the suspended particulate matter in most cases. This is especially true as the chain sections have increased in density in recent years.

Flue Gas Characteristics

The normal kiln will operate at low levels of excess air in the burning zone. It is not unusual to observe O_2 values of 0.5% by volume at the back-end of the kiln as measured by a probe installed about 2 to 3 ft ahead of the back-end seal. It is this seal location, at the junction of the rotating kiln and dust chamber, that provides the first major air inleakage amount and changes of 1 to 2% O_2 in the flue gas can often be measured in the breeching beyond a poor seal.

Release of CO_2 from the carbonate stone can provide upwards of 28 to 29% by volume in the kiln. Burner fuel comprised of oil or coal can add 1 to 2% to this value.

201

In a tight system, the decrease of CO_2 and increase of O_2 may only be about 4 percentage points between the kiln and precipitator discharge. More often, a CO_2 of 22% and O_2 of 6 to 7% is observed at the precipitator. In systems where kiln seals, hopper evacution seals, and expansion joints provide excessive in-leakage, CO_2 of 17% and O_2 values of 11-12% are possible. This condition will lower flue gas temperatures and can produce localized spark-over in the precipitator.

Moisture levels in the wet process gas system will closely proximate the H_2O percentage of the slurry in a tight system. Moisture in the dry process kiln will primarily depend on the quantity of water spray conditioning used to control back-end temperatures. Additional moisture is obtained by the combustion of the fuel.

Table 9—12 shows flue gas measurements obtained at a number of precipitator installations in the cement field. The quantity of air inleakage can be quite important since it not only increases the flow rate through the precipitator, but also works against optimum power inputs. Consider 7% of infiltration air on a dry basis for every 1% increase in O_2 measured downstream from the kiln as a rough estimate of this dilution effect. The actual effect on total flow will be less because of the H_2O portion and the decrease in flue gas temperature.

Table 9—12

Tabulation of Kiln Waste Gas Characteristics at Inlet of ESP—Wet Process Kilns

Kiln Size	Gas Flow Rate K acfm	Moisture % by vol.	% O_2	% CO_2	Gas Temp. ° F	Inlet Conc. gr/acf
11' x 175'	102	21	—	10.0	490	2.0
11'-3" x 425'	134	34	—	19.3	575	7.3
11'-3" x 425'	133	31	—	18.0	460	9.5
9' x 290'	75	24	12.0	12.0	535	3.6
11'-3" x 336'	122	32	9.0	15.0	470	4.5
11'-3" x 425'	180	35	9.5	15.0	540	5.7
10'-3" x 330'	90	37	7.2	18.3	600	4.9
11'-3" x 425'	119	42	6.0	19.0	400	1.8
11'-3" x 240'	90	32	9.0	15.0	420	7.6
10'-3" x 240'	110	26	7.1	15.1	600	2.6

Process Factors

A number of relationships can be of use in estimating process factors and flue gas factors in the cement industry. Some of these values are based on tests and observations on only about 25 kiln installations and may not encompass every field situation.

Production Rates

Production rates of rotary kilns are sometimes expressed in barrels of finished cement. Each barrel weighs 376 lb. Generally, the larger the diameter and length of the kiln, the larger the production rate as well as speed of rotation will be. A common kiln size of 11′ 3″ diameter and 425′ long may result in about 140 to 150 barrels/hr. It takes about 570 lb of raw kiln feed to produce a barrel of cement; most of the loss is in the CO_2 release.

Fuel Rates

The fuel rates to produce one (1) barrel of clinker will depend on a number of factors, but a good estimate for the wet process kiln would be 1,200,000 BTU per barrel. The dry process kiln will be 10 to 20% lower and the new long fuel efficient kilns may only require a nominal 700,000 BTU per barrel. Quantities of fuel required per barrel of clinker in the wet process kiln could show:

   70 to 120 lb of coal

    6 to 11 gal. of oil

   700 to 1600 cf of natural gas

Gas Flow Rates

A good estimate of gas flow rate at the precipitator of a wet process kiln would be approximately 1100 acfm per barrel/hr production rate. Table 9—13 shows the relationship between production and flow rate for a number of installations.

Particulate Emissions from Kiln

Based on a number of field measurements, the measurements of waste dust in the discharge gases of wet process kilns gave the following:

Table 9—13
Tabulation of Kiln Sizes, Gas Flow Rates, Production & Dust Emission Rates

Size of Kiln	Type of Process	Gas Flow Rate K acfm	Clinker Production bbl/hr	Dust Emission to ESP lb/hr
11'-3" x 336'	Wet	122	104	4,700
11'-3" x 425'	Wet	180	150	9,000
11'-3" x 400'	Wet	180	133	2,300
11'-3" x 425'	Wet	134	137	6,250
11'-3" x 425'	Wet	133	150	10,900
11'-3" x 425'	Wet	120	150	1,740
9' x 290'	Wet	75	62	2,550
10'-3" x 240'	Wet	110	80	2,900
11'-3" x 200'	Dry	—	210	9,500
12' x 380'	Dry	—	125	8,000

$$1.5 \text{ to } 6.0 \text{ grains/acf}$$

$$3.0 \text{ to } 11.0 \text{ grains/scf Wet}$$

$$1500 \text{ to } 7500 \text{ lb/hr}$$

A range of dust emissions from the wet process would be 30 to 50 lb dust per barrel of clinker. The dry process will normally emit 10 to 20% higher dust rates than for the wet process kilns of similar production rates.

Basic Oxygen Furnaces for Steel Production

One of the most interesting applications for the precipitator is the process for refining steel in basic oxygen vessels. At first sight, everything appears wrong for a successful match between the collector and the process. Production includes a batch operation usually lasting between 18 to 23 minutes and with a continuous change in flue gas and dust characteristics during the entire cycle. However, the electrostatic precipitator has proven to be an effective collector for this process if the design and maintenance knowledge is matched to proper operating conditions. I will list a number of areas in this section where some successes can be achieved.

Process Description

The basic oxygen vessel has surplanted the open hearth as the major steel refining process in this country (as well as in the rest of the world). The process is an extension of the original Bessemer Converter, except that the oxygen impingment occurs on the top surface of the charge rather than as a result of having air blown through the vessel bottom as it was done in the converter. All the vessel designs are similar, but variations occur in size, the type of auxiliary components (such as the fume capture hood), and the collector system. The high-pressure drop venturi wet scrubber has also been used with success on this process. This section will only cover the open hood design where sufficient air is drawn into the hood to basically consume all the CO produced from the carbon in the raw charge materials.

Early installations in this country were small in size, possibly 30 to 60 rated tons with pure oxygen consumption of maybe 3000 cf per minute. Recent designs have produced nominal 250 to 300 tons per heat and utilize 20,000 to 30,000 cf per minute blow rates of oxygen.

The process usually starts with a cold scrap steel charge (this may also include a pig iron portion) which is approximately 30% by weight of the full charge. The hot metal from the blast furnaces is poured on top of the scrap. When the vessel is turned upright, the oxygen lance is lowered through the hood and the impingement of pure oxygen begins when the lance is about 100 to 150 inches from the charge. There must be an ignition with heat evolvement from the bath during this initial period, and sufficient air should be available to minimize the presence of carbon monoxide in the flue system.

From the ignition to the conclusion of the heat, a number of transient events occur with relatively rapid changes in gas flow rate, gas temperature, carbon dioxide generation, moisture levels, and magnitude and character of the fume. These events effect the performance of the precipitator to some degree. Figures 9—3 and 9—4 show variations in these conditions for a typical heat.

The heat and material reactions in the vessel are beyond the scope of this book. In my observations, what is important to the precipitator is an effective input of moisture to the gas stream, and this is somewhat related to temperature levels. Figure 9—5 shows a plot of temperature measured at the inlet of the precipitator versus the quantity of spray water used in a 190 ton vessel.

Figure 9—3 Plot of Variation of O_2, CO_2, and Temperature of Waste Gases from BOF Process.

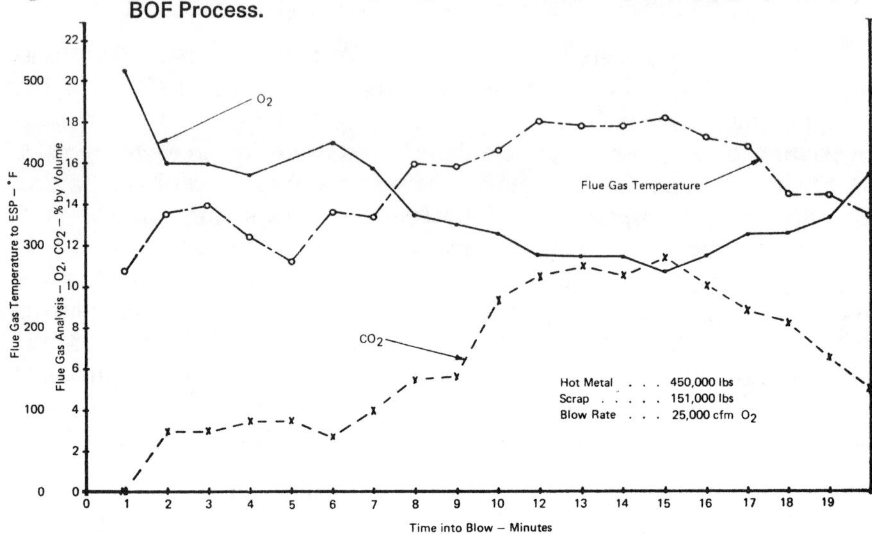

Figure 9—4 Variation of CO_2, Gas Temperature and Spray Water During BOF Heat.

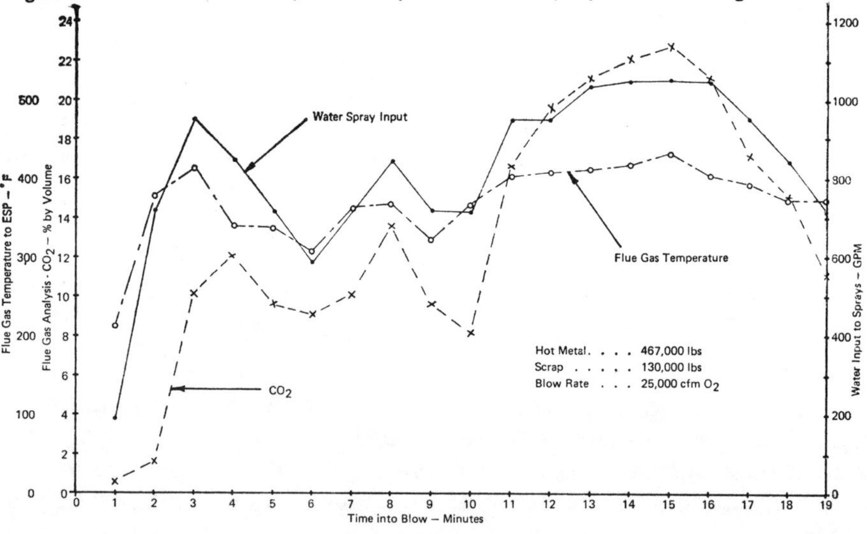

Figure 9—6 shows a much faster plot, at the same installation, of the spray action versus minor changes in flue gas temperature for the first 10 minutes of a heat. Another concern is for the close control of the gas flow rate through the system relative to the power input of the precipitator. A number of other pertinent process conditions will be discussed throughout this section and in the Case History Chapter.

206

Figure 9—5 Plot of Primary Voltage on Inlet Field, Spray Water Input, and Flue Gas Temperature versus Time into Blow for BOF Heat.

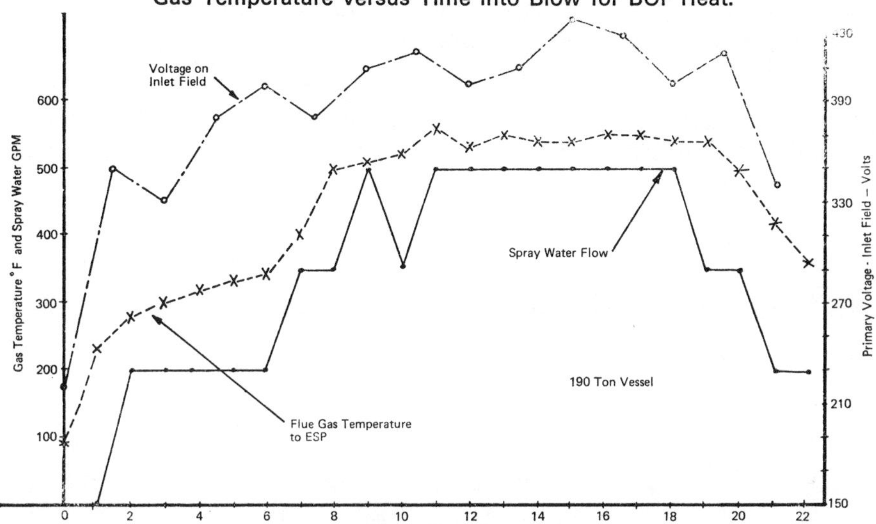

Figure 9—6 Plot of Flue Gas Temperature to ESP and Spray Water Flow for First 10 Minutes into BOF Blow.

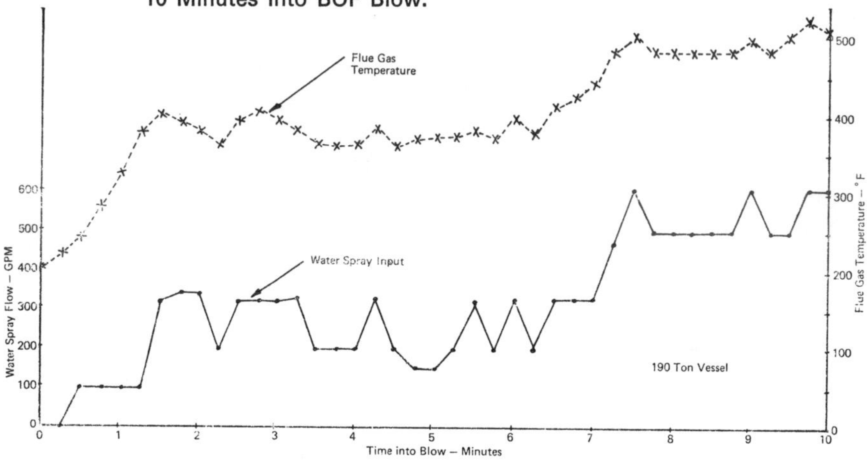

Raw Materials

There are several points that I would like to make concerning the raw materials fed to the vessel.

1. The hot metal supplied by the blast furnace has about 4 to 5% by weight of carbon which supplies the basic fuel in the process. The object of the refining period to to reduce this carbon to a low set point depending on the grade of steel desired.

2. If a major part of the scrap is pig iron, additional carbon is available that increases potential heat release and may require a cut-back in oxygen.

207

3. A minor amount of iron ore pellets is usually added soon after the start of the heat for control of the product. This addition can cause a brief disturbance.

4. Lime, dolomite, and spar are added as flux materials to promote a fluid slag formation, and some precipitator disturbance is likely at that time. This slag absorbs and dissolves most of the impurities removed from the bath. About 8% by weight of the the metal charge is comprised of these flux components.

Effect of Equipment Design and Operation

A number of factors in various installations can effect the levels and performance of the flue gas system. Briefly:

1. The size of opening between vessel and hood determines to some degree the quantity of air required to maintain an acceptable entry pressure drop. A small opening at this location may cause some operating and maintenance problems at the vessel, but is considered desirable to minimize the total air flow into the system.

2. Maintenance of a high temperature in the lining of the hot stack can be of help in promoting the use of more moisture during the initial couple of minutes of the heat. This requires a tight damper seal for the vessel that is down and frequent swing operation of the vessels if this is possible. It is usually better to run consecutive heats within short time spans than to stretch out the heats during low production periods. Much depends on the availability of hot metal from the blast furnaces.

3. A tight seal to minimize air flow through the non-operating vessel will also reduce the total gas flow handled by the precipitator.

4. At least 30,000 to 50,000 lb/hr steam flow would be useful for flue gas conditioning at the start and conclusion of the heat.

5. The banks of water sprays should be spaced and controlled so that a relatively smooth moisture input can condition gases in the hot stack. It is desirable to use high pressure water and nozzles capable of a good atomizing pattern. The control of the spray banks should adjust the quantity of water to the different temperature levels; fine spray controls should be used at the operation level that is sensitive to the precipitator. While not important during the middle portion of the blow, this

type of control can be quite important at the critical start and finish of the heat.

6. I like to see an effective fume deflector at the hood so that minimal air is required during the hot metal charge.

7. Automatic operation of fan control that is set to bring up the required air flow just prior to onset of the O_2 blow and to reduce flow immediately at the conclusion of the heat is quite important to minimize stack puffs during these periods.

Characteristics of Waste Dust

The fume generated in the basic oxygen vessel maintains a relatively uniform characteristic between the heats at any given installation. The prime differences as noted by the precipitator power supplies are related more to temperature and moisture conditions than to major changes in the chemical constiuents. Of course, differences in carbon content and wide variations in quality of the scrap can play a part in the characteristics of the heat.

Chemical Analysis

The chemical analysis of the fume in the waste gases certainly depends on the charge materials. I never asked to have an analysis made since performance goals can be achieved in BOF precipitators without the need to know the make-up of the fume. Table 9—14 shows an analysis from a 1966 V.D.I. publication.[11] Since high moisture content is present, the knowledge of resistivity has also been downplayed in this process. As can be seen by the analysis, the iron oxide portion dominates the make-up of the fume.

Table 9—14

Chemical Analysis of BOF Fumes at Inlet of Precipitator

Constituent	% by Weight		Constituent	% by Weight
Fe_3O_4	70 to 72		P_2O_5	2.50
Fe_2O_3	6 to 8		CaO	10.25
Fe	7		S	0.05
SiO_2	1.45		Others	0.75

From VDI Standards, Publication VDI 2112, June 1966.
Appears based on one set of data.

Particle Size

The bulk of the condensed fume particles entering the precipitator is below one (1) micron in size. There are some large particles of raw material carried up the hot stack but these are usually separated in a large dust drop-out chamber located ahead of the precipitator manifold. Large particles that get through the drop-out chamber can be found deposited in the manifold or will usually drop out in the low velocity conditions of the precipitator. About 85 to 90% by weight will be below one micron in diameter and many sub-micron particles attached or agglomerated in small groups will probably be less than the one micron size as they enter the precipitator. There is some further agglomeration as these particles pass through and are reentrained in the precipitator so that a larger size classification is usually found at the exit of the collector. As with the chemical analysis situation, particle size as such has never been considered important since ample power inputs exist during the times of major fume generation.

I would mention that the particle size distribution in the waste gases will vary during the cycle with larger particles existing at the beginning and end, with the finer fume probably occurring during the decarburization period that generally encompasses the mid 7 to 15 minute period of the heat.

Flue Gas Analysis

The transient conditions of the waste gases entering the precipitator was noted in the process description. Figure 9—7 shows the variation in moisture and flue gas temperature during a heat on an installation with a 16,000 O_2 blow rate on a nominal 190 ton vessel.

Excess air of 150% or more should be sufficient to convert the CO to CO_2 during most of the blow, but this is not always true, and excursions of CO will occur. Usually levels of CO above 7 to 8% by volume will cause the blow to abort because of the explosion danger. Levels of 1.5 to 2.5% may occur for short transient periods.

The most stable period of the process lasts for about 9 to 10 minutes and starts after about 7 to 8 minutes into the blow. This will generally show the highest moisture, temperature, and CO_2 levels of the cycle.

Figure 9—7 Plot of Flue Gas Temperature, Spray Water Flow, and Moisture in Flue Gas versus Time into BOF Blow.

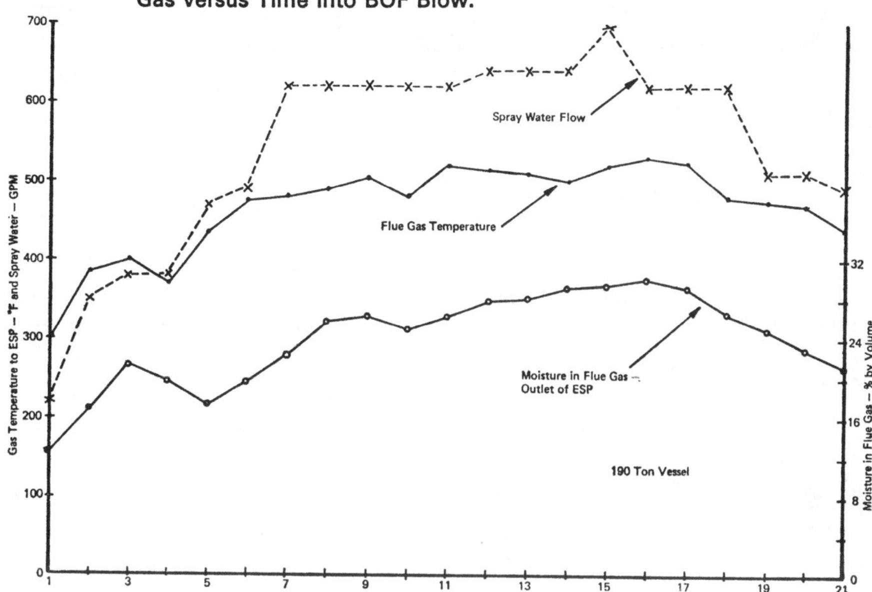

Process Factors

Variability is the normal condition in the oxygen steel making process, but several relationships have surfaced in the installations I have studied.

Raw Material Rate

The vessel takes about 20% by weight more of raw metal materials when compared to the final steel product. About 8% flux and slag forming materials are also added during the course of the heat.

Gas Flow Rates

About 3400 acfm average gas flow rate per finished ton is a good estimate for the gases measured at the discharge of the precipitator. This figure can deviate by ±10% at any time during the blow. The acfm through the precipitator ranges from about 35 to 40 for each scfm of the O_2 blow.

Dust Generation

Very little has been done in recent years to test or identify the variation in fume generation during the span of the heat.

I would judge an average of 6 grains/scf wet with levels of 9 to 12 gr/scf wet during the peak of the decarburization period as being reasonable figures. This would equate to about 110 lb entering the precipitator for every ton of steel processed during the time period of a heat.

Sinter Strand Application in Steel Industry

If there is an ultimate tricky and variable process for the application of a precipitator, it is the sinter strand or machine used for the preparation of blast furnace burden. The sinter mix, or burden, is comprised mainly of iron bearing ores, flux, waste materials from various steel mill operations, and fuel. The object of the sintering process is to increase the production rate of the blast furnace by supplying a material that is chemically and physically prepared for furnace charging.

The fact that the sinter process attempts to recover usable elements from many of the waste dusts and mill scale in the plant is a definitely positive factor. But the variability and chemical make-up of this wide assortment of raw materials makes for an interesting but difficult process. The precipitator installations in the early 1970's still did not recognize the process difficulties, and precipitator surface area was considered designed on the low side. The precipitator can do an effective job but the application must take into consideration all the variables of the system's design.

Process Description

The different materials are usually fed from separate bins onto a common belt. This material is mixed and moisture is added to form a handling consistency for feeding into the inlet of the machine ignition zone. The machine itself is a slowly moving group of pallets, each pallet about 6 feet wide and 2 feet long, and operated very much like the track of a military tank. Sizes vary, but a common machine would be about 95 feet long, 6 feet wide, and travel between 5 to 8 feet per minute.

Height of the bed varies between 9 to 12 inches and this level is controlled by an adjustable bar at the entrance of the ignition furnace. The burners use either coke oven, natural gas, or a mixture that raises the temperature of the top surface of the bed in order to ignite the coke or coal particles mixed in the burden. Then the burning process is maintained by air flow passing through the permeable bed as it travels the full length of strand. A desired bed condition might show about a 1″ band of burning still present at the bottom zone of the burden as it is dislodged from the pallet at the discharge end of the strand. The shed pallet moves over a sprocket for a return trip to the front of the machine.

The air as it passes through the bed picks up products of combustion and particulate matter in varying proportions throughout the length of the strand. These gases enter a series of hoppers called windboxes from which the gas is transferred to a bustle pipe. The bustle pipe usually carries the gases directly to a mechanical collector, I.D. fan, and precipitator, or, in some cases, the precipitator will be located after the I.D. fan.

Raw Materials

You name it, and you will probably find it in the sinter mix. Table 9—15 shows the most common sources of material that make up sinter. Not all the same sources are used at every installation but at least 9 to 12 different inputs can be anticipated. Various ores, blast furnace collector dust, hot and cold returns, dolomite, and cyclone precipitator catch returns are generally common to all in addition to coal or coke breeze. Table 9—16 shows the chemical constituents of the burdens at several plants. After tumbling in a mixer with a 3 to 6% H_2O addition, the feed material takes the form of small pellets about $\frac{1}{4}$ to $\frac{3}{8}$″ in diameter. Usually low sulfur bituminous, anthracite, or low sulfur coke breeze with sulfur contents under 0.7% by weight is utilized as the fuel mixed in with the raw materials.

Table 9—15

Sources of Raw Materials for Sinter Strand Operations

Constituent	Range % by Wt.	Constituent	Range % by Wt.
Iron Ores	33—38	BOF Slag Fines	0—6
Pellet Fines	12—24	Filter Cake	0—4
Blast Fce. Flue Dust	5—9	Cold Returns	6—18
Mill Scale	0—8	Hot Returns	0—6
Limestone	11—25	Sand	0—3
Coal or Coke	2—6		

Table 9—16

Typical Chemical Composition of Sinter Dust Entering Precipitator

Constituent	Range—% by Wt.	Constituent	Range—% by Wt.
Fe_2O_3	30 – 40	MgO	1 – 3
Al_2O_3	2 – 6	Alkali	1 – 4
SiO_2	4 – 15	Cl	1 – 6
CaO	6 – 14	Carbon	1 – 4

The make-up of the raw materials can be related to a basicity ratio desired in the final sinter product. The higher the basicity (or base) to the acid ratio of five key compounds, the greater the blast furnace's eventual production. Basicity ratios of 1.0 to 1.4 could be considered low and do not generally present resistivity problems in the precipitator. Recent trends to ratios of 3.0 or more have provided some difficulties with electrical characteristics as well as electrode build-up problems. The compounds used to calculate this ratio are:

$$\frac{CaO + MgO}{SiO_2 + Al_2O_3 + TiO_2}$$

Equipment and Process Effects

A number of design factors in the process equipment and control of the process will effect the performance of the precipitator. The sinter operation is a classic case where benefits to the production rate and quality of product will generally match precipitator benefits when attention and modifications are made in certain areas. Briefly:

1. Uniform blending of the raw materials is important in efforts to minimize process cycles.

2. Close control of the carbon level in the raw materials can minimize the need of frequent fuel changes to correct erratic burden characteristics. Changes made at the control floor are after the fact, and gas and temperature cycles can be appreciable before the proper fuel correction is finalized.

3. Any efforts to minimize machine stoppages will greatly decrease precipitator disturbances. With the machine stopped, the continued flow of air through the burden raises gas temperatures while decreasing moisture levels which combines to increase re-

sistivity conditions. This is a transient condition that may last for 10 to 15 minutes and cause disruptions in the precipitator.

4. Maintenance of good finger bars in the pallet will minimize the development of holes in the bed, that will short circuit air into the windbox zone.

5. Reduction of air leakage into the windbox through seal-bar openings and worn pallet ends is important. The seal-bar drops down at the sides of the pallet and seals the approximate ½″ opening between the side and wheels of the pallet. This bar can hang up with dust particles or warped pallet sides. If there are about 50 pallets riding the top side of the machine, it is not hard to see the amount of open area that can exist if 8 or 10 of the bars do not drop down on the rail. If the jammed seal-bars happen to occur in one area, the effect on system vacuum and gas temperature is often pronounced.

6. In fact, the high negative static pressure of the machine in the range of 20 to 26″ H_2O will also invite cracks or holes in the windboxes as well as a poor fit of flaps at the bottom of the hoppers. This will allow appreciable dilution air to be drawn into the system. Suction of 30″ H_2O or more at the I.D. fan provides a tough battle to maintain a tight seal of the overall system.

7. Too much water in the burden or excessive fuel used in the mix can blind the bed, and not only raise inleakage air, but slow the burn-through process. Any changes to the burden in these two areas should be of low magnitudes to help minimize the hunting or cycling of temperature and fume characteristics.

8. Reduce the circulation of condensed fines and alkalies by discarding the catch of the precipitator. While it may be satisfactory to return the mechanical collector catch to the burden mix, I have observed that by not wasting this relatively small amount of material, difficulties always arise.

Characteristics of Particulate Matter

It is difficult to characterize the dust and fume coming off the sinter bed by any means. There are three or four different populations of material and the size can vary from sub-micron to relatively large size rocks. Generation of material is different in each windbox with those closest to the ignition burners receiving some of the raw materials that drop through the pallet openings, while the iron oxide and alkali occur in various degrees throughout the strand length. Ignition furnace temperatures of 2500 to 2600°F initiate the melt down of the top of the mix, while the bulk of burden remains relatively untouched for a number of feet of strand movement.

Chemical Composition

The chemical composition of the material entering the precipitator is shown in Table 9—17. Table 9—18 shows a number of analyses obtained at the discharge of the precipitator. The type of raw materials and sources will determine the potential range of compositions observed. But the burn characteristics of the bed will also determine the magnitude of some of the ingredients.

Particle Size

With the usual mechanical cyclone collector in the system, about 98% by weight of the material entering the precipitator will be under 10 to 15 microns in diameter. About 5 to 8% by weight will probably be under one (1) micron. At the discharge of the precipitator, one might find any combination of particle size distributions depending on the reentrainment conglomerates that occur. Probably in pure form, the stack discharge will contain about 50% by weight below 5 microns in diameter.

Table 9—17
Typical Chemical Composition of Sinter Dust Leaving Precipitator

Constituent	Range—% by Wt.	Constituent	Range—% by Wt.
Fe_2O_3	40 – 55	Cl	2 – 10
CaO	6 – 16	SiO_2	2 – 6
Alkali	4 – 8	Al_2O_3	2 – 4

Table 9—18
Typical Composition of Sinter Strand Waste Gases to Precipitator

Constituent	Range—% by Wt.	Constituent	Range—% by Wt.
CO_2	1 – 2	N_2	76 – 78
O_2	16 – 19	SO_2	0.01 – 0.03
CO	0 – 2	H_2O	3 – 12

Flue Gas Characteristics

The composition of the windbox gases entering the precipitator can vary greatly but within a relatively tight band for any particular installation. Figure 9—8 shows the variation of moisture and temperature over a two hour period at one installation.

It can be seen that high levels of O_2 and low levels of CO_2 are characteristic of the process while the moisture band will generally vary between 4 to 10% by volume. Levels of CO are possible at times but is considered abnormal for most installations. Sulfur dioxide emissions are usually in the 200 to 350 ppm range for low basicity operation.

The temperature variation at the precipitator during a normal sinter run might cycle between 30 to 40°F and will be generally found within the 200 to 330°F range. The higher precipitator temperatures for any specific installation will be found with 550 to 700°F levels in the windboxes near the discharge of the strand. Operation at these higher windbox temperatures usually denotes acceptable sinter quality and a maximum contribution of iron oxide to the gas stream.

Figure 9—8 Plot of Flue Gas Temperature and Moisture for Waste Gases from a Sinter Strand Operation.

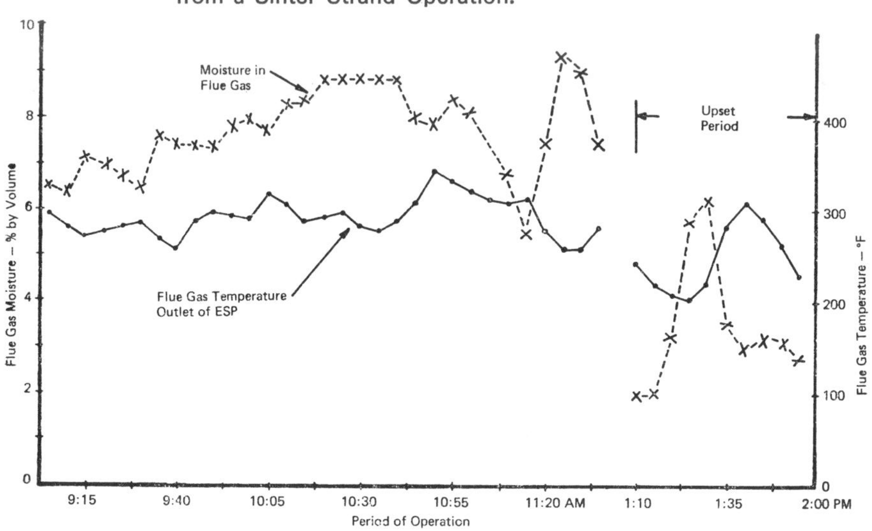

Process Factors

The sinter process variability has been stressed and this variability also makes it difficult to project dust concentrations to the precipitator. Measurements ahead of the mechanical cyclone have shown a range of 0.4 to 2.5 gr/scfd. The range of concentrations to the precipitator can vary from 0.1 to 0.5 gr/scfd for most installations.

A gas flow rate of 1800 scfmd for each ton per hour of sinter production is a reasonable figure for the precipitator. This indicates a nominal fume emission to the precipitator of about 4.5 lb/ton of production.

Chapter 10
Test and Other
Evaluation Techniques

Introduction

Correction of precipitator problems or the program for the upgrading of these collectors can only be as good as the background data collected. It is poor practice to rely on data that is a couple of years old. I intend to discuss some ideas on various evaluation techniques that have proven successful in my work. They are not all inclusive, but it is the purpose of this chapter to stress reasonable and practical tools. More involved techniques are sometimes warranted, but usually only to confirm and elaborate on preliminary results.

Indicators

Any simple method or technique is useful if it helps provide an indicator of performance or dust characteristic. The point is not to worry about achieving a true representation or magnitude, but to develop confidence in a short term indicator of conditions. This indicator can take several forms:

1. Memorize the optimum electrical characteristics of several critical fields as being indicative of good performance. Usually the stack appearance and precipitator performance will match the levels of these readings. In a small collector, the inlet field will be most critical. This technique is useful when there are a large number of power supplies and more than one precipitator.

2. The pattern of the length of time it takes to remove dust from hoppers can often provide valid background data. I would be interested in the chamber to chamber patterns for similar fields as well as the inlet to outlet patterns. This could pin down potential gas balance and distribution problems as well as excessive reentrainment.

3. Adhesive tape can provide a short time visual observation of dust in the flue gas at the discharge of a precipitator. This technique is especially important to detect combustible grit or provide a rough indicator for the size make-up of the effluent.

Figure 10—1 shows the simple way a test stick can be made up. I have used this technique for about 28 years and it can pinpoint differences between chambers or the locations of excessive dust loss within units.

The stick is inserted with the sticky part of the tape opposite to the direction of flow. White medical adhesive tape has been found useful for this technique. Quickly seal the opening of the test port. Turn the stick 180° for about 10 seconds and then pull it out for evaluation. The length of time the tape is exposed should be less than it takes to uniformly cover the adhesive surface. Sometimes the individual large particle can be clearly seen by eye, and other times, a magnifier will be needed. For purposes of comparison, coordinate with rapping cycles, or completely turn the rappers off for a short period. The tape approach can even evaluate the qualitative effects of rapping.

Sampling for Dust

The knowledge of effective testing for dust must be acquired if precipitators are to be evaluated properly. I have been extensively testing since my early serviceman days, and I am sorry to say that the test business has gone through the same hodgepodge of problems that the application of the precipitator has. Testing for a true representative sample of particulate matter from a process requires knowing the process. And testing a precipitator requires knowing the characteristics of this collector. Unfortunately, under the great haste of the Clean Air Act of 1971

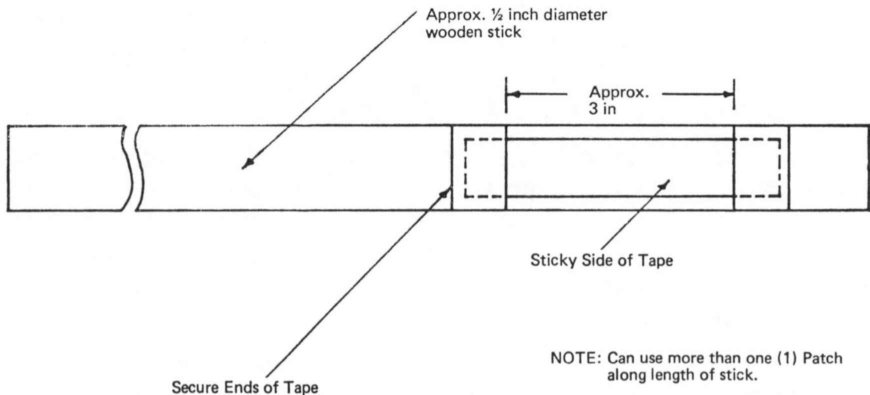

Figure 10—1 Suggestions for Diagnostic Probe using Sticky Tape for Collecting Particles in Gas.

to obtain numbers, a sizeable group of test results have generated some erroneous conclusions about the performance capability of the electric collector. In addition, some questionable test methodology has been promulgated since 1971.

Two basic methods are widely used in the field today. The EPA Method No. 5 draws a sample through a nozzle and long probe in the gas stream in conjunction with a filter placed in a heated box outside the flue or stack. The in-stack test, for years referred to as the ASME method, utilizes a nozzle and filter holder within the gas stream. In both cases, the quantity of gas drawn through the nozzle is at the same velocity as the main gas stream, and this procedure is known as isokinetic sampling. I am not going into a treatise on the pros and cons of each philosophy since this has been hashed over for years. While each method can provide valid test results, my personal opinion is that the in-stack filter can minimize a number of potential trouble areas present with the out-of-stack filter test. Specifically, the out-of-stack filter train has on occasion produced chemically reactive materials between the nozzle and filter that may not represent the actual material found in the gas stream. Again, it's a matter of the knowledge and caliber of the test group that may decide the final test result. Source measurements are tough to obtain, so any test method that may be questionable can only complicate the situation.

There are two groups of test measurements for particulate matter that the user of precipitators can experience:

1. Performance or emission sampling follows relatively tight test procedures. These tests will include simultaneous inlet and outlet samples for the performance result, and may only include the outlet of the precipitator for the emission measurement. The tests are characterized by long time periods and comprehensive sample traversing in the cross-sectional area of the flue. The outlet station test procedure will conform to regulatory acceptance for the emission measurement; this is also stipulated as the test procedure used for the guarantee of the specifications.

2. The purpose of the diagnostic group of tests is to define problem areas and evaluate modifications in the process or precipitator. Methodology of these measurements involves confidence and the reproducibility of results that can be substantiated by full scale tests at a later time. The measurements are generally of short time duration and usually involves samples from small sections of the flue.

The full-scale performance or emission test must be comprehensive, but even this test can produce diagnostic results if set up properly. Diagnostic testing, by its very term, denotes a searching approach, but its results can often be representive of true emissions. The following brief sections intends to outline some suggestions for both procedures.

Performance and Emission Tests

The performance test report should be a permanent record of all critical data obtained during the measurement period. It is amazing to me that thousands of samples are accumulated each year that become worthless in time because key process or precipitator data is not included in the report. One set of electrical readings from the precipitator during a test is usually not sufficient for about 98% of the installations. More than one set of readings is necessary where a variation in the process or fuel can change the electrical characteristics of the precipitator to effect the emissions. Therefore, without this knowledge that is gained from a number of readings, the test result could not be readily explained.

1. Make sure the process is not in a transient or upset mode prior to testing. I personally like about 1 to 2 hours of stabilization for a steady-state process.

2. Record all pertinent process operating data on a 10 to 20 minute frequency. Include this in the report.

3. Continuously record the precipitator electrical readings on a 10 to 20 minute frequency throughout the test period. Include this in the report. On critical jobs, a recording voltmeter and ammeter can be placed on an inlet field.

4. Break the full test down into the smallest sub-divisions commensurate with confidence in the handling and weighing procedures of the filter. This is as important at the inlet station as the outlet. This procedure may add about 5% more time to the test period, but the end results are more than worth the efforts. For example, with two precipitator units, use separate filters for each outlet flue instead of a common filter. In a large breeching or stack, use filters over shorter time periods to pin down differences in patterns or results that can be explained by process upsets or transient trends. The high flue with 7 to 9 test ports in the vertical wall can be broken down into 2 or 3 filters. The sampling train nozzle or probe need not be processed for each filter sample under this procedure.

5. If at all possible, operate simultaneous trains between different sections of the outlet flue or between separate flues.

Diagnostic Sampling

It is quite difficult to trouble-shoot or evaluate the precipitator with the full-scale comprehensive test. Aside from the time factor on the test site, extensive process time of the sample adds to the problem. At the same time, short term indicators (such as the adhesive tape) cannot be used with confidence for quantitative evaluation. I prefer a short time dust test that is obtained at isokinetic conditions. Test periods as short as 10 to 20 minutes can provide a representative comparison of conditions between parts of a flue or for the effects of rapping. Some key points:

1. Use small filter mats of glass fiber that range in weight from 100 to 130 milligrams. Look for a weight pick-up of at least 10 milligrams.

2. Have sufficient filter holders that field changes are held to a minimum. These tests are nearly always of the in-stack method.

3. Always attempt to apply a control filter, either at one spot while the second filter is moved, or in one precipitator duct while changes are being made in the other unit.

4. Experiment in the flue or stack to obtain a representative test location. From experience, a point of average velocity about half-way up a vertical flue or a point of average velocity about two (2) feet inside the stack wall will generally provide a valid sample. I suggest the use of about one day to ascertain patterns depending on the scope of the program.

5. In this type of testing, nozzle processing is not done, but I suggest running a flexible pipe cleaner through the nozzle at the conclusion of each filter run. I generally add 10% to the test result for this disregard of any nozzle deposition.

6. While these short time tests can be valid representations of emission, they are used primarily to spot areas of maximum dust losses. Usually the bottom of a horizontal flue is sampled to attain the greatest rapping losses, because changes in rapping will have the greatest effect at this location.

7. In most diagnostic testing, learn to go through several cycles of process or precipitator modifications to gain further confidence in the results.

8. This method of testing will help to pinpoint effects of small changes in the process, while the comprehensive full scale test will mask these variations.

9. Frequent recording of process and precipitator electrical readings are important in the diagnostic program.

Visual Observations

Visual observations play an important part in the evaluation of precipitators, but they should be applied with care. Any optical device, including the eyeball, produces more of a qualitive measurement in terms of what may be existing in the stack, but that indicator is useful. Three basic visual methods exist.

Stack Discharge by Eyeball

The plume appearance can tell you whether rapping puffs or some other factor is present. Check the frequency of puffs against the frequency of the outlet field rappers. In some cases, it may only be one chamber that is causing the puff. Of importance is the visual uniformity of the plume. I get concerned when the outer zones of the plume is much heavier than the center.

Plant personnel involved with the precipitator can develop a strong feeling for early morning plume appearances relative to operating levels and collector difficulties. Variations in atmospheric and sky backgrounds sometimes make for difficult visual conditions, but time will help sort out some of these factors. I suggest taking a look at the stack appearance sometime between 11:00 A.M. and 1:00 P.M. and recording this observation against conditions and load levels.

After years of looking at the plume compared to emission tests, I can on occasion guess close on the final test result, but frankly, it's mostly luck because the misses are just as frequent as the hits. In some cases, the density of plume by eye indicates 3 to 4 times the actual test measurement. Fine-particle size, gaseous condensation at the top of the stack, angle of sighting, diameter of plume, height of stack, hygroscopic factors and chemical composition all play some part in this viewing process. However, each industrial process has a spread of dust concentrations that can often match up with a range of approximate opacity viewings. For example:

Fly Ash	5 to 10% opacity	0.01-0.015 gr/scfd
Cement	5 to 15%	0.005-0.015
BOF	5 to 20%	0.005-0.015
Sinter Plant	5 to 20%	0.01-0.02

Opacity Meters

Modern opacity meters can be costly both in installation and upkeep maintenance, but these meters can be important tools if reading confidence is gained by the plant personnel. Therefore, any problems with reliability should be attacked quickly after installation. I also suggest that a tabulation of process variables be coordinated against the chart so that the operator has a feel for stack effects caused by some of the operation changes that are made. Thus, the meter can be a useful training aid. Trends of the opacity chart can point out changes in fuel as well as rapper effects. Is the chart a true indicator? In many cases the outside visual opacity will show lower plume density than the chart, but a well-maintained meter and optical system appears most useful for relative patterns. Figures 10—2 and 10—3 show some chart patterns of opacity meters. The high peaks are indicative of rapper reentrainment caused by a poor gas distribution.

Internal Precipitator Observations During Operation

The best observation tool is the ability to visually look into the precipitator proper or the ductwork during operation periods. This can determine areas of dust losses, gas sneakage, location

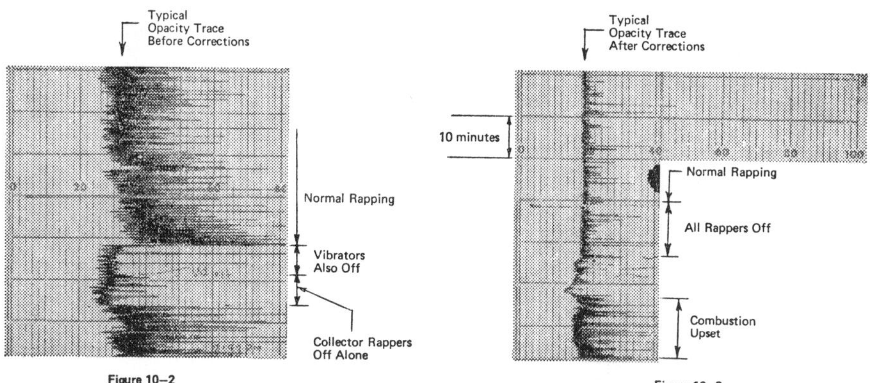

Comparison of How Opacity Tracings Would Look at Discharge of Precipitator under Poor and Good Conditions—with and without Rapping.

and severity of spark-over, and rapping effects. A few methods are used:

1. In the outlet test ports, a light placed in one port, and a safety glass placed on an adjacent port will allow viewing levels and patterns of dust in the flue. Intensity of the light should be changed to determine the best light scattering effect. I suggest concentrating on shadows to gain the best viewing. The light should be in a porcelain socket with high temperature leads. Wrapping the leads with glass fiber tape will help. Seal off the light port to keep the lamp from oscillating. The ports can be either in a vertical direction or along the horizontal plane. Light and glass are moved from port to port to gain an overall picture of the patterns. This procedure can be done with and without rapping. If rapping effects are desired, it helps to have an associate with a walki-talki call out the rappers in operation to the viewer. It is much easier to proceed with this evaluation when the port location is under negative pressure, but special air locks can be installed, such as those used for sampling pulverized coal from the transport pipes, to overcome a positive pressure installation.

2. Special observation doors can replace existing doors for short evaluation periods. These doors are best installed during outages and safeguarded in some cases with interlocks or other means for safe securing. Figure 10—4 shows one design that can be used with a special light apparatus. In some cases, obser-

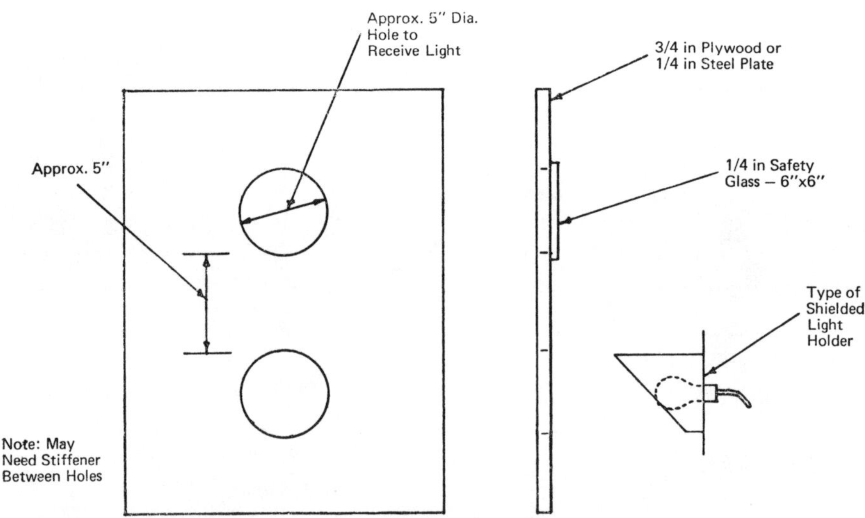

Figure 10—4 Suggestions for Observation Door for Use in Viewing Internal Conditions of ESP during Operating Periods.

vation ports can be placed in a spare door for installation. The use of 4″ pipe couplings as observation ports in doors or walls should be about 1½″ deep to help achieve a wide viewing span. Ports can also be placed in hopper doors for observation of gas sneakage or effects of air inleakage. These ports are usually placed so that couplings are parallel to the floor.

Gas Distribution

There are several techniques used to determine the magnitude and scope of the gas distribution patterns at the inlet face of the precipitator. I suggest that an internal inspection be made of the plenum to detect dust patterns along walls, vanes, and distribution plates. In many cases, these visual patterns can often show the direction of the gas flow. But actual measurements are often necessary to gather a quantitative pattern to help implement solutions to a poor distribution pattern. Briefly:

1. The traverse of a Stauscheibe or double-S pitot tube, through ports placed across the inlet face of the precipitator, during gas opration is the method I have used extensively. This technique can also be used through test ports in the vertical wall of the inlet plenum. Gas velocities are low, but the object is to detect relative flow patterns and their direction. Use of a sensitive draft gauge, capable of recognizing 0.001″ H_2O changes, is part of the procedure. The double-S pitot tube tends to amplify the differences between low flow and the 15 to 25 ft/sec velocities often measured in poor distributions. A true magnitude is not as important as the differences measured. An important factor of the pitot tube is that it can detect the direction of gas vectors which also can be as important as the magnitude.

Since the pitot tube sometimes has to traverse depths of 30 feet, a number of designs can be devised to effectively handle this length of proble without bending it. One good design used a 1¾ to 2″ diameter aluminum tubing as the support through which the connections are carried.

A commercially purchased 0.25″ H_2O draft gauge has been found sufficient, but even finer measurements can be achieved by using apparatus such as a Hook Gauge. The long length of tubing used with the pitot tube helps dampen out some of the oscillations in the draft gauge. The average reading of the slight swings of the gauge is satisfactory. A sensitive magnehelic gauge has also been used in these measurements, but I believe the elongated type draft gauge provides more stable readings.

I have stressed that the temperature pattern at the inlet face of the precipitator is an important measurement because of effects of gas temperature on electrical characteristics, but it is not necessary for the pitot tube traverse. However, it is still often convenient to attach the thermocouple wire to the long proble and obtain these temperatures at the same time. Effect of 40°F on the velocity measurement is small. For example:

Vel. Head	At 340°F	At 380°F
0.01″ H$_2$O	8.2 ft/sec	8.4 ft/sec

The above velocities are uncorrected for the pitot tube factor which reduces the values by 15%. I usually plot the velocities in the uncorrected form since relativity is what is considered important. Variations of 100% or more in flow areas should stand out, and the examples of Chapter 11 will show these differences clearly.

2. The hot-wire anemometer is also used for low velocity measurements in the field. It is easier to handle and appears to be more accurate, but it does not have the capability of determining direction of the gas vectors. For that reason, I have stayed with the pitot tube approach because the direction of gas movement is considered most important. Whether the hot-wire or pitot tube measurement is obtained, the spacing of the test ports and the number of readings per port stays the same. I look for centers of equal areas across the inlet face with widths of 5 to 6 feet being generally satisfactory for large installations. If the width of the precipitator chamber is less than 20 feet, about five ports can be used. Either 1′ or 1½′ spacings between readings are usually satisfactory. Both the port location and number of readings per port must be based on the design of vaning and the obstructions in the plenum. It may require more readings to nullify the shadow effect of splitter vanes on the analysis of the pattern.

Just where to place the test ports always presents a problem. I suggest an internal inspection to help evaluate obstructions that might interfere with the probe. Blue prints can also provide this information, but there are sometimes differences between the print and what actually exists. Figure 10—5 shows some ideas on where to locate these ports relative to gas distribution plates. In every case, it is recommended that the inlet field adjacent to the test area be deenergized during the time of measurements.

3. Although I prefer the measurements obtained during normal flue gas operation, there are times and certain installa-

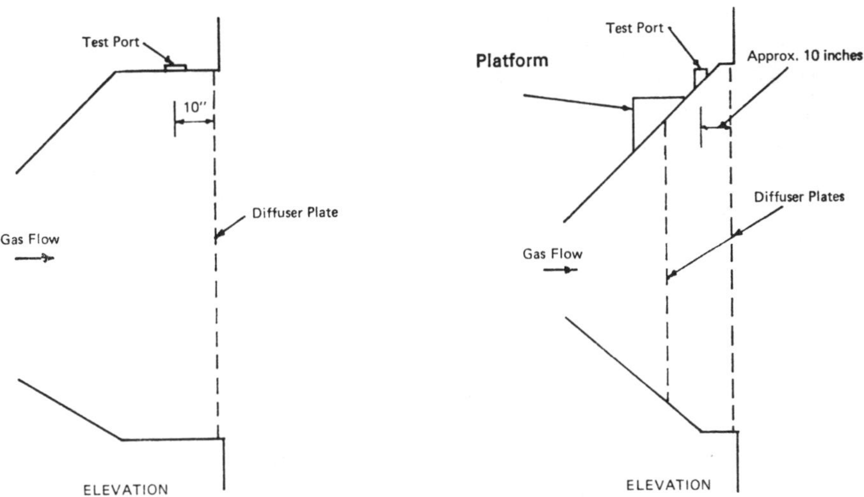

Figure 10—5 Suggestions for Location of Test Ports to Obtain Gas Distribution Pattern at Precipitator.

tions where flow patterns can be best obtained on air flow by personnel stationed inside the plenum. Instruments, such as the vane anemometer, can be used for quantitative results although indicators such as long silk ribbons attached to a stick can also show the directions of flow.

Several safeguards should be employed with this method. I suggest the fans of the system be operated at maximum levels for about 10 to 15 minutes prior to the lowering of speeds to allow access into the plenum. A full internal light system should be present. Radio communication as well as a back-up contact system should exist. I always like two (2) persons inside. The fan dampers can then be slowly opened so that about 95% of the maximum fan ratings are attained. The persons inside the flue should be faced toward the outlet with face protection until the air flow stabilizes after several minutes. After that, it is usually safe in many cases to remove the face cover, but I do suggest retaining a pair of dust-tight glasses. Usually the only dust problem occurs by disrupting build-ups in the plenum. It is, of course, mandatory that the precipitator, rappers and other associated equipment be safely locked out prior to this procedure. There is usually no danger involved in obtaining this set of measurements, but it can be uncomfortable. Speed of fan operation should be ramped upward slowly to minimize the variation in ear pressure. The low velocities of 5 to 20 ft/sec do not impose significant wind pressures.

One problem that arises with this measurement is that the

total air flow rate sometimes cannot match the flow obtained during gas operation because of fan limitation. Usually about 80 to 90% of the velocities are achieved when compared to the higher gas temperature measurements. Another factor to consider is that the higher density of the cool air does not react the same as the low density flue gas. But direction measurements are considered valid, and any bulge in flow detected during the air check may be considered to be worse with normal gas operation.

Flue Gas Analysis

Methods to arrive at the composition of the flue gas have been observed and evaluated for many years and a basic number of factors are offered for your information. Equations and background data can be obtained in chemical handbooks or other literature. I intend to relate the most common tools used in the field.

O_2 and CO_2 Measurements

These measurements are the most commonly obtained in most of the basic industries. The CO_2 chart was used for years in boiler house control rooms, but the O_2 has gained prominance in the last 20 years. It has only been relatively recently that the industrial O_2 analysis methods have given me confidence for most fly ash boilers, but they still appear as high maintenance devices. One of the measurement problems lies in getting a representative sample at the economizer outlet.

Portable O_2 heat-provers are used and are considered reliable. The Orsat apparatus containing the O_2, CO_2, and CO chambers has been used extensively in the field, but is considered subject to many handling problems. In recent years, the individual bottles of O_2 and CO_2 absorbing solutions, distributed as Fyrite apparatus, have been found reliable and useful field devices. Extensive cross-checks with the normal Orsat has produced variations of $\pm 0.5\%$ during comparison readings.

I always suggest using both the CO_2 and O_2 Fyrites in a fossil fuel installation to achieve added confidence in the readings, since the two values should complement each other and add up to $18\frac{1}{2}$ to 19% in most instances. Allow the solutions to come to temperature equilibrium at the site, but use sufficient tubing to keep the solutions from treating high flue gas temperatures. Solution temperatures below 110°F are considered to be in a good range.

While the flue need not be traversed as extensively for CO_2 and O_2 as for dust, there should still be a well spaced set of measurements depending on the size of the flue and the upstream equipment. For example, the test location downstream from a rotating air heater should have at least 8 to 10 pairs of measurements, spaced equally across the width and depth of the flue. Be especially concerned with the outer wall areas.

Temperature

The seemingly simple reading of the flue gas temperature has caused as many problems as any other measurement. Single point measurements can be far off the average especially if obtained near the side or top surfaces of the flue. I usually suggest readings no less than 12″ from an outer flue wall. The thermocouple and either a direct reading device, or potentiometer, should not be relied on in the field without frequent checks against an accurate glass thermometer or bimetallic dial thermometer that has been previously calibrated.

Temperature patterns can be used to substantiate thermocouple measurements in some cases. Gas temperatures normally form trends in the vertical and horizontal direction of the large boiler flues. Whenever large variations to these patterns begin to occur, I suggest a recheck of the instrumentation. In most of the process flue or stacks where gas mixing occurs, the temperature deviation from average may only be 10°F. A single point confirmation of the thermocouple lead is usually considered satisfactory with this condition.

Moisture Evaluation

During sampling periods for dust, the moisture determination is usually based on water condensed in the train and related to the gas volume that passed through the test apparatus. This method generally provides a reliable measurement. Wet and dry bulb measurements can also provide a valid moisture determination when the dry bulb reading can be kept in the saturated range of 212°F or below. Below 212°F the thermometers could be inserted directly into the gas stream. Above 212°F, the wet and dry bulb could be used in a hygrometer through which the flue gas is drawn. Conversion formulas can be found in handbooks.

SO₂ and SO₃ Measurements

Adaptations of the Shell Technique for the determination of SO_2 and SO_3 contents of the flue gas have been well documented. EPA has published techniques for SO_2 sampling that are acceptable. SO_2 measurements have gained confidence by a number of cross-checks against sulfur balances of the system. Actually, if the sampling of the fuel is done carefully, the SO_2 emission can be calculated quite close by the relationships discussed earlier.

However, determination of the SO_3 content is quite difficult, and I would place a $\pm25\%$ confidence level on any measurements of this constitiuent.

Particle Size Analysis

Determination of the valid size characteristics of the dust was noted earlier to be quite difficult. Not that the sizing techniques alone are questioned, but the ability to obtain a representative sample from the flue gas stream is the main problem. I do not have any quick and easy accurate sizing method, but I can supply a number of comments and ideas in the various techniques that have been obtained over the years. Briefly:

1. Discussion of sizing must refer to either projected size diameters or representative aerodynamic diameters that are based on the density and shape of the particles. A projected diameter value can be obtained by microscope, Coulter Counter, or sieve methods. The aerodynamic size can be obtained in the centrifugal Bahco classifer or cascade impactor techniques which are based on the inertial characteristics of the particles. The stated particle size can be vastly different depending on which method is used.

2. Weight of a particle goes up as the cube of its diameter, so it is clear that the large particles will constitute the bulk of the weight for materials such as fly ash.

3. Practically all dust particles found in industrial gases will follow a log-normal distribution if the particles are part of the same population. This relationship will generally provide a straight line on a log-normal graph as shown in Figure 10—6. This graph shows a comparison of a sieve, microscopic, and Bahco

Figure 10—6 Comparison of Sieve, Microscope, and Bahco Sizing on Same Fly Ash Sample.

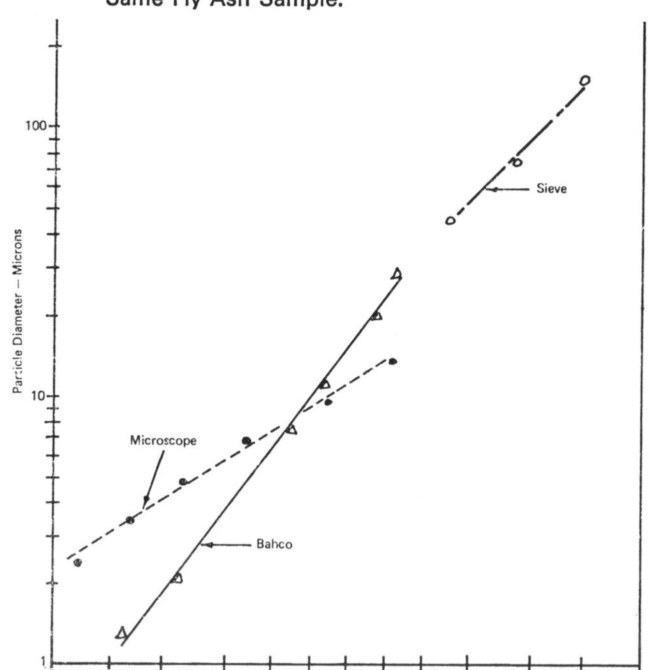

sizing of portions from the same ash sample obtained at the inlet of a large utility precipitator. Actually, a slight curvature often occurs in the very fine-particle zone. Figure 10—7 compares a microscopic sizing to a Bahco determination on parts of the same fly ash sample from an industrial boiler flue gas system.

4. Sieve and Bahco samples are normally obtained by traversing a precipitator inlet flue; at least several grams are required. An outlet sample is difficult to obtain in bulk form, although large volume sampling trains may collect sufficient samples over a number of hours. Both of these evaluation methods are described in the ASME publication P.T.C. No. 28, "Determining the Properties of Fine Particulate Matter."

I prefer the use of the 3″ diameter sieves with 100, 200 and 325 mesh screens comparable to the 149, 74 and 44 micron cuts. I suggest that you experiment with each type of material that is to be sized to help gain confidence in the results. For example, you might vary the time period, quantity of material, and the severity of the tapping of the sieve, and then compare the results of the size fractions with each change in procedure. Based on

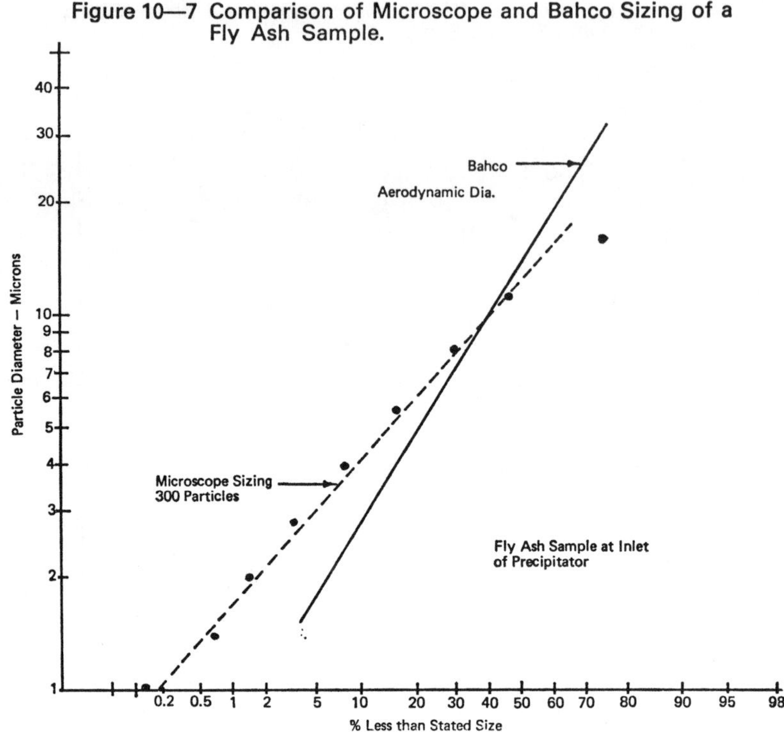

Figure 10—7 Comparison of Microscope and Bahco Sizing of a Fly Ash Sample.

the results of this investigation, future analysis on the same type of dust should maintain the same procedure.

The Bahco has been found to be reproducible and provides a reliable aerodynamic sizing of fly ash, especially in the range of 1 to 20 microns. One potential problem in both the Bahco and sieve technique is determining whether the agglomeration that may take place in the bulk sample is effectively separated back into the size fractions that existed in the flue gas stream. Some exploration of the size fractions by microscope will help gain confidence in the procedures. The finer the material (it may have hygroscopic tendencies), the more the situation may be aggravated.

5. The microscope technique is considered a primary standard to which many of the other sizing methods are checked. I cannot cover all of the ramifications of this technique, but some of the following comments should be of interest.

a. The optical microscope can be used for sizing particles down to about $\frac{1}{4}$ micron diameter, but confidence is less for the $\frac{1}{4}$ to $\frac{1}{2}$ micron range than for the larger sizes.

b. Magnification up to about 1000 is used for ultra-fine particles. Use oil contact between the objective lens and cover glass of the samples for best viewing.

c. Preparation of the sample is a key factor. One method to retrieve a representative sample for viewing involves using a toothpick to pick up a small sample, and then stirring the particles for about a minute within a drop of immersion oil placed on a slide. Then place a cover glass over the mixture.

d. To gain confidence in the size distribution, several slides should be prepared from the same bulk sample and the counts compared.

e. A measuring scale placed in the eyepiece lens housing is viewed as superimposed over the sample, and the calibrated scale is matched to the projected area of the particle to estimate the approximate diameter. The number of particles matched to each size range constitutes the final size distribution by number that can be further converted to the weight fractions by the cube relationship. You can see that the experience of the analyst and the number of particles counted will determine the final confidence factor in this technique. However, the microscope can provide a reliable sizing method even though it is a tedious routine and lacks the benefits of aerodynamic sizing.

6. Any attempt to capture the particles as they exist in the flue gas stream is desirable. The best method I have ever used involved placing a membrane filter (made from cellulose ester gels) in the flue gas for a matter of seconds and capturing the the particles on the filter surface. The characteristic of the filter is such that its index of refraction is practically the same as balsam oil, so that a drop of this immersion oil actually spreads to make the filter optically transparent. Thus, the particles can be viewed and sized in the state in which they deposited on the surface. Unfortunately, to the best of my knowledge, the membrane filter cannot be used above 230 to 240°F, which means it has to be adapted to this temperature limit instead of being used in-situ for many installations.

7. Recent efforts to arrive at valid particle sizing has concentrated on various designs of cascade impactors that are placed within the flue gas stream. About 4 to 6 size fractions is normally obtained based on the inertial properties of particles. A

small cyclone can be used for a pre-collector cut of large particles so that the stages will possibly cover the range from 15 microns down to as small as a ½ micron diameter, with a back-up filter for the finer-sized portion. The weight of material collected on each stage provides a distribution that is referred to the size fraction that should have been deposited on the specific stage at a given gas flow through the impactor. One problem that arises with large particles is that they may bounce from one stage to another, which will tend to distort the true size curve. As with any other sampling system, part of the problem in attaining a true representation is that large flue areas of 300 to 600 sq ft often have to be traversed. However, if you are interested in a specific size fraction alone, such as the 1 to 5 micron range without regard to the above 20 micron weight fraction, then the impactor begins to even look better. Because the device does look at the areodynamic characteristics of the particle right at the test site, the future should see much more activity in this type of instrument.

8.   Other classifiers that separate particles into weight fractions, based on aerodynamic principles, involve the use of cyclones, or the use of settling chambers with closely spaced plates. Difficulties in sample recovery are present to some degree because of the large surface areas of these devices. There is always concern with a sampling device that contains appreciable surface area between the nozzle and the receptor, because wall deposition can produce a sizable error in the final analysis of the distribution depending on the quantity of material sampled.

Gas Flow Determination

It would be remiss on my part not to discuss the measurement of gas flow rates in greater detail since this quantity is frequently discussed in precipitation. The same type of pitot tube used for gas distribution is also extensively used for flow rate measurements. The double S tube provides a higher draft gauge reading than required to attain a true velocity. Therefore, each double S tube must be calibrated against a standard pitot tube to determine a factor which is used to reduce the indicated velocity. A number of comments should be of help:

1.   The standard pitot tube is designed to provide a true reading of velocity head with a factor of one (1). The double S tube factor can vary between 0.82 to 0.86 depending on the design of the tube terminations. If designed to close tolerances with

sharp edges around the openings, the factor can be judged accurate at 0.85. However, most practical tubes used in the field have squared-off edges which tend to give factors of about 0.84. The use of an 0.84 factor will be considered conservative in favor of reporting a slightly higher flow.

2. Practically all of the past history of gas flow measurements for the application of collectors are based on the pitot tube. There have been questions in that the velocity measurements have sometimes shown gas flow rates to be 5 to 10% greater than calculated by stoichiometric calculations of the process. Part of the problem lies in gas tuburlence, insufficient traverse points, or errors in arriving at a true area determination at the plane of the measurement. When the traverse is obtained under the right conditions, the flow rate will usually check the calculated value closely.

3. The best traverse location provides for 50 to 70 ft/sec velocities in a straight section of flue for at least a couple of diameters downstream from any physical disturbance. The number of traverse points should often be 2 to 3 times the number of positions chosen for the sampling of dust. It is especially important to make the number of equal areas great enough in order to provide a traverse of the zones near the wall surfaces of the flue. Valid temperature and gas composition measurements, including moisture content, are necessary to arrive at a true gas flow rate.

4. It is desirable to periodically rotate the pitot tube 45° from the center line of the judged direction of impact to help determine if eddy current exist. The two side readings should be equal and slightly lower than the center line readings.

5. Frequently check for leakage in the pitot tube system or for dust pluggage in the end of the tube.

Chemical Analysis

Since the resistivity of some dusts can be judged at times from the chemical make-up, it is often prudent to obtain a chemical analysis. Some problems arise in arriving at a representative sample and being able to judge the variability of the raw materials or fuel over a given time period. In almost all cases, I suggest analyses of material which is collected at isokinetic sampling rates in the flue gases. Analyses of materials removed from electrode surfaces can also be important in evaluating corrosion as well as performance characteristics.

The easiest analysis results usually involve x-ray diffraction and other spectrographic methods. Unless the lab has proper standards, these results might be questionable. In critical jobs, it may be wise to send parts of the same sample to two different labs. The use of wet chemistry methods can be quite accurate for most of the constituents. Always ask for the expected deviation of the analysis results. This is especially true for the x-ray diffraction method. In some cases, the analysis results may be considered more qualitative than quantitative. Always ask the lab for how much sample is required for the desired tests and this will help determine the length of the sampling periods.

Loss On Ignition

I have just a couple of points on the laboratory analysis of the loss on ignition, which in the case of fly ash primarily means the combustible fraction. Procedures for this test are also included in the ASME PTC 28 publication. Since the combustible particles are larger, porous, and less dense than the ash segment, any bulk sample left for a few minutes will allow the combustible particles to rise toward the top of the container. This is why I frown on hopper samples as the means to check out combustible values.

A gram sample is sufficient for accuracy in the LOI analysis. The problem arises when 10 to 15 grams are collected. I suggest an approximate one (1) gram sample be taken in pie shape form from the whole sample as it is dumped into a pryamid mound. In other words, if a 10 to 15 gram sample is mixed in a container which releases the mix into a loose pile, the mound will take the shape of a pyramid with segregation of the large combustible immediately taking place. This part of the preparation is sufficiently difficult so that two samples should be obtained for comparative analysis on critical jobs.

Another point that I would like to emphasize is that if an alumina or any other porous thimble is used to collect the sample, the true combustible (this also holds true for the chemical analysis) may be distorted somewhat if the total sample weight collected is only about 1 gram. Approximately 0.2 to 0.3 of a gram could be left in the pores of the thimble, and since this residue will constitute the finer-ash segment, any loss of ignition on the loose sample can be biased on the high side. There are a couple of ways to get around this potential error. With the alumina thimble, the thimble including the sample can be combusted for loss on ignition. Another approximate method is to remove all the loose sample and also weigh the residue left in the thimble. If you consider the residue weight to have 25% of the loose sample combustible, then a calculation can be made for more confidence in the analysis. By the way, this possible error must also be considered in the sizing techniques. This is why pre-blinded thimbles are used in some cases to further reduce the residual in the surface pores. Look at an example for combustible:

Combustible analysis of a 0.7 gram loose sample $= 6\%$

Residue weight left in thimble $= 0.3$ grams

$$
\begin{array}{ll}
0.7 \text{ x } 6 \ \ = 4.2 \\
0.3 \text{ x } 1.5 = \ \ .45 \\
\hline
1.0 \qquad\quad\ \ 4.65
\end{array}
$$

$$\frac{4.65}{1.0} = 4.6\% \text{ specified combustible}$$

If only the loose material was considered, the error could be appreciable. The weighted calculation is considered much truer for this case with a $\pm15\%$ confidence factor.

Large combustible particles have the tendency to bounce off some surfaces or stick to others depending on the residual electric charge. In some containers, the combustible tends to cling to the walls, so that care must be exercised in not leaving this portion during any transfer of the material.

Chapter 11
Gas Distribution Factors

Introduction

A good portion of the "Art of Precipitation" must be devoted to the area of gas distribution. There are well accepted principles of fluid flow, but there is also a gap between this knowledge background and what has been observed in the field. A number of reasons exist for this situation, including what I believe to be a lack of field evaluation being fed back to the designer. I have spent years testing for gas distribution patterns and applying corrections, and I still use a trial and error approach. I have a healthy respect for the strange happenings that can occur in flow patterns, and this chapter will discuss many of the field experiences and corrections that can be helpful in your programs.

Model Studies

Since proper gas distribution is known to be a difficult area, the use of small scale modeling to pre-judge the best design of baffles and vaning has been extensively employed. Some questions come up about whether the model, usually at a 1/16 scale to the full size unit, is effective in predicting the actual gas flow patterns. I would place the model in the same category as the resistivity measurement. It should be used as a tool, and the confidence expressed in its results should be tempered with caution and previous experiences. Where the duct-work does not fit standard designs, the small scale air flow model can be a solid starting point. The caliber and experience of the modeling team is considered an important factor for a measure of success in effectively applying the model results.

My first disillusionment with model studies happened in the early 1950's, soon after I became a serviceman. At that time, a critical installation produced a very satisfactory model study, but as it turned out, its implementation in the field was unsuccessful. However, after a 8" section was removed from a distribution plate to eliminate a build-up of dust, the collector efficiency rose from 85 to 94%. What I have learned in my field experience is that mounds of dust, poor flow patterns at the inlet nozzle of the precipitator plenum, biased gas vectors that tend to crowd gas flow non-uniformly, and floor and wall obstructions cause problems that models cannot anticipate. I believe that it is sometimes difficult to close the disparity between the model and the field results.

Key Observations

While I will discuss an assortment of field observations and directions throughout this chapter, there are certain key bits of information that I want to stress. Briefly:

1. The gas flow vectors in a dynamic system will tend to keep going in the direction pointed, until striking another obstruction. Now that statement will not become as memorable as Lombardi's, "Run to Daylight," but it is fundamental to the understanding of why some installations have problems.

2. The velocity level of the gases entering an expansion plenum will determine the final patterns at the face of the precipitator. If there is a poor vector pattern at the entry of the nozzle, then higher flow rates will aggravate the distribution by the time the gases reach the precipitator.

3. The 40 to 50% open area diffuser plates will provide little correction of a poor gas pattern if the pressure drop across the plate is less than $0.5''H_2O$. However, these plates will reduce the rolling action of the gas, and most of the kinetic energy of the gas will be transferred into smaller jets. Generally, in the 10 to 15 ft/sec range, only minimal benefits will accrue in the gas spreading effects of the low pressure drop diffuser.

4. With a 40 to 50% open area diffuser plate, which is commonly used, any gas vectors striking the plate at 45° or more from the prependicular will have a sizable fraction of that gas flow slide across the plate.

5. Any flue expansion with more than about a 8° slope will generally have some separation of gas from the surface. The common practice of 30 to 45° plenum expansions tend to present distribution problems for that reason.

6. Any process whose flue gases contain particles over 30 microns in diameter could get into distribution troubles by the settling of dust in the expansion plenum. This condition becomes worse during long periods of reduced process operation with its attendent low gas velocities.

7. I have some concern that the emphasis on using the hot-wire anemometer, rather than a directional velocity device, may

have added to the type of gas distribution problems observed in the field. Direction of the gas vectors may be just as important a measurement as the magnitude.

Goals of Distribution Design

The quality of a gas distribution pattern, in a practical sense, is based on the layout of the flue system. More efforts and money must be allocated for poor flue designs. Poor designs will require more steel in terms of vanes and diffuser plates in addition to possible field modifications. A gas pattern has been stated to be acceptable in past years by the INDUSTRIAL GAS CLEANING INSTITUTE with the following guideline:

> "Uniform gas distribution shall mean that a velocity pattern five feet or less ahead of the precipitator inlet flange shall have a minimum of 85% of the readings with ±25% of the average velocity in the area with no reading varying more than ±40% from the average."

I believe the above goal is reasonable, even though it may be difficult to achieve in most field measurements. Part of the trouble lies in the shadow effects of splitter vanes in the plenum, or a multiplicity of problems involved with dissipating the high kinetic energy of the incoming flue gases. I have seen a number of installations that never met the guideline and still gave acceptable performances of the precipitator. Deviations of 100% or more from the average velocity begin to concern me, especially if the precipitator performance is considered marginal. However, where the aspect ratio is high and precipitator design velocities are low, I usually do not place initial emplasis on improving distribution patterns.

I do place emphasis on some improvement of the gas distribution of precipitators with aspect ratios of 1.0, or less, and with poor power input characteristics. If the precipitator is underperforming, then gas distribution is attacked as a fundamental factor, and a goal of 70 to 80% improvement in the existing pattern is usually attempted.

Proper Input to Flue Designs

There are a number of flue and vane designs that I have observed and evaluated that tend to minimize gas distribution problems. Uniformity of design is a key objective in practically all cases, especially in terms of the inlet and outlet nozzles of the precipitator plenums. The following layouts of designs are considered to represent good designs, but do not include exact dimensions of either the flue or internal vaning.

Figure 11—1 shows a solid design, if the installation allows this type of flue configuration. I usually do not take measurements for the distribution pattern of this design. The object is to make sure uniformity of gas velocity exists ahead of the elbow. Even one (1) perforated diffuser plate of 40 to 50% open area is satisfactory. This type of design can have a elbow entry from the top, but this is less desirable than the bottom entry because of potential dust build-up in the lower corner.

The common type of a nozzle expansion will provide a fair distribution pattern, if means are provided to guide the incoming gases uniformly in all four sections. Either vanes or diffuser plates can be used, but sufficient steel must be used to keep the gas vectors from crowding the extreme zones. Many of these problems will be seen in later sections of this chapter. Figure 11—2 shows a couple of design types generally considered

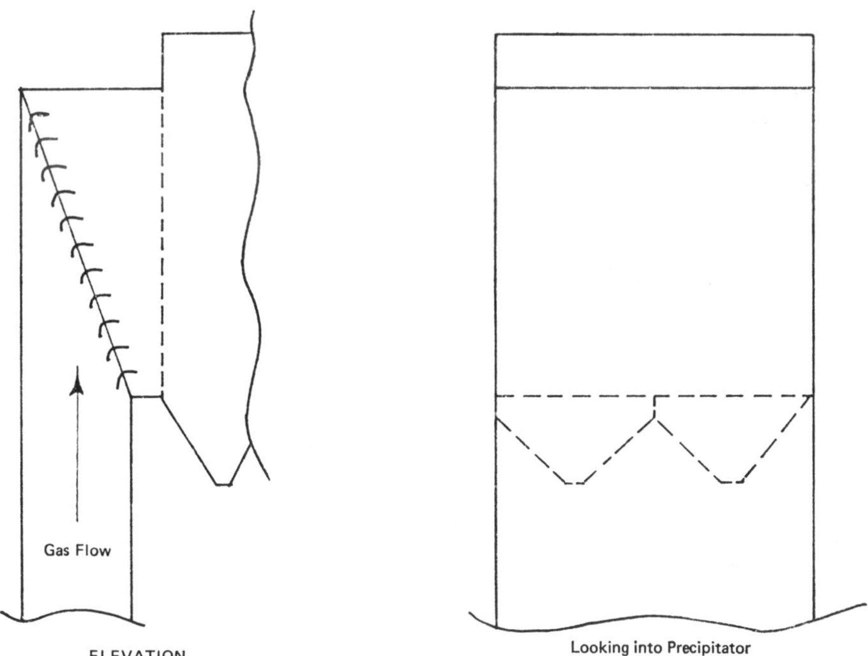

Figure 11—1 Layout of a Good Inlet Plenum to ESP Showing a Large Number of Closely Spaced Vanes.

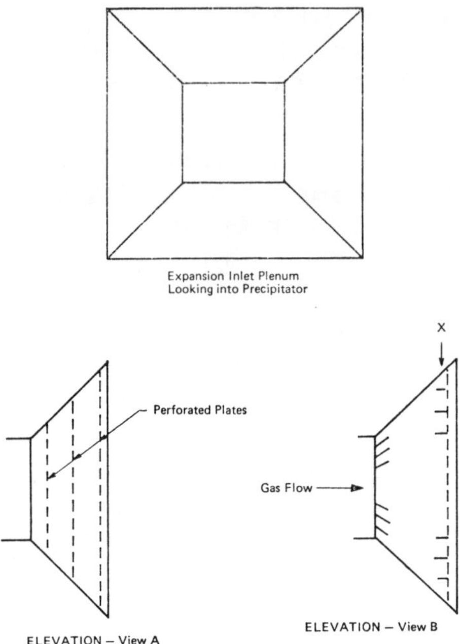

Figure 11—2 Expansion Inlet Plenums Showing Two Methods of Spreading the Gas Pattern.

acceptable. The side elevation shown as (A) uses at least three (3) perforated plates, sized and spaced to provide a satisfactory spreading of the gases. The second approach shown inside (B) utilizes two dimensional deflectors at the nozzle to direct the gas outward. Then apply horizontal and vertical shelves shown at point X to straighten out the vectors before the gas enters the precipitator. The second approach will probably do the better job, but it is more complicated and costly. Again, it will be the integrity of the gas flow pattern at the entry of the nozzle and dust build-up on the bottom slope of plenum, that determines the eventual success of this design.

Poor Inlet Flue Designs

Examples of poor inlet flue designs are shown in Figures 11—3 and 11—4. The downward direction of gases in the flue, shown in Figure 11—3, will generally cause performance problems. In Figure 11—4, we see a common two-way nozzle expansion design that will produce distribution problems even if dual diffuser plates are placed at the inlet face of the precipitator. I am especially concerned when the distance X is less than the distance Y of the incoming flue.

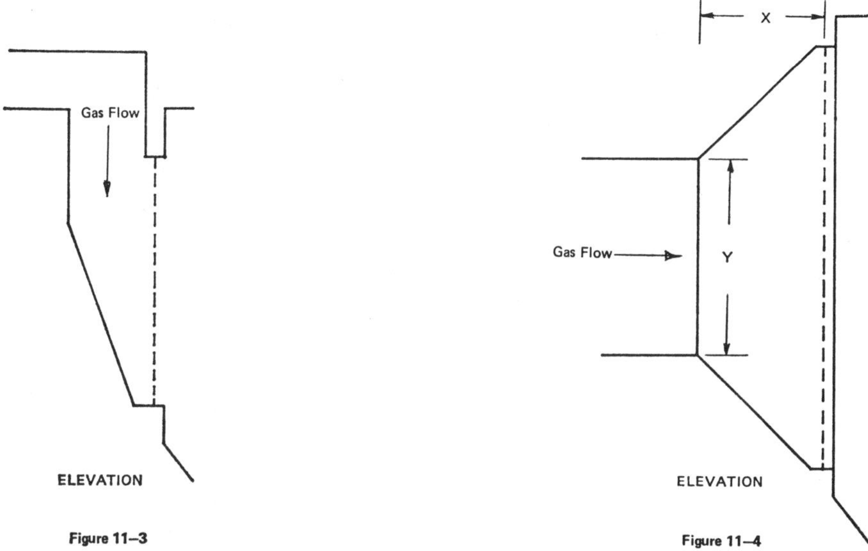

Examples of Two Inlet Plenum Designs that Generally Cause
Gas Distribution Problems.

Importance of the Outlet Flue Design

The main point in good outlet flue design is not to distort
the static pressure at the outlet face of the precipitator. This will
often occur when the outlet flue nozzle discharge is not uniformly
placed in relation to the collector. If an expansion outlet flue is
used as noted in Figure 11—5, I have never been concerned with
the absence of deflectors in the plenum. However, about a 40%
open area perforated plate may be desirable if the nozzle discharge is relatively close to the outlet face of the precipitator.

One outlet flue design that should be averted, if possible, is
shown in the Figure 11—6 (A) and (B) views. The problem of
pulling gas flow through the precipitator on a bias is present,
even if a perforated plate is placed at the outlet face of the
collector. There are different approaches used to overcome these
problems and I will cover them later in the book.

Manifold Problems

The manifold inlet and outlet flues for multiple parallel
precipitator chambers are primarily used in the BOF, open hearth,
and some cement installations. Flexibility in the isolation of
chambers is desirable, but there is no doubt that the manifold
design has caused gas balance and distribution problems in these
systems.

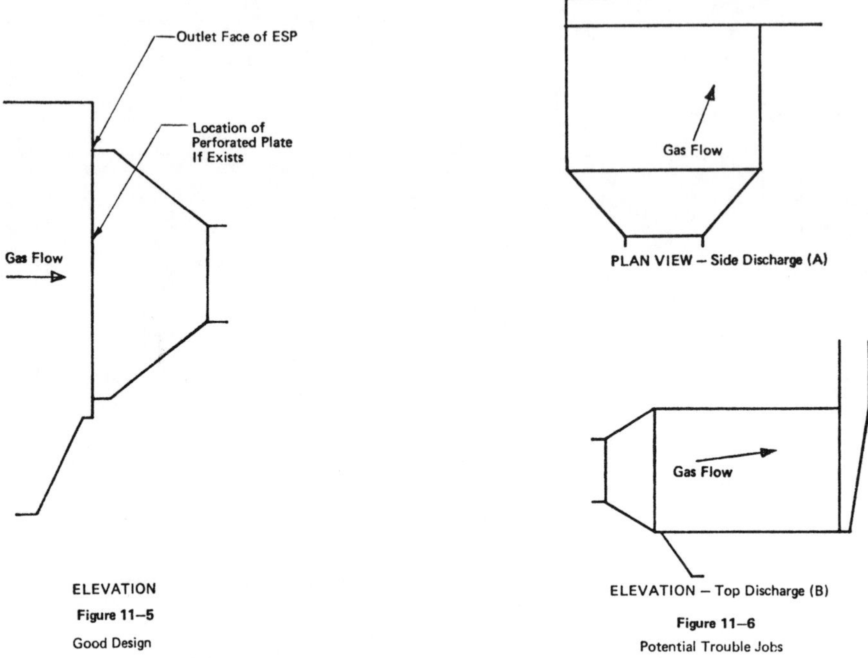

Examples of Good and Potentially Poor Designs of the
Outlet Plenums of Precipitators.

Figure 11—7 shows one type of manifold commonly used in which the incoming gas enters near the center of the main and allows gas to split toward both ends. Turning vanes at the center entrance and in front of each chamber are normally employed. The manifold is generally tapered down in area after each chamber entrance to maintain sufficient gas velocity in the manifold,which tends to keep particle deposition to a miminimum. Two other manifold types are shown in Figure 11—8, in which the gas flow can either enter at one end of the main as shown in view (A), or from either direction as shown in view (B). Turning vanes are possible in the main at each chamber entrance of view (A). However, no turning vanes can be used in the design of view (B), usually applied with gas flowing alternately between the two ends of the manifold. A number of points should be made concerning manifold systems:

1. A minimum of four (4) chambers usually exists, but six to eight are not uncommon. Each chamber can be isolated with guillotine dampers at the inlet, and sometimes with a guillotine at the outlet nozzle discharge in addition to a louver control damper. The louver damper is normally used to balance gas flow rates between chambers.

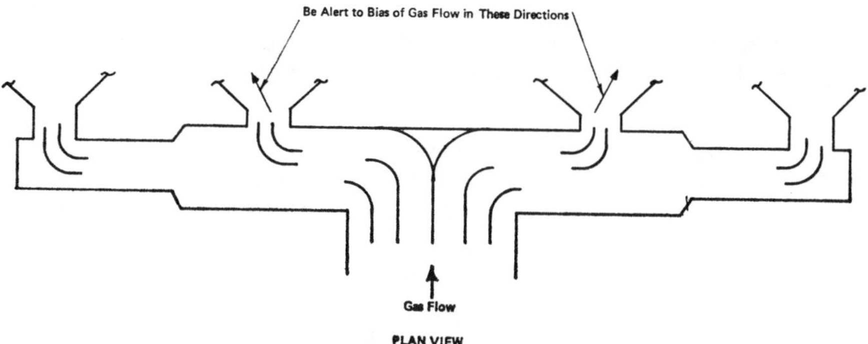

Figure 11—7 Layout of Typical Manifold Plenum with Gas Entry at Center.

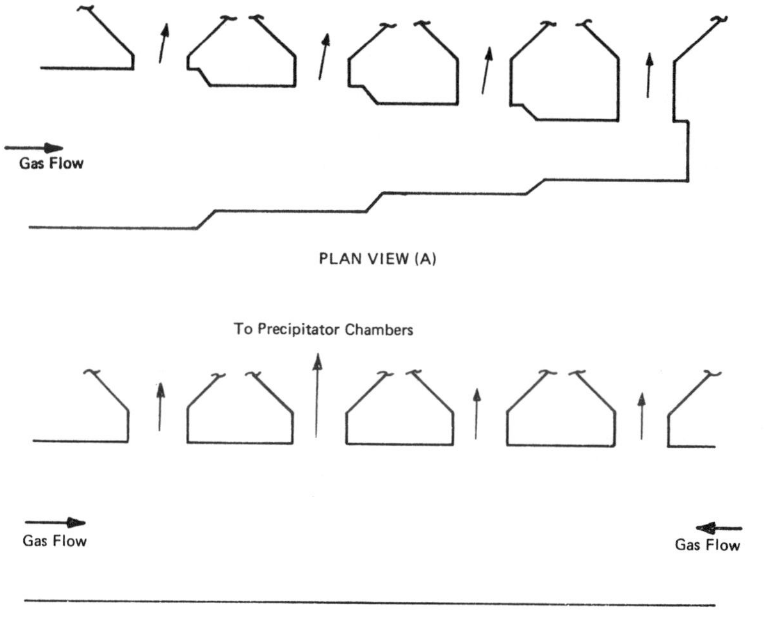

Figure 11—8 Two Methods of Inlet Manifold Design Depending
on Layout of Process.

2. The guillotine gate damper at the inlet should not be used for any balancing, since this can produce a mal-distribution in the chamber. In fact, always check to see that these guillotine dampers are completely out of the gas stream. Pressure drops of $1\frac{1}{2}''$ to $3''$ H_2O across the outlet louvers are quite common to achieve proper gas balance. The more chambers, the greater the difficulty for equal flows.

3.   One of the problems in gas balancing occurs when one of the chambers is isolated. Depending on chamber location, and the degree of balancing sophistication, a short period of disturbance could exist.

4.   Unless a sizable pressure drop is taken at the nozzle entry of the inlet plenum, it is difficult to achieve good gas distribution within each chamber. Even turning vanes in the manifold has been found lacking in this regard. Part of the problem lies with the ratio of areas between the nozzle entrance and the face of the precipitator. This allows any biased gas vector to magnify into sizable flows toward one side or the other. The directions of gas flow vectors are basically shown on the above figures. Unless a well designed set of turning vanes are installed (and this is not easy to achieve), some bias tends to occur at the entrance. That is why I suggest a grid or high pressure drop perforated plate at the entry of the plenum designed for about 2.0″ H_2O drop at maximum gas flow rates.

5.   Any obstruction between the manifold and the nozzle entrance can deflect gas vectors appreciably. In fact, the location of the transition opening in relation to the curvature of the manifold can effect the directional flow of gas into the plenum of the chamber. The effect of these biased flows will be nullified to some degree by the use of a high back pressure grid at the nozzle entrance.

6.   Once a uniform flow pattern is obtained at the nozzle entry, multiple perforated plates or deflectors will be required in the expansion plenum.

7.   Proper gas balance between chambers can be best achieved by valid measurements of these flows. Common practice utilizes the pressure drops across the complete chambers by use of taps in the inlet and outlet plenums. This method has not proven successful, because of the low pressure drop usually experienced between these two points in the system. I suggest the drop be taken across a high pressure grid at the inlet, or an orifice plate placed at the outlet duct. The object is to work with drops over 1″ H_2O.

Gas Sneakage Through Hoppers

The prime reason for the major sneakage of gas flow through hoppers is related to poor gas distribution patterns. Some sneakage can occur above the active electrical zones of each field, but I usually do not get concerned about this portion, especially in a three or more field precipitator.

Expansion plenums and top entry plenums ahead of the precipitator are the usual culprits, as these designs tend to direct gas vectors toward the hopper area. The bottom slope of the expansion chamber is tailor-made to guide gas flow downward, and multiple perforated plates can aggravate this directional flow. For example, the bottoms of the perforated plates have often been cut-out to prevent excessive build-up of dust. Unfortunately, these openings provide a clear path to the hopper area. Figure 11—9 shows the direction of these gas vectors. Note that the center baffle of the inlet hopper will do little for this downward flow, except help guide it toward the apex to further reentrain collected material.

Another cause of gas sneakage into hoppers could occur with excessive flow striking any collector surface baffle at the leading edge of the collector plate. This can happen more when gas velocities of 10 ft/sec or above exists. Once the gas flow enters the open hopper area, there will be nothing to redirect the flow back into the collection zone except the hopper wall or baffle.

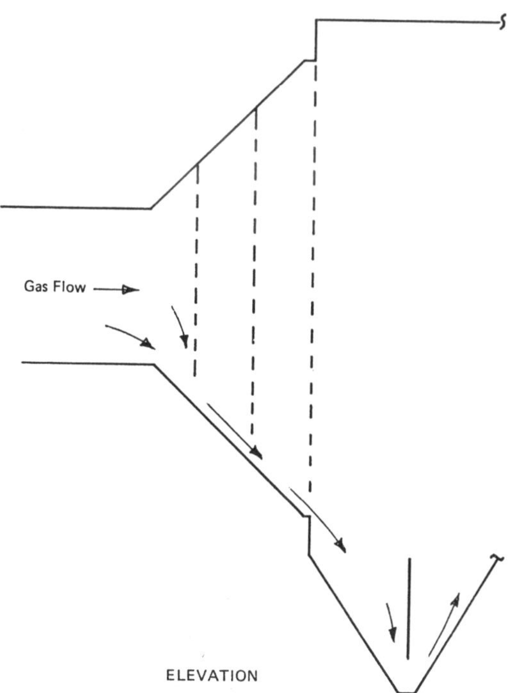

Figure 11—9 Gas Flow into Hopper Caused by Excess Opening at Lower Portion of Diffusers.

Field Evaluations and Corrections

I believe the best way to explain what happens to gas flow patterns in certain flue and plenum designs is to show a number of typical field cases. There can always be several methods used to correct field problems. I generally suggest the simplest approach with refinements implemented as required. In most cases, a perfect gas distribution is not needed. As I said before, for most modern installations, a much less than perfect modification of poor distribution will generally be satisfactory.

The example discussed in the model study section is shown

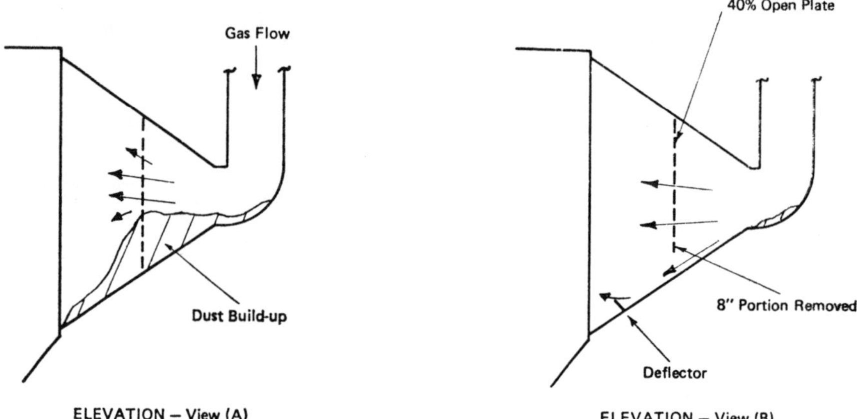

Figure 11—10 Elimination of Dust Build-up Helps Gas Distribution Pattern to Precipitator.

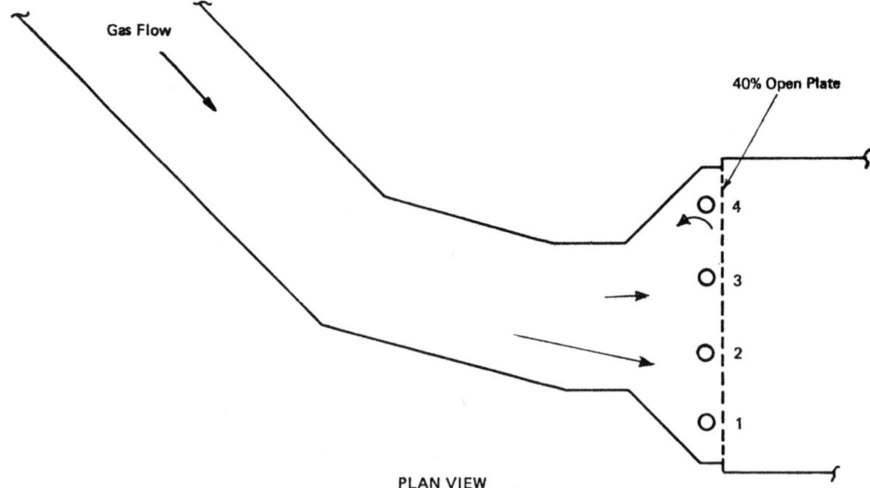

Figure 11—11 Effect of Layout of Flue without Proper Vaning on Gas Distribution in ESP.

in view (A) of Figure 11—10. Removal of the bottom 8″ of the perforated plate with a small deflector added on the slope shown in view (B) provided an appreciable gain in performance, even though the modification was considered relatively simple.

An example of how the contour of the incoming flue can effect the gas flow distribution at the precipitator is shown in Figure 11—11. Uncorrected pitot tube measurements showed:

Port	1	2	3	4
Measured Vel. ft/sec	9.9	10.8	2.8	—1.5

Another interesting flue design is shown in Figure 11—12. High gas velocities occurred in the upper portion, with a tendency for low flows through the center of the precipitator. The layout of the vanes indicate that a center shadow effect might occur. Simple correctional steps can include modification to the vanes (as shown) and deflectors (shown at points A and B). The performance of the precipitator was marginally satisfactory with the original gas distribution when power input was relatively high. However, rapping puffs were detected.

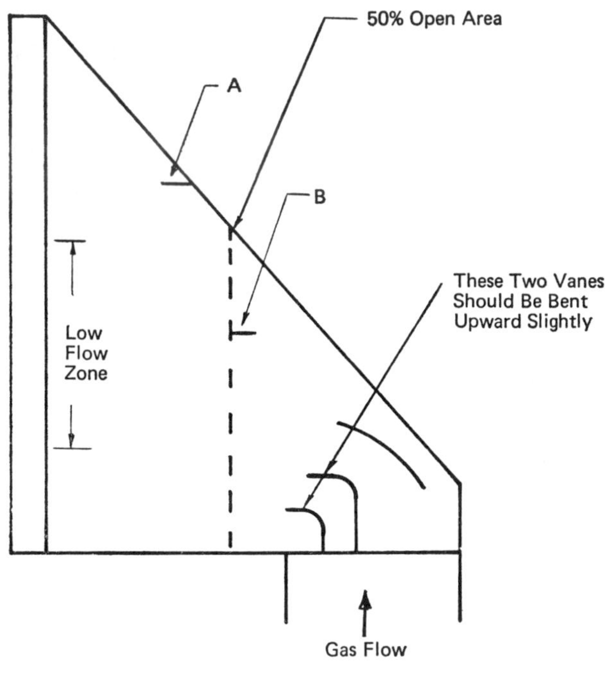

Figure 11—12 Minor Modifications to Gas Flow Pattern at Precipitator.

Although not common, the flue layout shown in Figure 11—13 has been observed in the field. Rapping reentrainment is usually high because the bulk of gas flow will remain in the upper strata, especially in a short precipitator. Possible correction would take place at the entrance of the plenum, in addition to horizontal shelves placed across the first perforated plate, as shown in view (B). Actual pitot tube measurements about 1 foot in front of the first perforated plate for the as found condition of view (A), showed the upper half to have about 2 to 3 times the bulk gas flow rate as did the lower half. Actual rapper reentrainment was high at this installation.

A common flue design observed when the process discharge elevation was higher than the precipitator is shown in Figure 11—14. I recall that the hanging baffle was installed by the user to minimize the large dust build-up that occurred on the floor of the transition duct. Before the baffle, I have no doubt that the upper half of the precipitator received the bulk of gas flow. After the baffle, it can be seen that the bulk flow picked out the lower third of the precipitator as shown by the gas vectors, and a sizable part of this flow entered the hopper area. Simple corrections suggested include a small floor deflector shown at (A), some perforations in the hanging baffle, and a series of small deflectors at the discharge of the perforated plate shown at (B). A number of different ways could have been implemented, but the object is to reach a relatively uniform flow across the plane of point (C), and prevent gas sneakage into the hopper.

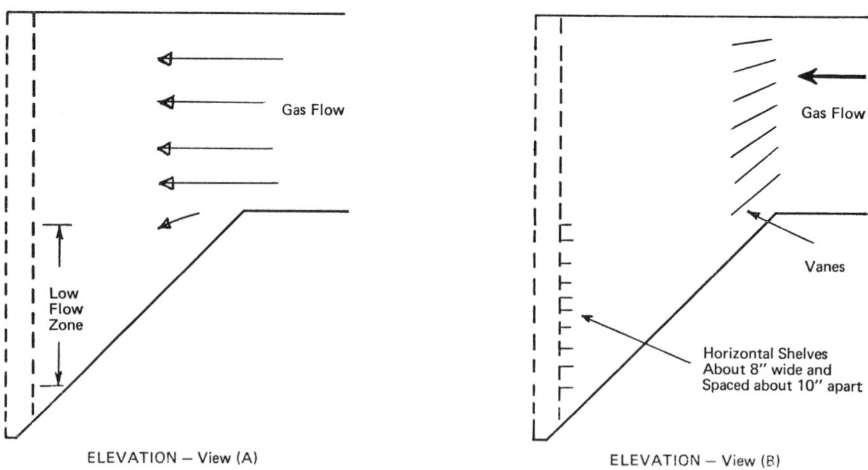

Figure 11—13 One Method to Correct a Biased Gas Flow Pattern to the Precipitator.

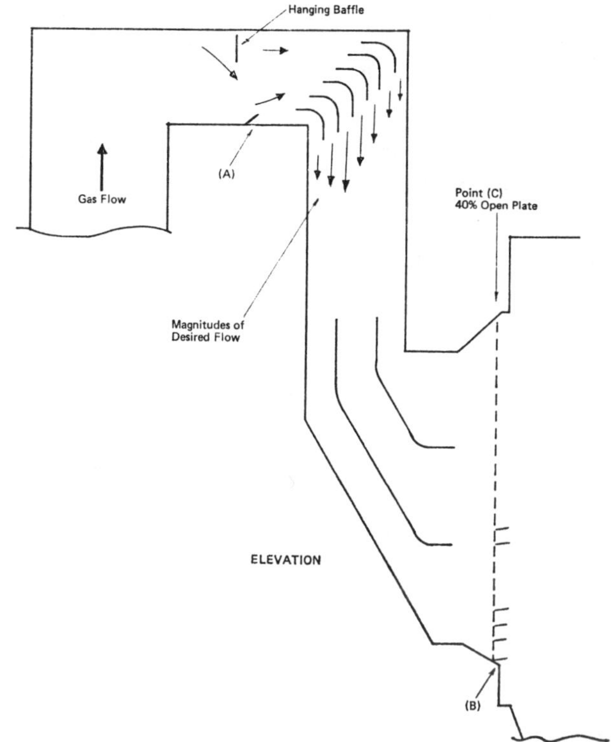

Figure 11—14 Difficult Inlet Plenum Where Gas Balance to Three Verticle Zones Should be Biased Slightly toward Center Zone.

There is a tendency to believe that the gas flow fills up the available space in a satisfactory manner. Nowhere is that more pronounced than in retrofit jobs, where some strange flue arrangements occur. Figure 11—15 (A) shows the basic pattern

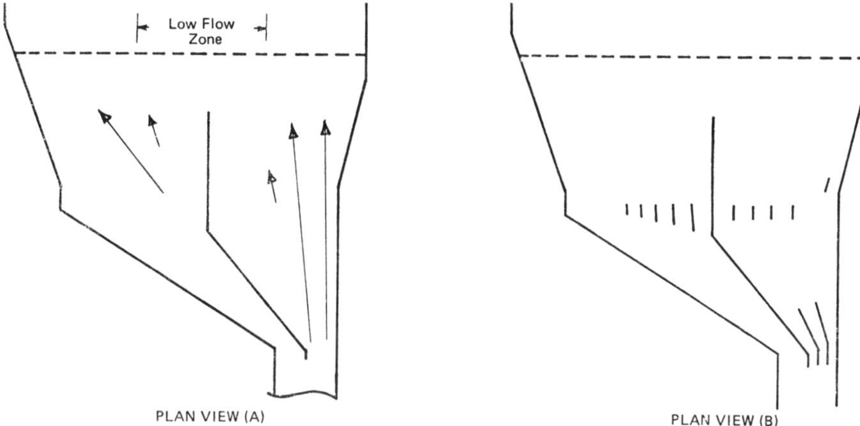

Figure 11—15 Suggested Modifications to Minimize Low Gas Flow Zone in Precipitator.

observed in one such retrofit, which produced low gas flow in the center zone of the precipitator. Corrections could take the form of splitters shown in view (B).

Figure 11—16 shows a common design that attempts to do a good distribution job, but is less than optimum in a marginal installation because of the wide spacing between the horizontal shelves at the face of the collector.

Another expansion design that poises problems is shown in Figure 11—17 (A). The appreciable distance between the splitters and the inlet of the precipitator allows the biased gas vectors to crowd the outer zone, while some reduced flow is observed in the center zone. While this can be corrected by several methods, I suggest additional splitters as shown in view (B). Improvement in performance occurred with this minor change.

A plan view of another side expansion is shown in Figure 11—18. Correction of this pattern can be helped by even moving two of the splitters. A model study had given high marks to the original distribution.

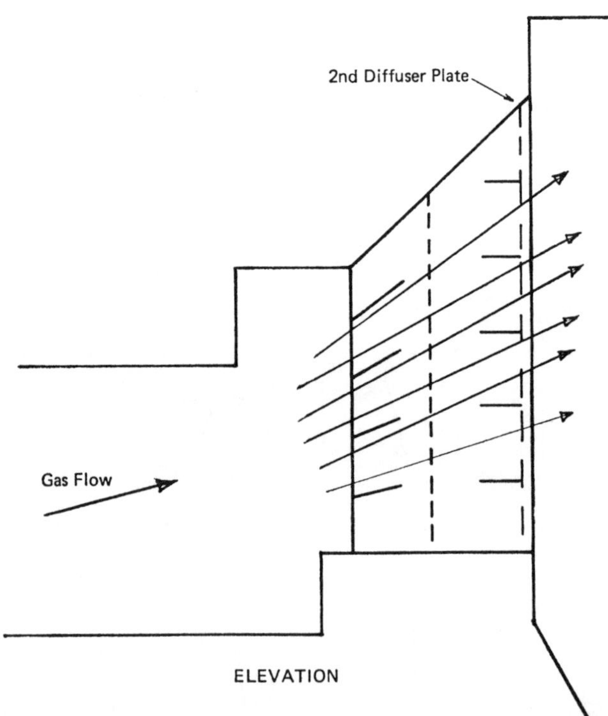

Figure 11—16 Upward Biased Gas Pattern into Precipitator Because of Wide Spacing of Horizontal Shelves.

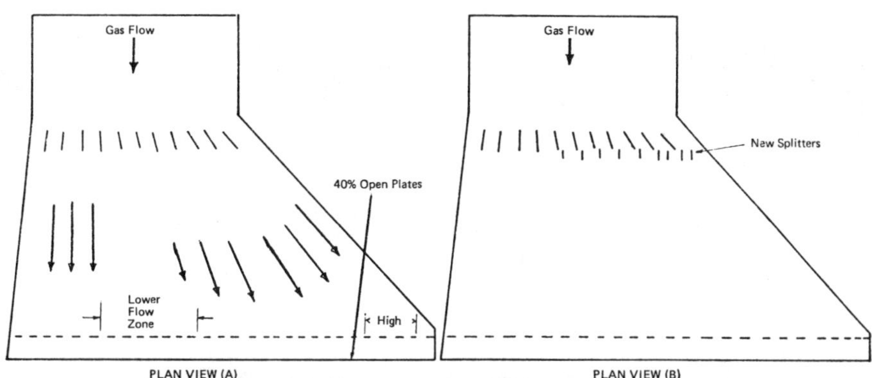

Figure 11—17 High and Low Gas Flow Zones Caused by Use of Splitters Placed Far from Inlet Face of Precipitator.

Another distribution problem occurred in what would be considered an ideal expansion flue as shown in Figure 11—19. The sparse vaning in the elbow, coupled with drop out of dust noted in view (A), produced the following patterns:

Zone	% of Mass Flow Rate	Gas Velocity Ft/Sec
Top 4'	25.1	14.5
2nd 4'	19.2	11.1
3rd 4'	22.4	12.9
4th 4'	18.9	10.9
5th 4'	14.4	8.3

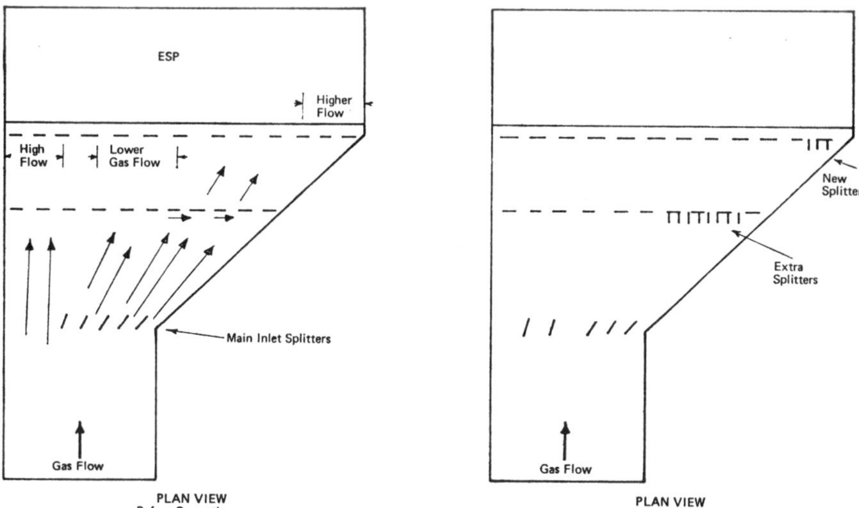

Figure 11—18 Shadow Effect Seen Caused by Main Splitter Vanes at Inlet of Expansion Plenum.

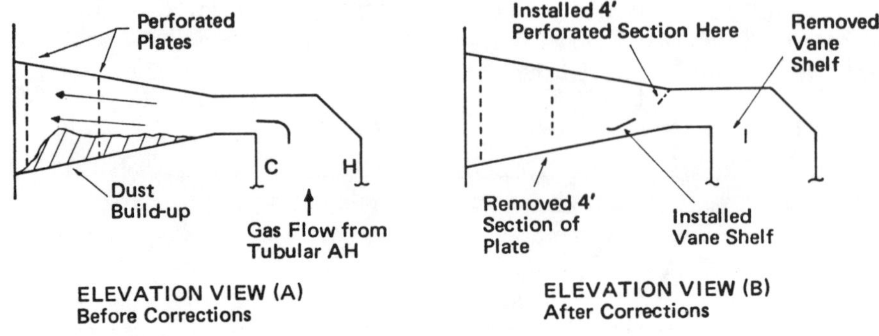

Figure 11—19   Modifications to Flue Gas Patterns by Dust Build-up and One Corrective Method that can be Used.

About a 30°F gas differential existed between the top half and bottom half of the inlet face area of the precipitator, due to the tubular air heater of the boiler system. In order to correct both the velocity and temperature patterns in this marginal installation, corrections were suggested as shown in view (B). The repeat evaluation showed a more even flow pattern from top to bottom with a resultant 15°F gas temperature variation between extremes. This change, among others, brought the precipitator within compliance.

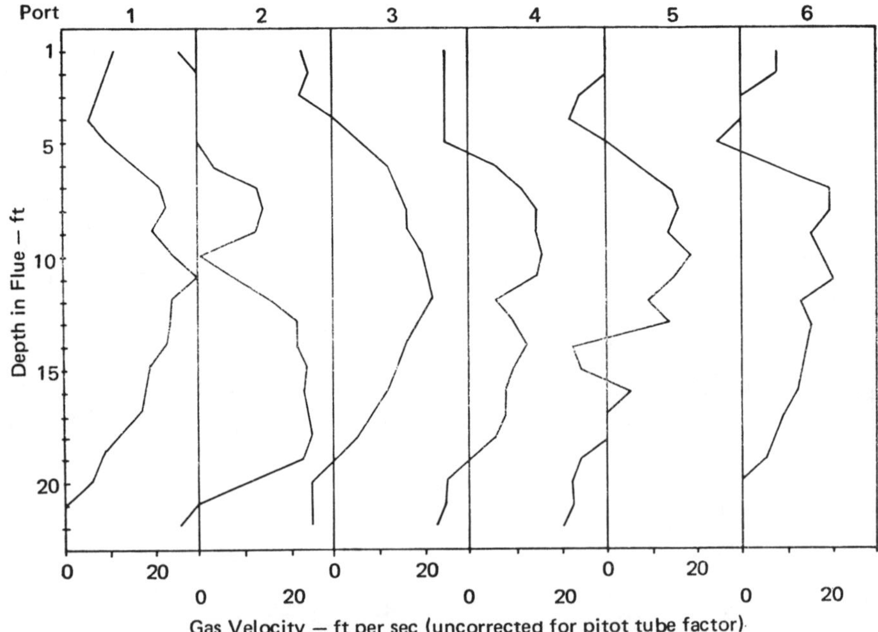

Figure 11—20   Gas Distribution Pattern at Inlet of Precipitator Before Blocking Off Part of Perforated Plate.

I indicated earlier that the commonly applied expansion plenum shown in Figure 11—4 will provide poor flow patterns into the precipitator. Figure 11—20 shows a distribution pattern, obtained by pitot tube inserted after the first perforated plate. Rather than utilize turning vanes, the center zone was restricted with ⅛″ flat bars spaced to cover much of the existing open area. The upper and lower zones were opened to a greater degree by notching out some of the plate between the holes of the perforated plate. Figure 11—21 shows a repeat traverse after the modification. The correction was acceptable and no further work was contemplated. Increased pressure drop was negligible across the modified plate.

The basic inlet plenum design as shown earlier in Figure 11—3 had been observed at a number of installations. Most of these precipitators were of older vintage, but some are still in existence around the country. In some cases, a single vane might have been installed, as shown in Figure 11—22. For one installation, I had six ports placed up the vertical wall and the direction of the gas velocities were averaged as shown. Note that the upper zone had an average velocity of 20.4 ft/sec for 38% of the precipitator volume while a 9.5 ft/sec average existed for 62% of

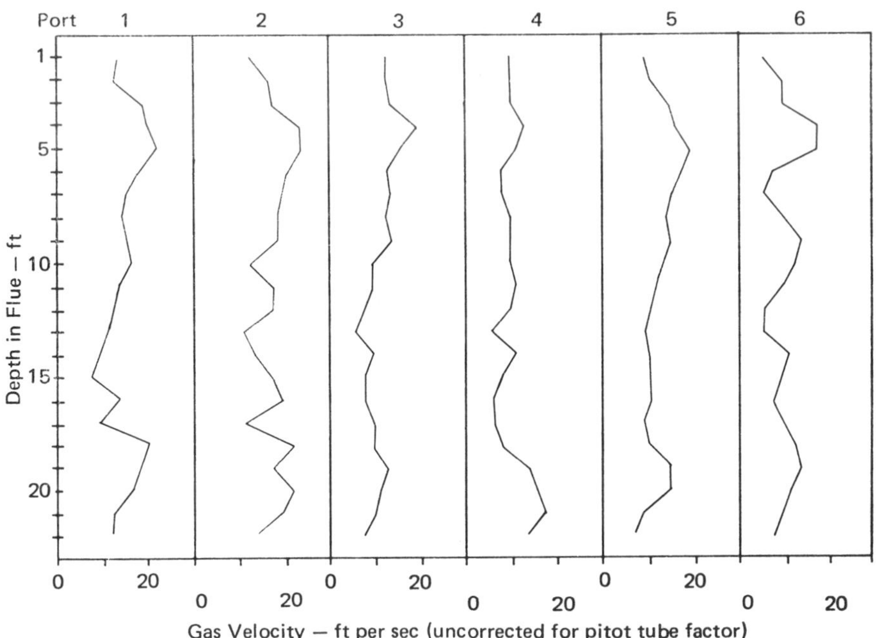

Figure 11—21   Gas Distribution Pattern at Inlet of Precipitator After Modifications to 40% Open Perforated Plate.

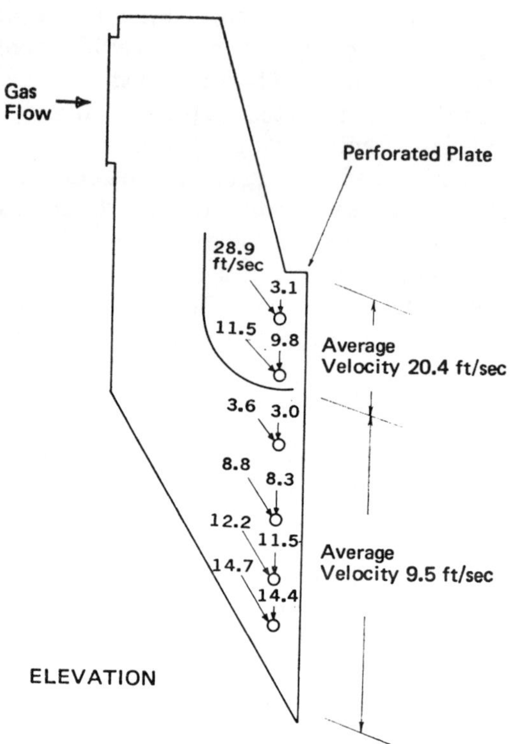

Figure 11—22 Distribution of Gas Flow and Direction of Flow Vectors in Inlet Plenum.

the lower area. Even though a perforated plate existed at the inlet face, downward vectors would prevail into the precipitator hopper. In other top entry designs, the sweep of the inlet hopper was common, as shown in Figure 11—23. Field corrections for this type of plenum design usually include a number of differ-ent step shelves, as shown in Figure 11—24. The large opening at the top of the vane (A) was usually closed down by half. The simple three shelves as shown in view (C) was applied as part of a program on a four parallel chamber precipitator that reduced emissions from 1140 lb/hr to 140 lb/hr. It usually does not take much correction to see immediate effects on designs of this type.

One example of insufficient turning vanes in an elbow pre-ceding a plenum is shown in Figure 11—25 (A). Correction was made with small deflectors at the trailing edge of the vanes as shown in view (B).

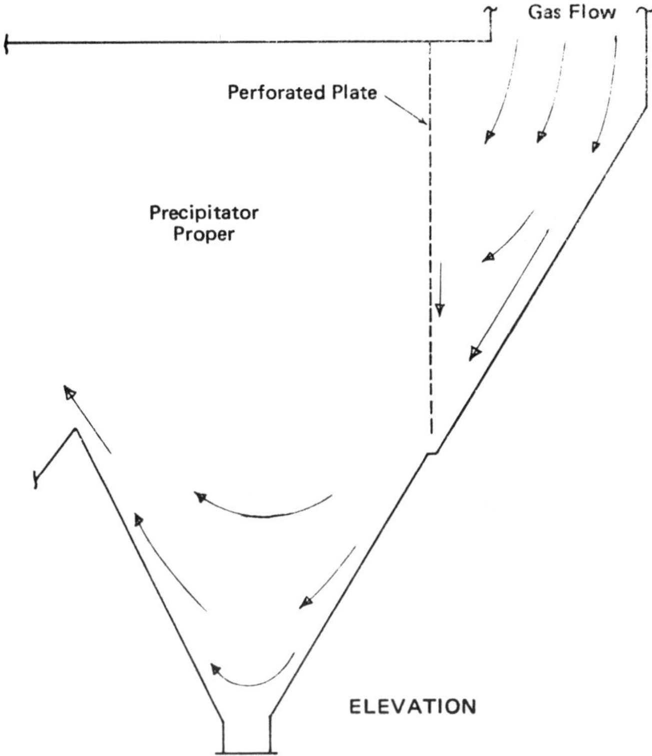

Figure 11—23   Gas Flow Sweep of Hopper Caused by
Absence of Vanes

Figures 11—26 and 11—27 show pitot tube readings obtained before and after correction of a poor distribution pattern which was caused by a biased flow into an expansion chamber that was part of a manifold system. Figure 11—28 shows the plan layout of the plenum, and the gas vectors that existed before a corrective grid was placed at point (X). The tabulations shown in the figures are representations of the average velocity measured by placing the head of the pitot tube into the expected direction of flow, and at 45° angles to each side of that initial position. If you follow the pattern of Figure 11—28, it becomes a little clearer what is occurring between the t wo perforated plates. The flow can be seen heading toward the east, based on the readings of Ports 7 and 8. A slight amount of gas appears to have turned back toward the west in Port 6, based partly on the turbulence caused by the east wall of the transition duct. Ports 4 and 2 show a decided bias of gas flow toward the west. Port

259

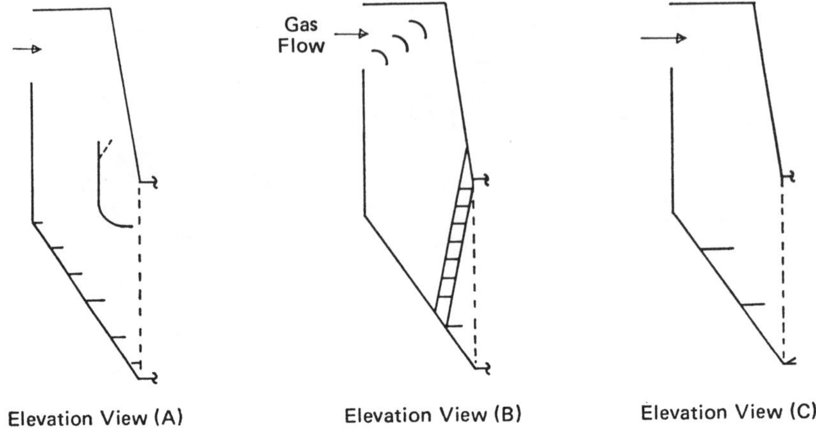

Figure 11—24   Three Modifications in the Field that can Overcome
Some of the Effects of a Top-entry Flue Causing
Poor Gas Vector Patterns.

No. 1 shows a balanced flow, even through much smaller than Port 8. There is still a higher flow tendency toward Ports 7 and 8, which would require another corrective step. But the ability of the angle grid, shown in Figure 11—29, to minimize the bias gas vectors can be clearly seen.

The use of the perforated target plates in an expansion plenum can be dangerous if the gas vectors striking the plate is not perfectly perpendicular to the surface. Deflection of the gas flow toward the bottom or sides of the precipitator can occur, as shown in Figure 11—30.

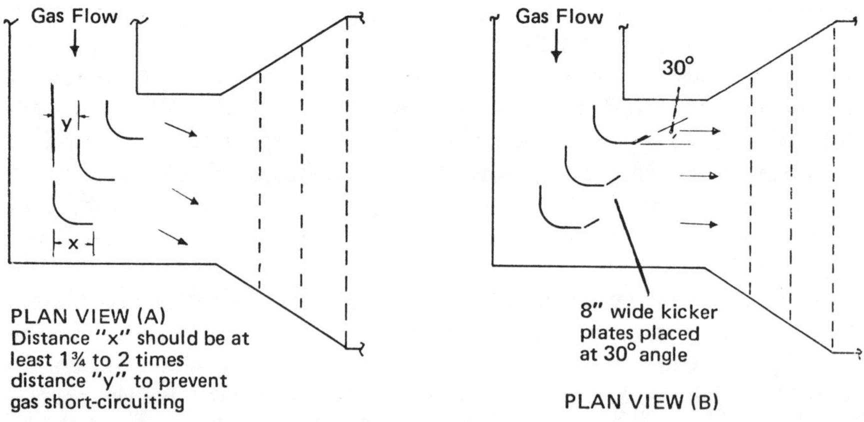

Figure 11—25   Use of Kicker Plates to Change Gas Flow to Desired Directions.

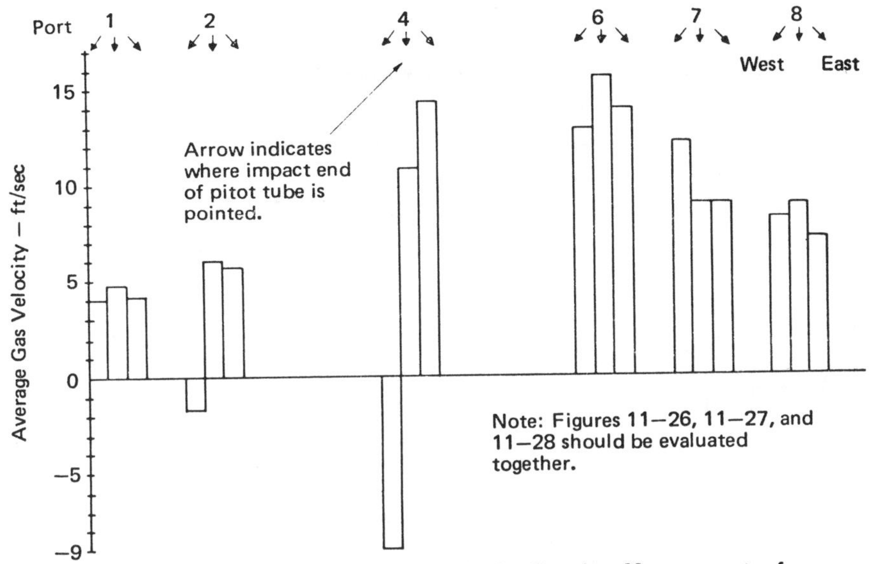

Figure 11−26 Pattern of Gas Distribution Based on Measurements of Pitot Tube Rotated in Three Directions — Prior to Corrective Grid.

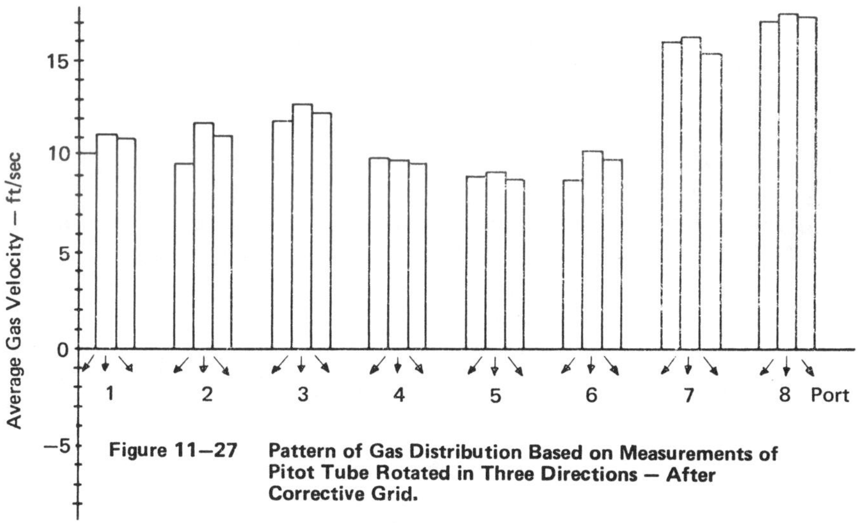

Figure 11−27 Pattern of Gas Distribution Based on Measurements of Pitot Tube Rotated in Three Directions — After Corrective Grid.

Effect of Mechanical Collector on Distribution

The use of a close-coupled mechanical cyclone collector can adversely effect the gas flow patterns into the precipitator. While the 2 to 3″ H_2O pressure drop of the mechanical collector will generally provide a uniform gas discharge from the cyclones, the

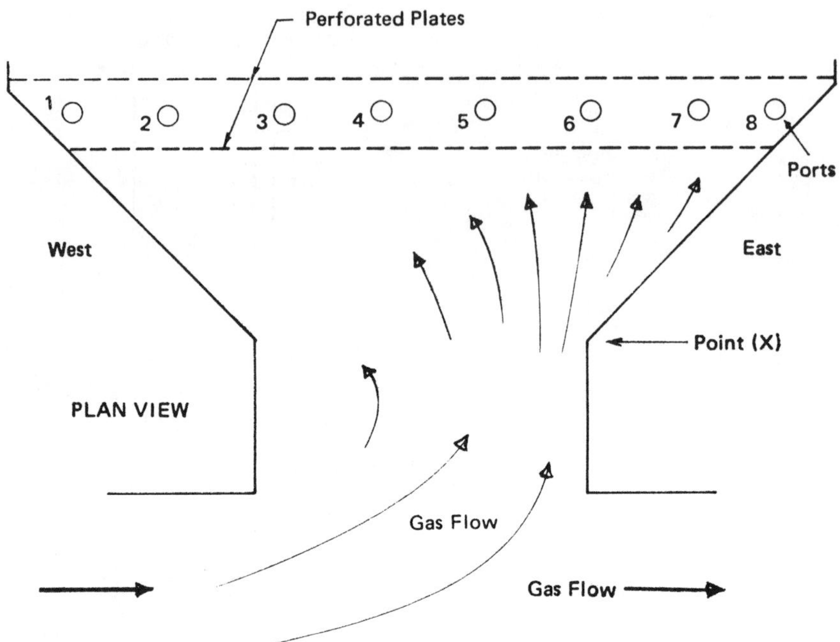

Figure 11—28   Gas Flow Vectors into an Expansion Plenum of a Manifold
System Prior to Installation of a Grid at Point (X).
(Refer to Figures 11—26 and 11—27)

physical position of the outlet tubes tends to direct the flow
upward. Perforated turning vanes and hanging baffles are used
extensively, but a slight bias upward is more the rule than the
exception. If the precipitator is performing in a marginal band,
some effort in the gas distribution area between the cyclones and
the precipitator should be explored. See Figure 11—31 for several
design problems that have been observed, and where the corrections produced improvement in the precipitator performance.

Dust Build-Up Effect on Distribution

A question frequently asked over the years has been concerned with the dust build-up on turning vanes, as shown in
Figure 11—32 (A). Since these build-ups are primarily caused by
the gas separation and resultant eddy currents at the 90° bend,
I usually do not even recommend cleaning during outages. In
other words, this build-up is not considered to have a basic effect
on distribution. The wider the spacing between vanes, the higher
the build-up. I do evaluate the uniformity of build-up between

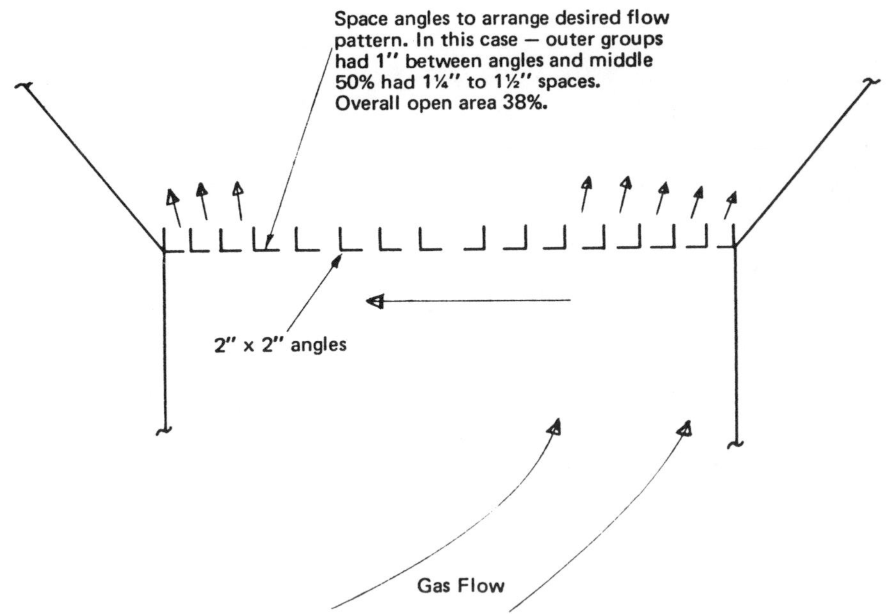

Figure 11—29   Suggestion for Angle Grid that can Redistribute a
Biased Incoming Gas Flow.

the different vane ledges to help identify a poor incoming gas
flow pattern at the elbow. When wide horizontal ledges exist
which are normally applied to prevent short circuiting of gas, the
excessive mound of dust could be minimized by burning slots in
the ledge as shown in Figure 11—32 (B). But I generally do not
recommend these efforts unless the weight of the dust is distorting
the horizontal ledges of the vane.

As pointed out earlier, the dust build-ups on the horizontal
or sloped floor of an outlet plenum can be detrimental to gas
distribution in many cases. Efforts to eliminate the build-up has
caused hopper sneakage in other cases. Some attempts at dust
blow-off , or aeration by steam or air nozzles in the floor of the
plenum, has enjoyed some success, but has also caused enough
problems over the years that it is generally not recommended on
modern installations. In some marginal installations, I have gone
to a chute design in the expansion flue to help promote a dust
cleaning action, as well as direct more flow into the bottom zone
of the precipitator. The downward gas flow must be re-directed
into a horizontal pattern as well as split up by the method shown
in Figure 11—33.

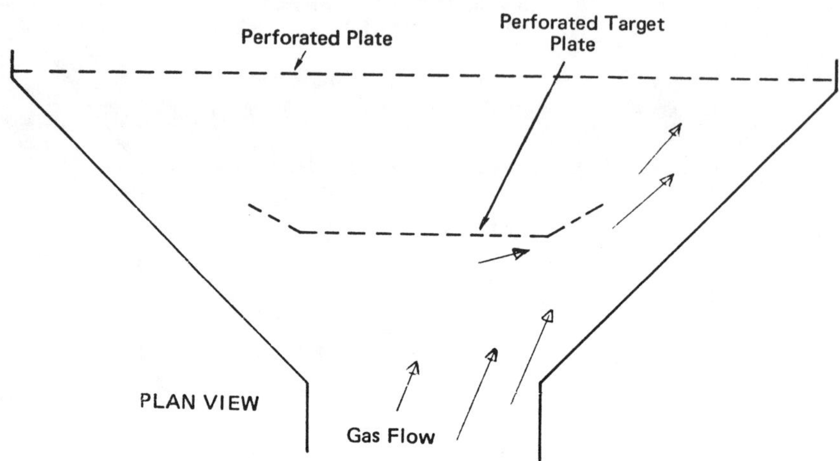

Figure 11—30   Perforated Target Plate Can Produce a Poor Gas Pattern at Precipitator if Incoming Gas Flow is Biased.

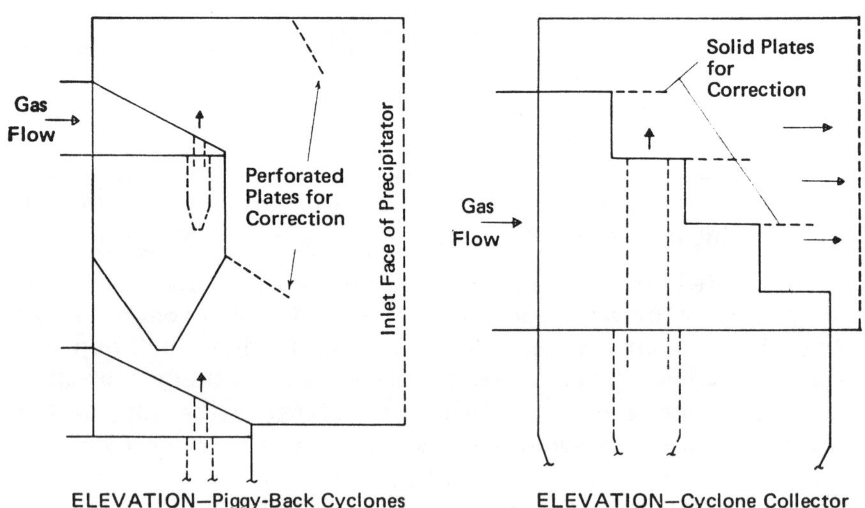

Figure 11—31   Two Methods used to Improve Gas Flow Patterns from Mechanical Cyclone Collectors placed immediately Ahead of the Precipitator.

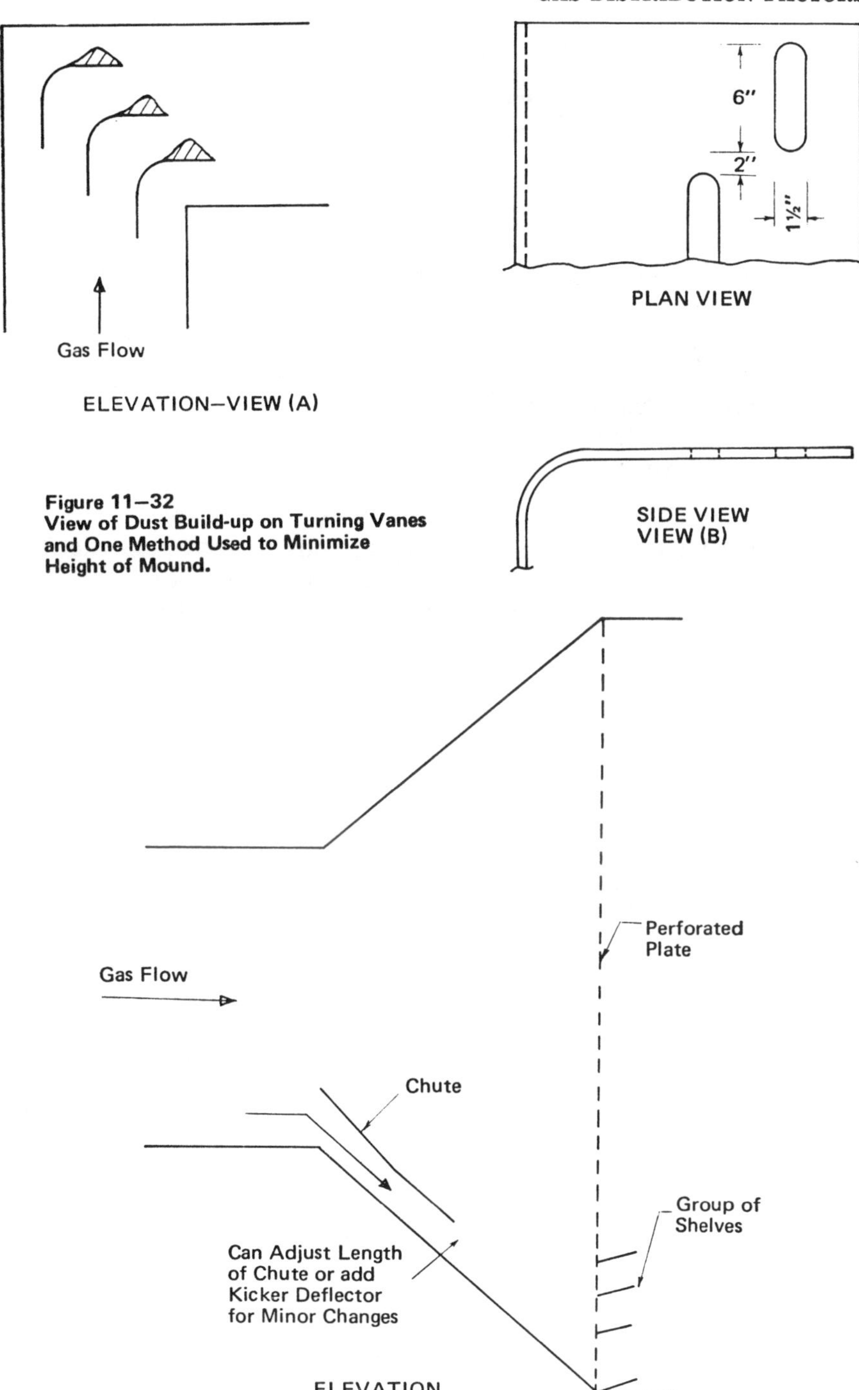

Figure 11—32
**View of Dust Build-up on Turning Vanes
and One Method Used to Minimize
Height of Mound.**

Figure 11—33 Use of Chute to Gas Sweep Bottom Slope of Expansion Plenum.

Chapter 12
Gas Conditioning

Introduction

The preferred method to improve the performance of existing precipitators involves the use of higher power inputs. Poor resistivity levels can be overcome by the modification of the flue gas characteristics. These modifications are the subject of this chapter, and for the following discussions, I will disregard all the effects of the precipitator components on possible spark-over and concentrate on the base chemical reactions in the process.

The term "Gas Conditioning" normally refers to the various methods used for injection of chemical constituents into the flue gas stream, primarily to help alter resistivity levels in the precipitator. I would like to enlarge the term to include any method, whether or not it is inherent in the process or supplied from an external source.

The use of gas conditioning is usually considered on an economic basis. In practically all cases, the precipitator can be enlarged sufficiently to overcome a reduced power input, but the cost would be prohibitive. In other cases, the useful life of the process facility might be limited. All options are usually weighed before the best economic decision is made.

Conditioning of the flue gas is not a new technique. The use of water spray cooling towers, or some other form of temperature control, is an integral part of many industrial operations. Dr. White mentioned in his book that the use of anhydrous ammonia in the cat-cracking gasoline process enhanced the collection of the catalyst. I heard about the use of SO_3 injection into gas streams soon after I became a precipitator serviceman in 1950. High resistivity problems began to surface in fly ash precipitators with the increased use of strip mine coal. Soon after this period, the term "gas conditioning" became almost synonomous with fly ash.

However, even though some of the gas conditioning techniques were known in the United States, little field activity occurred in fly ash until the late 1960's and early 1970's. By this time, problems with low resistivity became as important as the high resistivity installations. More stringent emission regulations for particulate matter, coupled with the need to reduce sulfur contents of coal, began to proliferate the need for systems to help modify the resistivity of fly ash in the mid-1970's.

These systems took different forms and utilized various chemicals to achieve the desired effects. Separately, or in some combination, sulfur and ammonia have been the most common conditioning substances. Recent work with sodium carbonate (soda ash) to overcome high resistivity has shown some promise. The object is to improve electrical characteristics in the precipitator, and anyone who can come up with a new approach or method that is simple, economical, and will not produce detrimental effects to the process will be applauded.

Conditioning Concepts

The complex interplay between the flue gas characteristics and the particle is the area of interest. Each process has an infinite number of variables, but there are two main concepts:

1. Uptake of moisture and gaseous constituents on the particle surface is the prime mechanism below about 400°F in which the electrical characteristics of the precipitator are effected.

2. Above about 400°F, subtle changes begin to occur in the bulk resistivity of the particle. This becomes the prime mechanism that controls the power input above about 700°F of the flue gas. Surface factors are still important in the temperature range of 400 to 600°F, but a reduction of resistivity appears to occur more rapidly by the rapid rise of temperature than by other possible factors.

The process should be first explored to determine if inherent changes in operation equipment can modify resistivity levels. I have often observed net improvement in the resistivity by this approach. Some of these techniques include:

    a. Substitute, blend, or prepare some of the bad actors in the raw materials or fuel in a manner conducive to precipitation. This may require some modification of the handling equipment.

    b. More efforts on the maintenance of moisture levels in the flue gas is important. Just the elimination of inleakage air will have this net effect.

     c. Awareness of the temperature effect on resistivity must be uppermost for any process change. Even the elimination of high to low gas temperature zones may help moderate a poor performance to one that is acceptable.

Probably the use of additional moisture in the flue gas by way of water sprays or steam injection can be considered a primary method of gas conditioning. Water addition to many process gas streams is often part of the operation. Water forms part of the raw material in some cases, or in others, is primarily used to control gas temperature levels at the discharge of the process. Fuel supplies another source of moisture. As previously covered, moisture contents over 20% by volume tends to nullify resistivity problems depending on the gas temperature range at the precipitator. The use of steam is much less utilized because of the cost factor, but is useful on a short term basis where water may present condensation problems.

The use of chemical additives offers a secondary approach if the time factor or economics dictates that any modification of the process is not a satisfactory route. Fly ash collection has been the greatest area of implementation for this method in recent years.

Conditioning of Fly Ash

The areas of high and low resistivity has been identified and discussed in previous chapters. Chapter 9 discusses a number of process factors that relates to the final ash characteristics in the precipitator. I have spent most of my time evaluating ways to overcome resistivity problems by juggling these process factors, so that my experiences with conditioning systems has been infrequent. But, there are installations based on design, economics, and environment that defy relatively simple corrective actions to attain satisfactory performances of the precipitator. Usually these installations will involve one or more of the following factors:

1. High ash contents of bituminous coal above 18% by weight, matched with sulfur contents below 1.3% by weight.

2. High ash contents of bituminous coal above 18% by weight, matched with sulfur contents above 3.0% by weight.

3. Average gas velocities of 7 ft/sec or more through the precipitator.

4. Flue gas temperatures at the precipitator inlet between 330 to 350°F, especially matched with the above No. 1 point.

5. Flue gas temperatures at the precipitator inlet between 240 to 280°F, especially matched with the above No. 2 point.

When either points 4 or 5 are matched with high gas flow rates, the probable need for conditioning in the flue gas is quite good. Just which conditioning method to use, and how and where to use it, are the concerns of the following sections. I will discuss some of the factors common to all the conditioning systems.

1. Provide an effective safety and operation training course for all personnel who may come in contact with the system. Even though some chemical systems already exist in a power station, the average person who will be working on the system will probably have little chemical background. Set-up the safeguards and mechanisms to completely train new people, as they are assigned to monitor and maintain the system. Any hesitancy of personnel, because of a fear of the system, can gradually lead to a deterioration of operation.

2. Provide the means to isolate each probe, and install devices to monitor the flow through each probe of the injection system. This measure increases the cost of the system, but is considered necessary to detect plugged nozzles, as well as to provide accessibility for probe removal and its clean-out during operation.

3. Provide a practical method to support a long probe inside the duct or chamber. I have seen several cases where it was quite difficult to return the probe after clean-out to its support receptacle. To remedy this problem, Figure 12—1 shows a design that was used at the bottom of a 12 foot high duct, in order to keep a ½″ injection pipe from oscillating in a 60 ft/sec gas velocity. As you can see the flared top allows relatively easy access for returning the probe to operation.

4. Provide an inert-gas purge system, if at all possible, when the chemical flow is not in operation. Eddy-currents of flue gas around the probe can cause some blockage of the nozzle holes.

5. If the injection nozzles are installed in close proximity to the expansion plenum of the precipitator, as often happens in retrofit cases, it may be more desirable to provide a small turbulance zone than rely on a greater number of nozzles to achieve proper mixing. Much depends on whether fan capacity is available for the ¾ to 1″ pressure drop that may be required to mix the gas.

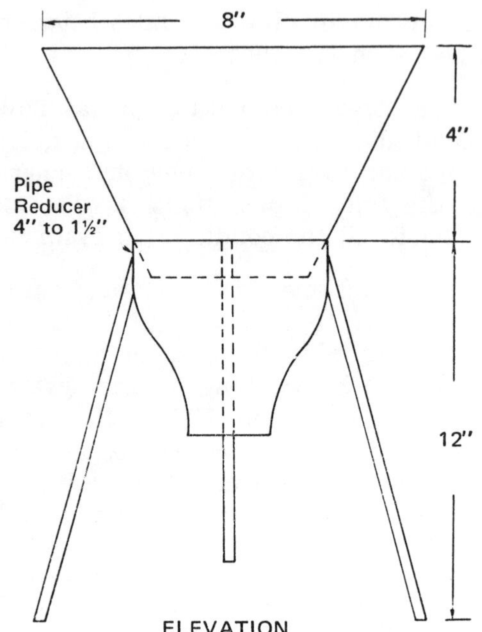

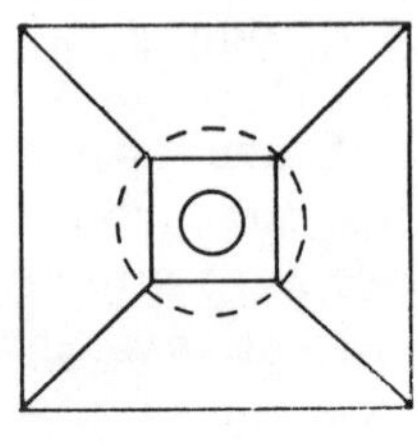

Figure 12—1
Suggestion for Floor
Receptacle to Restrain Long
Probe from Excessive Side
Motion yet Allow
Free Access.

6. If at all possible in large precipitator systems where different gas temperature zones exist, attempt to inject the conditioning agent according to need. Of course, this may not be possible on all installations; however, it is a technique that minimizes chemical consumption, with the benefit of preventing over-conditioning of part of the flue gas. There is an example of this later.

7. The effectiveness of the gas conditioning system will be directly related to both the uniformity of the process (including coal characteristics) and whether or not it is a steady-state operation. Conditioning can be applied on a cycle boiler, but it is more difficult to match the correct agent to the ash conditions in a transient state.

8. Consider the use of two conditioning methods on the same application, if the flexibility and benefits outweigh the disadvantages. For example, the use of small quantities of water vapor and low levels of chemical injection may be better than the use of large levels of chemical input alone. In some cases, the flue gas temperature has to be brought into a relatively narrow range for the chemical agent to be effective at reasonable injection rates.

Moisture Addition

I sprayed low pressure water into the flue gas of a highly resistive fly ash installation back in 1953, and increased collection efficiency about 15% to reach the guarantee point of the precipitator. In 1968 at a power station installation, I experimented with spray water for fly ash conditioning on a marginal collector. The use of 0.8 to 1.0% by volume water vapor reduced the exit flue gas temperatures by about 10°F at the precipitator. Some gains in power input occurred, but since the flue gas temperature range was between 285 to 295°F, further work was halted, especially since only 85 psig water pressure was available.

Early in the 1970 to 1974 period, when I invited RECON SYSTEMS, INC. of New Jersey, to join me in various conditioning projects, we worked with several installations that used spray water injection. It was felt that the proper application of small droplet water spray still offered a viable conditioning method for high resistivity ash. Several experimental projects, utilizing small capacity nozzles, with either steam or compressed air used for the atomization of water in the nozzle discharge, showed some positive and negative results. In Chapter 13, I will cover one project in detail, but the following brief comments should be helpful.

1. Utilize a vertical duct for spraying if at all possible. If a horizontal duct is available, it is desirable to provide at least a ¾ second residence time before the gases enter the inlet plenum of the precipitator. In horizontal ducts, drainage connections should be considered as moisture check points, in the floor of the spray zone.

2. Make sure that the spray patterns do not strike any wall surface or overlap with adjacent sprays.

3. If steam or compressed air is not used for effective atomization at the nozzle, high water pressures of 400 psig may be required. Water outputs of less than 2 gpm per nozzle are desirable, regardless of the atomization principle. I believe it is much better to use more nozzles of smaller capacity in the gas temperature range of 290 to 330°F, especially for purposes of temperature control.

4. Do not control the water sprays from thermocouples at the discharge of the precipitator, although the outlet flue gas temperature should be monitored to operate safely above the dew point level. The heat sink capacity of a precipitator can be seen

by a series of measurements obtained on the installation, shown in Figure 12—2. The normal temperature drop measured between the inlet test ports and the precipitator outlet was 18°F. Average inlet gas temperature was 291°F, before the start of a 7.2 gpm water injection into the 250,000 acfm gas steam, at the location noted on that figure. The average gas temperature measured at the inlet dropped about 35°F to about 255°F, corresponding to the zero point of Figure 12—3. The decay °F temperature curve shows that only a 18°F drop occurred at the outlet after one (1) hour of operation at the low spray rate. The final gas temperature at the outlet test ports dropped to about 250°F near the end of the 95 minute spray period. After the sprays were turned off, the average temperature almost immediately returned to 291°F at the inlet ports, while the average temperature at the outlet was still 266°F after 34 minutes.

This example is meant to show the span of gas temperature that can occur with spray conditioning between different points of the system. These values will change depending on some of

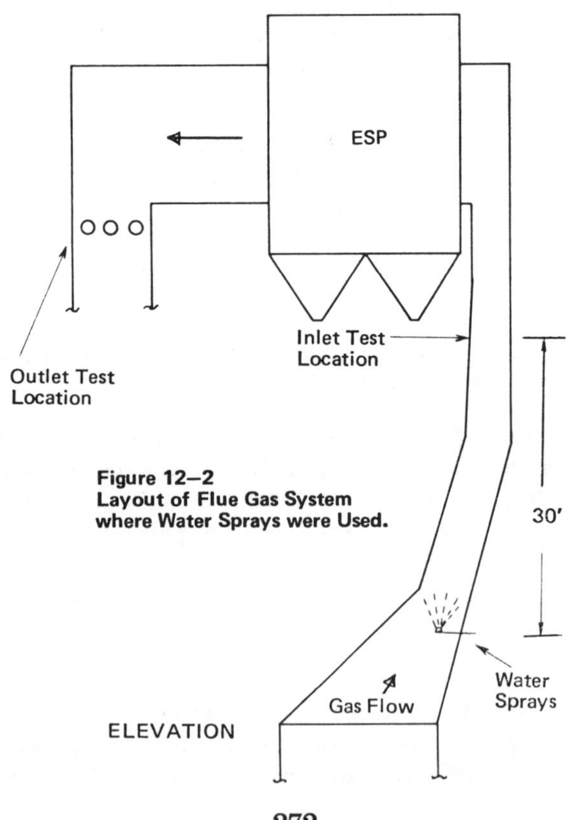

Figure 12—2
Layout of Flue Gas System
where Water Sprays were Used.

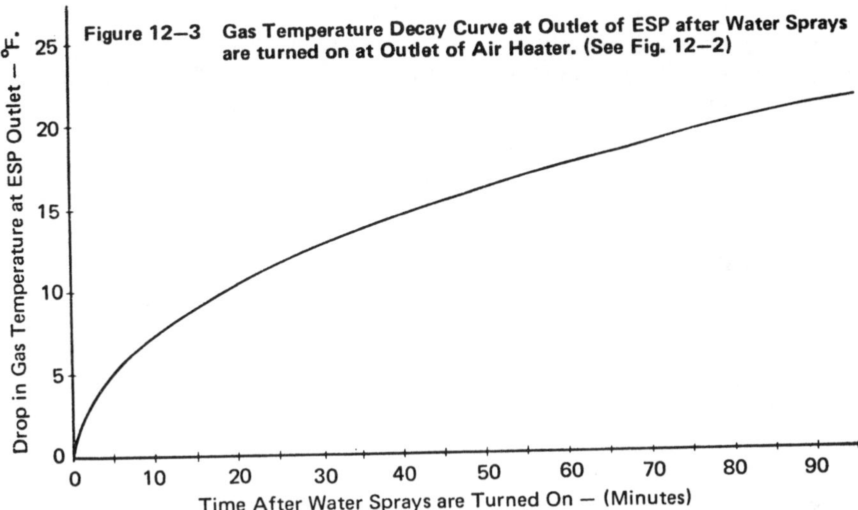

the variables, but the principle remains the same. Consequently, any system must consider the time-lag effect in the monitor and control features. Another factor is to position the control thermocouples far enough downstream from the sprays, so that complete water evaporation is assured between the nozzles and the temperature measurement.

5. The amount of water vapor required to cool the gases down to a desired level can be obtained from engineering handbooks. A rough estimate for cooling gases from 340 to 270°F is one (1) gallon per minute of water for a 4.5°F reduction for every 100,000 scfm of flue gas flow. The range might be 3 to 6 gpm, depending on the starting level of the gas temperature. Levels about 400 to 450°F might require as much as 6 gpm for every 4.5°F drop per 100,000 scfm.

Steam conditioning is used primarily for its moisture effect rather than cooling. A rule of thumb would indicate 5,000 to 10,000 lb/hr of steam flow (at 250 to 400 psig) is required for every 100,000 scfm gas flow rate to make an appreciable effect on a high resistivity dust collection. This becomes a fairly expensive method of conditioning.

6. When steam is used as an atomizing medium in water spray nozzles, amounts of 300 to 600 lb/hr of low pressure steam is usually required for each gpm of water flow. Expensive, yes, but I believe this is the most effective conditioning method, since the steam is also adding moisture to the flue gas.

7. In practically all cases where a highly resistive ash must be brought into a reasonable operating range by sprays, one must work toward flue gas temperatures of about 290°F. However, below 300°F, the monitor and maintenance of the system assume more critical proportions in terms of nozzle integrity to prevent wetting or condensation problems. I am only talking about total moisture contents in the flue gas of 9 to 13% by volume, so that the gas temperature reduction is the prime mechanism to help reduce resistivity in that case. If water sprays are used in conjuction with a chemical additive, then temperature reduction to the 310 to 320°F range might be sufficient.

8. Some minimal help in the reduction of resistivity was achieved when we added water through lances in the combustion zones of small boilers. The technique was simple and eliminated the problems of spraying water in the cooler portion of the flue gas system; the effect on boiler efficiency did not warrant further work in this area.

9. Maximum heating of hopper wall and other ash evacuation features discussed in Chapter 5 should be present if spray water conditioning is considered. Control of air inleakage is very important throughout the system after the location point of water injection.

SO$_3$ Injection

Early studies, as noted by Dr. White, indicated that the condensation of SO$_3$ on the particle surface was a main conditioning agent. This led to the experimental use of liquid SO$_3$ as a means to reduce resistivity levels of fly ash. The difficulties of handling the liquid SO$_3$ was a major problem since it had tendencies to solidify and cause pluggage unless temperatures of the solution was closely controlled. I have had no experience with this particular conditioning method, but evidently it has worked effectively in the past.

Sulfuric Acid Conditioning

Probably the simplest transition used to eliminate the problems of liquid SO$_3$ involved liquid sulfuric acid as the source of SO$_3$. I evaluated this method on more than one installation and it showed satisfactory results, but some difficulties were encountered. Only about 15 to 20 ppm of SO$_3$ or H$_2$SO$_4$ may be required on 1% sulfur coal applications in the 300 to 320°F temperature range. Upwards of 30 to 40 ppm may be required in the range of 360 to 400°F because of the difficulty of gaseous uptake on the ash particles at these higher gas temperatures. Concentrations required can also depend on the chemical composition and content of the ash in the coal. Some key points are:

1. The vaporization technique for sulfuric acid does not appear complex, but probably offers the area of greatest difficulty. Hot clean air is the carrier medium that transports the vaporized SO_3/H_2SO_4 to a nozzle system in the duct work, usually immediately after an air heater.

2. Although carbon steel can be used for handling the liquid acid, the material used for piping between the vaporizer and nozzles must be applied quite carefully to minimize corrosion. Even glass lined piping has been used, but temperature levels must still be kept high enough to prevent acid condensation in the main manifold. Based on normal acid feed rates, gas temperatures above 550 to 600°F may be required to eliminate condensation problems. All piping and manifolds should be adequately heated and insulated to keep wall surface temperatures above 550°F.

3. Nozzle design can be as simple as drilled holes in a manifold pipe, but the layout of the holes and the number of nozzle probes can be critical. Sufficient back-pressure in the probes will help in the maintenance of equal discharge velocities. One problem with an insufficient number of probes is the lack of mixing between the vapor and the flue gas. Excessive probes may cause an overlap of vapor that can produce a localized acid condensation problem.

Condensation of SO_3 to H_2SO_4 on the surface of the particle is the mechanism desired. It is well to maintain the nozzle discharge in vapor form for as long a period as possible, in order to obtain a wide spread distribution of SO_3 through the flue gas prior to condensation. I have one concern with a premature condensation of acid and that is H_2SO_4 adsorption on the ash particle might be inhibited, leading to higher levels of free acid in the gas stream.

The immediate area around the discharge holes in the probe should be clear of obstructions so as to minimize the formation of hard build-up. Actually, eddy currents of gas can cause a crusty build-up at the edge of the nozzle hole, which could distort the spray pattern and aggravate the conditioning problem. The ability to withdraw the probe for periodic cleaning can be important.

4. Sometimes a problem occurs with a cycle boiler where the system is turned off during half-load operation. The rate of power improvement after the start of acid vapor injection depends to some degree on the thickness of the collecting surface dust deposit and the rate of spark-over. Since each spark sort of blows out a small pocket of material from the layer, a higher spark rate will tend to show the fastest response to a new coating of conditioned ash. I prefer to see a spark rate in the 100-150 per minute range prior to injection, especially on the inlet half of the precipitator. Another technique is to slightly over-condition for a short period, as little as $\frac{1}{2}$ hour, and then return the system to its normal feed rate for the high gas flow rates.

The electrical characteristics will often improve during these low-boiler load periods. The transient period from the low level to full load can be short, and some variation of the above methods may be required to promote a high precipitator performance during the change in gas flow rate.

5. The first sign of conditioning should occur within 5 to 10 minutes with sufficient H_2SO_4 injection rates of 15 to 20 ppm. The outlet fields will probably show increased power first on most installations. Approximately $\frac{3}{4}$ hour after injection, one should see at least 70 to 80% of the desired power input levels, with the remainder of improvement occurring within several hours. I keep hearing about conditioning systems that are supposed to take many hours before changes in power characteristics occur. In my experience, I have found if a conditioning system is successful, it shows a relatively fast response soon after the initial injection. On the other hand, the overall electrical characteristics may hold up fairly well for 1 to 2 hours after the acid system is shut down. Within the first hour, the inlet fields show the fastest decline in power, whereas the outlet field may take twice as long to decay.

6. The speed of response in the electrical characteristics from acid injection may vary between similar fields, and this situation might identify differences in the distribution of acid vapor between chambers. Compare the power characteristics of the adjacent fields before and after injection.

Conversion of SO$_2$ to SO$_3$

From the standpoint of the precipitator, the more SO$_3$ produced in the boiler system from a given sulfur source, the better it is for high resistivity conditions and the worse it is for the conductive jobs. The more excess air, the more the chance for conversion for SO$_2$ to SO$_3$ from the combustion operation itself. Unfortunately, higher excess air levels means higher exit gas temperatures and higher gas flow rates for a given boiler load. Even at that, benefits often accrue to the precipitator performance with the higher SO$_3$ production and resulting improvement in power input.

I generally recommend a minimum of 30% excess air in the combustion zone of low sulfur installations. Even more may be required at times. If the fan system is limited, then a slight reduction of steam flow is indicated. This type of operation is usually considered as a stop-gap method while other options are being explored. It often amazes me at how important a small change in the air-fuel ratio is on the electrical characteristics of a precipitator, especially in boiler sizes below 600,000 lb/hr steam flow ratings. Some precipitators will show an immediate drop in power as soon as a slight deficiency of air occurs.

Just where all the conversion of SO$_2$ to SO$_3$ occurs in the boiler system is not well understood. No doubt part of the reaction occurs in the combustion zone itself, but partial conversion probably continues as the flue gas carries through the boiler tube sections. The intimate mixing of fly ash, iron surfaces, and gas temperatures in the 700 to 900°F range may also be conducive to some conversion. I remember an economizer tube replacement with one of a finned-tube design that appeared to add another degree of conditioning to an already high sulfur oxide installation.

The knowledge that the iron element can help the SO$_2$ to SO$_3$ conversion once led me to consider blowing iron oxide particles into the combustion zone, but I never followed through with this method. I still prefer the pulverizers that can grind up most of the pyritic sulfur segment on low sulfur applications, while others should have a high reject rate on the high sulfur coals.

Conversion of SO$_2$ to SO$_3$ is a viable conditioning technique and several ideas have been generated over the years. The use of a slip-stream of flue gas passing through a catalyst bed is one method, while others have talked about using the catalyst in-situ, that is within the gas duct. I do not know of any commercial application of these methods, and while they should be effective, I guess the basic success of the vaporization and injection of the sulfur chemical agent has produced some resistance for trying other and possibly more novel conditioning techniques.

Another method that appears practical and is used commercially is the catalytic conversion of SO_2 to SO_3 from either an external SO_2 source or a sulfur burner. The sulfur burner appears the better method from an economic standpoint. The injection system is similar to other methods of SO_3 production.

The use of trioxide injection to help combat high resistivity ash conditions in the precipitator is effective, and can be economical compared to other methods to improve the performance of the collector. There are inherent problems, mostly from the vaporization step to injection nozzles, but these problems can be reduced with proper system designs. With good intimate mixing of the SO_3 vapor and ash particles, a correct rate of injection should minimize the existence of acid mist particles in the stack plume. The method is quite reliable in its ability to alter resistivity levels of the fly ash as well as other particulate matter.

Use of Soda Ash

Recent studies by various groups on the effect of the chemical make-up of fly ash on resistivity properties indicated that the deficiency of sodium oxide (Na_2O) may aggravate the problems from low sulfur coals. This led to field trials of adding soda ash (sodium carbonate) as the source of sodium to the boiler system in attempts to improve resistivity levels. Success has been shown for some installations, while others show little help from this technique. About 0.5 to 1.0% by weight of soda ash per ton of coal consumed is the general range of application for the additive. The higher rates may cause fouling of the boiler tubes and must be monitored carefully.

There appears to be several methods for adding the soda ash to the boiler system.

1. By feeding the soda ash onto the conveyor belts that transfer coal to the burners.

2. The soda ash can be fed directly into the bunkers, even in a batch-type style. Methodology of coal feed from multiple bunkers of a large installation tends to smooth out the cycle

effect. The pulverizers also provides mixing benefits between the soda ash and coal burden.

3.   Another injection method of the dry soda ash would be at the coal feeders, just before entry to the pulverizers. A rotary feeder is one way this method can be implemented.

4.   A sodium carbonate solution can also be spray injected into the boiler flue gases in the 1000 to 1200°F temperature zone.

As covered previously, less than 0.5% by weight of Na_2O in the fly ash coupled with low levels of SO_3 generally indicates trouble with the electrical characteristics of the precipitator. There appears to be some relationship between the sodium segment and the sulfur content of coal in that both tend to be higher or lower at the same time. Where sulfur levels are relatively high, about 2% by weight in the coal fed to the boiler, the sodium quantity does not appear to be a critical ingredient. If an ash happens to have a sodium oxide analysis over 0.5% by weight with low sulfur, it does not mean that optimum power inputs will exist in the collector. I doubt that additional sodium oxide will be beneficial here.

Ammonia Conditioning

The use of liquid ammonia (NH_3) as a conditioning agent has found wide use in recent years, primarily for low resistivity ashes that exhibit excessive reentrainment from the collecting surfaces. A paradox occurs in that NH_3 has also been used to condition high resistivity ashes. The mechanisms of this method of gas conditioning, while effective, are still not well understood. A common theory indicates that the ammonia reacts with the H_2SO_4 to form an ammonium sulfate substance, and that has the same effect as increasing the resistivity of the ash to a desirable level. Another theory talks about the change in space charge without a change in resistivity. Whatever the mechanism, injection of the ammonia vapor, at rates of 15 to 50 ppm by volume, has produced marked improvement in high sulfur, low flue gas temperature applications.

The benefits of ammonia with the relatively high resistivity ash must lie with the adsorption of the vapor on the particle surface, and the subsequent uptake of moisture. Fly ashes that are acidic in nature appear more apt to benefit from the ammonia in these cases. Usually, additional moisture by use of water spray or steam injection is needed to gain the desired improvement in power characteristics. Nevertheless, I do not believe the use of ammonia in high resistivity cases has produced the same magnitudes of improvement that have been observed in the low resistivity cases, at least with eastern bituminous coals.

One facet about the use of ammonia conditioning is that it is simple to handle and relatively safe compared to some of the other chemical agents. However, NH_3 is highly soluble in water and will become irritating to the sensitive membranes of the eyes, throat, and lungs. Concentrations of 50 ppm in air will generally be detected by people, but this concentration is safely below the level that might cause any difficulties with an over-exposure to NH_3.

The base system for the vaporization of liquid NH_3 usually consists of a large storage tank or a parallel group of smaller tanks connected to a common manifold. The nice part about this method is that if the liquid temperature is kept above 60 to 70°F, the vapor pressure in the gas space of the tank will generally be sufficient. Thus, tank heaters are only required for part of the year. The driving force of the tank gas is usually reduced to 10 to 15 psig for transport to the injection nozzles. Other than the elimination of brass, copper, galvanized or cast iron fittings in the piping system, the material of construction is not critical. Several comments on this conditioning method follows:

1. The nozzle probes are usually of simple design, possibly 1/2″ or 3/4″ pipe size, with 1/8″ or 3/16″ holes spaced about 10 to 12 inches apart. I prefer a single row of holes in the probe, rather than holes spaced within a 60 to 90° arc, to help minimize pluggage as well as provide more of a directional pattern. I find that in most installations where NH_3 is used to overcome low resistivity problems, a uniform spray pattern in the flue gases is not what is desired. For example, NH_3 should be concentrated in the temperature zones below 290 to 300°F with very little NH_3 above 300°F, if indeed any.

2. Ammonia is usually used after the air heaters. The regenerative heaters would be very sensitive to pluggage, but even the tubular heater and possibly mechanical cyclones might be affected.

3. The effect of NH_3 on the electrical characteristics of a low resistivity precipitator should be observed within minutes after the start of injection. The inlet fields should show a rise of 20 to 30 volts on the primary meters in about 10 to 20 minutes with a possible slight drop in current. Some spark-over may occur, but it should only be occasional in this early period. With

that type of response in the inlet fields, the primary voltages of the center fields should also gradually rise, possibly 5 to 10 volts by the end of the 20 minute period. As the NH_3 injection rate continues, one might even observe the inlet spark rate increasing to where a power reduction occurs, with the center fields sparking at a low rate at some reduction in current, and the outlet field just showing a small rise in voltage. This set of meter observations should correspond with what I would now consider a good performance level in the precipitator. Stack puffing or spikes on the opacity chart should have been minimized or eliminated for all practical purposes.

4.   About 70 to 80% of the change in the electrical characteristics should occur during the first hour after the start-up of the system with the remainder of the electrical change occuring during the next few hours. The gradual electrical change may even span a longer period, but the effect on precipitator performance will be minimal after that initial main response. The gradual decay in electrical characteristics, that is, a reduction of spark-over and a return to lower voltage-higher current condition, may take 30 minutes or more after a shut-down of the system. In fact, a full return to the original electrical conditions may not occur for hours under uniform operation of the system.

5.   Ammonia vapor injection into precipitator zones of moderate resistivity will usually produce electrical characteristics simulating a higher resistivity condition. I have observed power reductions by a factor of 5 or more by over-conditioning with NH_3.

6.   The danger of over-conditioning with NH_3 is the presence of dust build-ups that is difficult to remove from both electrode surfaces and hoppers.

7.   A rule of thumb would indicate about 15 ppm by volume of NH_3 is needed for every percent sulfur of the coal, but this quantity also varies with the method of distribution and flue gas temperatures. At least 20 to 25 ppm would be a reasonable rate for a 2% sulfur coal with ash operating in the wrong temperature range. Injection rates are also expressed in lb/hr of NH_3, and about 15 to 30 lb/hr are reasonable quantities for 2,000,000 acfm flow rates.

8.   Another suggestion is to use clean compressed air as a continuous purge during the non-conditioning period. It may be desirable to use an inert gas as an initial purge prior to the normal air flow used to keep the system nozzles clear.

9. Because of low transport system pressures (and low NH₃ flow in many cases), it can be difficult to obtain equalized gas distribution between the different probes of a manifold system. This will generally take some investigation and correction after the initial operation. Observe the potential pluggage pattern of the nozzle openings. I also suggest that small pressure drops could be taken across orifice plates in the individual nozzle piping and which could serve as flow indicators and effective maintenance tools.

Other Conditioning Methods

The quest for improved methods of gas conditioning to modify the electrical characteristics of the precipitator is still active. Any method that is simple, economical, safe, and has no detrimental effects on the process should be considered, and I am sure new methods will be developed.

Aside from the techniques discussed earlier, attempts to use fine-sized carbonaceous material for electrical help have produced some minimal benefits. The use of oil burners for partial load in coal-fired boilers is thought to offer some benefits, but I have observed little evidence of conditioning help. The coating of high resistivity fly ash particles with carbon from the incomplete combustion of oil was tried about 25 years ago by Research-Cottrell, Inc. Results indicated that benefits would accrue with relatively large concentrations of the carbonaceous material, but the method did not appear commercially practical.

Other methods have used sulphamic acid (NH_2SO_3H), calcium chloride ($CaCl_2$), and the ammonium sulphate products. These solutions have shown conditioning benefits at times. Several companies have developed proprietary products from variations of the above chemical solutions, and performance success has occurred depending on the application.

Hot Side Precipitators

Since it is known that flue gas temperatures above 650°F improves electrical properties of fly ash particles regardless of surface characteristics, the use of the hot-side precipitator to combat low sulfur coal problems has met wide acceptance in recent years. High temperature precipitators had been used extensively in the process industries, but most all fly ash precipitators up to the early 1960's were found in the 300 to

420°F temperature range. This range has already been identified as a critical zone for fly ash precipitation with resistive dusts.

A reasonable strategy for low sulfur coal precipitators was to consider placing the collector between the economizer and air heater where flue gas temperatures of 650 to 730°F can be found. This approach is better applied on new designs rather than for retrofits. The rise in use of western coals, low in sulfur and high in basic constituents, promoted the concept of the hot-side precipitator.

The hot-side precipitator has proven successful, but not without problems. Expansion difficulties occurred in a number of the early designs, especially with the relative growth between the main structure and the upper roof plate and penthouse. Most of these design problems have been overcome. But elimination of the resistivity problem by reaching a high temperature over 650°F was not that easy to attain. Some dust characteristics appeared to require higher temperatures than existed at the economizer outlet, and good power inputs still escape a number of installations. Low magnitudes of sodium and iron in the ash of the western coals have provided a sensitive factor in this application. I will discuss some features of the hot-side precipitator in the design section, but I have had little personal experience in this area.

Chapter 13

Case Histories

Introduction

The bottom line in precipitation is its performance level. Can the precipitator meet its required guarantee as stipulated by its design criteria? Some original design precipitators do and others do not. In many cases, as noted earlier, the raw materials or process modifications could alter a reasonable design of the collector. The purpose of this chapter is to draw together in case histories some of the information covered previously.

I believe there is no better way to learn precipitation than through involvement in an actual field experience. But the second best way is to study case histories which describe a number of causes, effects and results of problems with precipitators. The 13 case histories I have chosen for this chapter will encompass a variety of inputs from the fly ash, cement, and steel industries. I may refer to installations throughout the chapter, but these 13 should specifically illustrate the areas of concern to the precipitator users as well as pinpoint methods to improve performances in the field. The information in this chapter will be most useful to you if you pay attention to the broad methodology rather than concentrate on the details of these case histories.

Whether your problems involve fly ash, paper mill, smelter, or some other application, the corrective steps taken to upgrade the performances of precipitators basically follow similar paths. Although the prime paths of improved power input, gas distribution, and temperature control are explored, be on the alert for all the subtle effects of the process on these variables. I have been extensively involved in all aspects of the installations discussed for many years, and I have a high degree of confidence in the internal conditions of the precipitators as well as the test results. The identification of these companies is really immaterial to this discourse, although I am sure they appreciate the use of their past histories to help promote efforts to acquaint and train new people in precipitation. For discussion purposes, I assigned capital letters to each of the 13 installations, and will refer to tables in the text concerning pertinent design and operating data.

The first three histories, A, B and C, are fly ash installations where gas conditioning methods were employed to enhance precipitator behavior. The purpose of these descriptions will be to identify electrical trends and show why gas conditioning has met acceptance on critical installations. Cases D through I will present a variety of fly ash installations which will be discussed for their specific characteristics. Cement Cases J and K should prove interesting. Case L will discuss a sinter plant installation, while Case M describes the evaluation and upgrading of the performance of a BOF precipitator.

Case A

This fly ash installation is considered small by recent standards, but was a common installation 25 to 30 years ago. The rated steam flow of the boiler, as well as pertinent design data of the original precipitator, can be found in Table 13—1 along with Cases B and C. The precipitator size was capable of achieving its original efficiency of 98% at the design gas conditions.

One early problem occurred in the precipitator when the four 25 KVA power supplies had to be replaced because of maintenance problems. Four 15 KVA supplies were made available from another installation. Unfortunately, these T-R sets were placed in a hot environment, and the limit on the transformer oil temperature prevented full use of the available power. Test A1 in Table 13—2 shows an efficiency of 96.5% during this period under a low power input to the precipitator. Note that a flue gas temperature of 420°F existed at the precipitator. While this is rather high, the design temperature of 301°F appears too low for this vintage boiler.

An upgrading of the power supplies occurred both in size and rating. The original T-R sets energized 8 bus sections in the precipitator. Therefore, 8 new supplies were installed, the inlet fields of 32 KVA capacity, and the outlet fields of 64 KVA. The effect of being able to utilize this much power can be seen in Test A2, where the normal coal consumed during that particular period was in the 5 to 6% S range. Here was a relatively high velocity, two field precipitator doing 99.3% with a SCA of 135. This seems hard to believe with our current thinking of large-sized collectors, but believe it! A contribution to the high collection efficiency could lie in the Vee plates of the outlet field, since little rapper reentrainment was observed. This type of collector plate was used years ago and consisted of close coupled fins protruding from the surface which created a zone to prevent

TABLE 13–1
OPERATION AND PRECIPITATOR DESIGN DATA
TYPE OF PROCESS – FLY ASH

PLANT IDENTIFICATION	A	B	C
Rated Steam Flow K lb/hr	400	894	6000
Precipitator Design Data (original)			
No. of Units	2	2	2
No. of Fields	2	2	4
Total No. of Gas Passages	Inlet 46 Outlet 40	76	312
Size of Passage H'xL'	20x18	17.5x18	30x30
Spacing of Passage in	Inlet 8¾ Outlet 10	8¾	9
Type and Size of Disch. Electrode	Nom. 0.1"	Nom. 0.1"	Nom. 0.1"
Total Length of Disch. Elect. Lineal Ft	20,640	31,920	374,400
Total Collecting Surface Sq Ft	30,960	47,880	561,000
No. and Size of T-R Sets	4-25 kva	4-15 kva	16-64 kva
Total kva installed	100	60	1024
Guarantee Efficiency %	98	95	99.5
Gas Flow Rate K acfm	215	366	2700
Rated Gas Temp. °F	301	305	285
Gas Velocity ft/sec	5.3	6.3	6.4
Net Aspect Ratio	0.9	1.0	1.0
Design Migration Vel ft/sec	0.45	0.38	0.42
Watts/1000 acfm	302	106	246
Watts /sq ft Coll. Surf.	2.10	0.81	1.19
SCA— sq ft/1000 acfm	154	131	208
Density— ma/1000 sq ft Coll.Surf.	48	21	28

deposited material from being reentrained. In my experience, these finned collector surfaces were best applied to the lower range of resistivity. On high resistivity dusts, the finned design appeared to be more sensitive to electrical break-down, possibly due to an edge effect.

Tests A3 and A4 were obtained during a later time span when low sulfur coal was being consumed. Even though there was over six times the available power for Test A3 when compared to A1, the actual corona power was less because of the spark-limited conditions of high resistivity ash. This is a graphic case where the actual power consumed in a precipitator can

TABLE 13–2

TEST RESULTS AND PRECIPITATOR PERFORMANCE DATA

TYPE OF PROCESS — FLY ASH

PLANT IDENTIFICATION			A	A	A	A	A
TEST NUMBER			1	2	3	4	5
Steam Flow Rate at Test		K lb/hr	410	409	430	435	375
At Outlet of ESP							
Gas Flow Rate		K acfm	231	230	250	245	234
Gas Temperature		°F	420	375	390	295	380
O_2		%	6	7	–	–	9.5
Moisture		%	6	6	8.2	16.0	6.8
Concentration		gr/acf	0.08	0.007	1.13	0.086	0.02
Concentration		gr/scfd	0.145	0.011	2.02	0.15	0.033
Emission		lb/10^6 BTU	0.29	0.05	4.8	0.34	0.08
ESP Efficiency		%	96.5	99.3	58.0	97.2	98.1
Test Gas Vel.		ft/sec	5.7	5.7	6.2	6.1	5.8
Test Migration Vel.		ft/sec	0.42	0.61	0.12	0.46	0.50
Test Average		KVp	34	54	26	50	42
Test	Watts/1000 acfm		53	320	50	182	149
	ma/1000 sq ft Coll.Surf.		12	67	13	41	36
Coal as Rec.	Ash	%	13.77	13.44	17.35	18.11	7.77
	Sul.	%	1.80	6.10	1.00	0.95	0.79
	BTU/lb		12,305	11,913	11,592	11,386	13,701
Ash Analysis—Inlet	AL_2O_3	% }	38.32				
	Fe_2O_3	% }					
	SiO_2	%	49.04				
	Na_2O	%	0.67				
	SO_3	%	0.20				
	L.O.I.	%	6.80	5.6			
Conditioning Method			None	None	None	H_2O + Steam	H_2SO_4
Amount Used			–	–	–	23 gpm 6600 lb/hr ST	13.5 ppm

widely differ from the installed power capacity. The 97.2% collection efficiency of Test A4, compared to the 58% of Test A3, was due to water spray conditioning of the flue gases for a temperature reduction of about 100°F at the inlet of the precipitator. As noted in the table, 23 gpm of water and 6600 lb/hr atomizing steam was required to achieve this reduction. Five sprays were used in a cross-sectional area of 64 sq ft, and the spray nozzles were placed in the vertical duct approximately 40 ft from the precipitator. No evidence of mudding was noted during a later inspection.

Figure 13—1 shows a plot of 8 emission tests obtained at the outlet of the precipitator before and after water spray conditioning. A couple of the water spray tests were obtained in the 300 to 310°F gas temperature range, and lower power inputs produced higher dust concentrations as noted on the graph. The power characteristics during Test A4 were not as good as Test

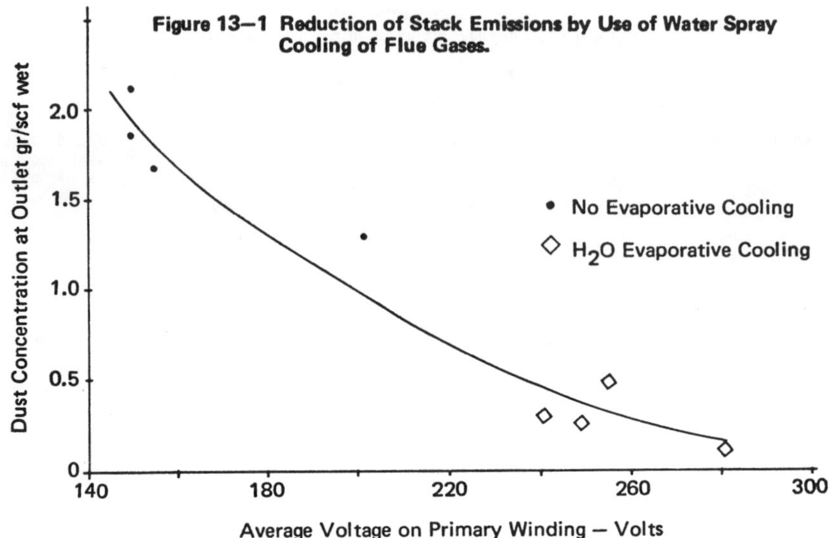

A2, but note the comparison of power characteristics between tests A3 and A4 as shown on Table 13—2. The higher ash content for the low sulfur coal of these tests placed a further burden on this small precipitator.

The results of another gas conditioning test program is shown by Test A5 in which a low ash-low sulfur coal was burned. Only minimal levels of sulfuric acid injection were required to overcome poor power characteristics and bring the collector into compliance. The acid was injected into the duct in the same vicinity as the previous water spray nozzles. Similar performance results were also observed with 12 to 15% ash coals, but acid injection rates of 30 to 40 ppm by volume were needed to obtain similar power levels to that of Test A5. It appeared that higher levels of acid injection were required because of the higher flue gas temperatures that existed at this installation. Probably half the injection rate might have had the same effect if flue gas temperatures of 300 to 320°F had existed for the higher concentrations of ash.

An interesting relationship of this group of tests obtained over a five year period is shown in Figure 13—2. The plot of the precipitator peak voltage versus collection efficiency follows an expected pattern regardless of the conditioning method used to attain the high power, but with some minor scatter based on the ash content of the coal. The key point to remember is that the precipitator cannot distinguish what method is being used to

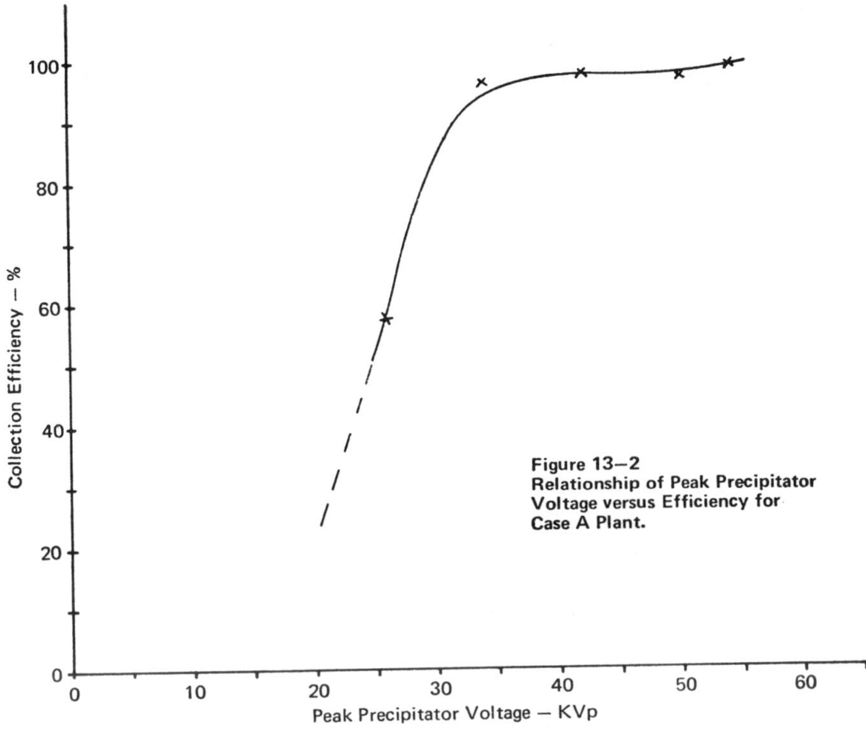

Figure 13—2
Relationship of Peak Precipitator
Voltage versus Efficiency for
Case A Plant.

condition the ash, but that its performance is based on the end result of high peak voltages and "Effective Holding Power".

During this total test evaluation period, no change was made on the rapping, gas distribution, or alignment of the precipitator. The rappers were of a impact type, and the number and location was considered poor by present standards. Approximately 3870 sq ft of collector surface was struck by each rapper, side mounted at the lower third of the collector. The gas distribution was considered good, but minor improvement was still possible, mostly in a vertical direction. Alignment of the electrodes was considered good, with no wire discharge electrode more than ½" off the center line of the gas passage. The wire electrodes were shrouded, and no localized break-downs were detected.

Case B

This precipitator of the early 1950 vintage had a common design of two fields, high velocity, and an aspect ratio of 1.0. Especially note-worthy was the small size of the T-R sets and the original design of low power density. Low power densities of 80 to 120 watts per 1000 acfm on fly ash installations were

289

capable of achieving design efficiencies prior to the occurrence of the higher sulfur-low flue gas temperature situations that began to surface in the 1950's. Along with exit gas temperatures of 280 to 290°F and about 2.2% sulfur coals, this installation was also troubled with high levels of combustible grit in the 10 to 14% by weight of ash at boiler full load, because of air starvation and a burner design problem. The 4 original power supplies gave way to 4 — 50 KVA power supplies, which was further upgraded to 6 supplies, totaling 366 KVA by the mid 1960's. As you will see, even this six-fold increase in power capability could not overcome the reentrainment problem caused by operating in the acid-dewpoint range at high gas velocities, especially with the scouring action of the combustible particles.

The 5 tests shown in Table 13—3 were obtained during a comprehensive evaluation period, and graphically show the effects of gas temperature on the precipitator performance. All five test measurements were obtained on one precipitator unit so that the 9.4 lb/hr and 26.8 lb/hr ammonia rates shown in the table are for flow rates of 260,000 and 240,000 acfm for Tests B2 and B4 respectively.

Test B1 was a base test showing that the low gas temperature combined with the other detrimental characteristics to produce a collection efficiency of 63.4%. A repeat Test B2 was obtained with a low level of NH_3 gas injected between the air heater and the precipitator. Note the 10% increase in peak voltage while the precipitator current reduced somewhat. Although the injection rate was less than desired, the rise in voltage and reduction of current must be observed if the beneficial effects of the NH_3 gas is going to be realized in the collector. A later test with 26.8 lb/hr of liquid ammonia at a 290°F gas temperature produced a collection efficiency close to 99%.

Tests B3 and B4 was another comparison set of measurements obtained at higher gas temperatures in the precipitator. To raise the temperature from 282°F to 325°F, the F.D. fans were biased between the units. This necessitated a slight drop in boiler load to maintain similar combustion characteristics to those of Tests B1 and B2. The high efficiency of 98% measured in Test B3 can be primarily attributed to the effect of the higher gas temperature in nullifying the low resistivity ash reentrainment problem. A slight rise in voltage, along with a reduction of current, can be seen in the power characteristics of this test compared to Test B1. Ammonia gas was injected into the flue ahead of the tubular air heater for Test B4 to determine its

Table 13—3
Test Results and Precipitator Performance Data
Type of Process — Fly Ash

PLANT IDENTIFICATION			B	B	B	B	B
Test Number			1	2	3	4	5
Steam Flow at Test — K lb/hr			860	860	800	800	800
At Outlet of ESP							
Gas Flow Rate — Kacfm			510	510	480	480	480
Gas Temperature — °F			283	282	325	330	246
O_2 — %			—	—	—	—	—
Moisture — %			—	—	—	—	—
Concentration — gr/acf			0.87	0.35	0.036	0.064	0.50
Concentration — gr/scfd			—	—	—	—	—
Emission — lb/10^6 BTU			3.30	1.34	0.13	0.23	1.85
ESP Efficiency — %			63.4	84.0	98.0	96.5	71.6
Test Gas Velocity — ft/sec			8.8	8.8	8.2	8.2	8.2
Test Migration Velocity — ft/sec			0.18	0.32	0.65	0.56	0.21
Test Average KVp			50.3	55.4	50.7	48.9	49.5
Test — watts/1000 acfm			190	176	175	163	304
ma/1000 sq ft Coll. Surf.			59	49	54	52	90
Coal as Received	Ash — %		11.97	11.97	11.63	11.63	11.63
	Sul. — %		2.80	2.80	2.34	2.34	2.34
	BTU/lb		12,163	12,163	13,032	13,032	13,032
Ash Analysis—Inlet (Partial)	Al_2O_3	%	30.21				
	Fe_2O_3	%	16.39				
	SiO_2	%	43.56				
	Na_2O	%	0.40				
	SO_3	%	0.78				
	L.O.I.	%	—	—	12.0	10.9	—
Conditioning Method			None	NH_3	HiTemp	HiTemp NH_3	—
Amount Used			—	9.4 lb/hr	—	26.8 lb/hr	—

effect on the high temperature conditions of Test B3. On other occasions, an injection of NH_3 into a flue gas situation where the ash was already in a moderate resistivity range usually initiated or increased spark-over with a subsequent lowering of power input. Test B4 substantiated previous observations in that increased spark-over did cause a reduction of 2 KV and current compared to Test B3. Collection efficiency deteriorated to 96.5% as shown in Table 13—3.

Another comparison for the effect of low flue gas temperature of the precipitator performance can be seen in Test B5. This measurement was obtained on the cool gas temperature precipitator unit under the same boiler conditions of Test B4. Even though the peak voltage fields were similar, Test B5 almost doubled the current density because of the reduction of resistivity. Therefore, even though the power input practically doubled, the collection efficiency was only 71.6% compared to the 98.0% and 96.5% values of the higher resistivity conditions. Now, the use of NH_3 for the conditions of Test B5 would have probably brought the collection efficiency into the high 90's, again with a 4 to 5 KV increase in peak voltage and some drop in the current density of the precipitator. Additional knowledge on those effects can be gained by a study of the next case. Figure 13—3 shows the results

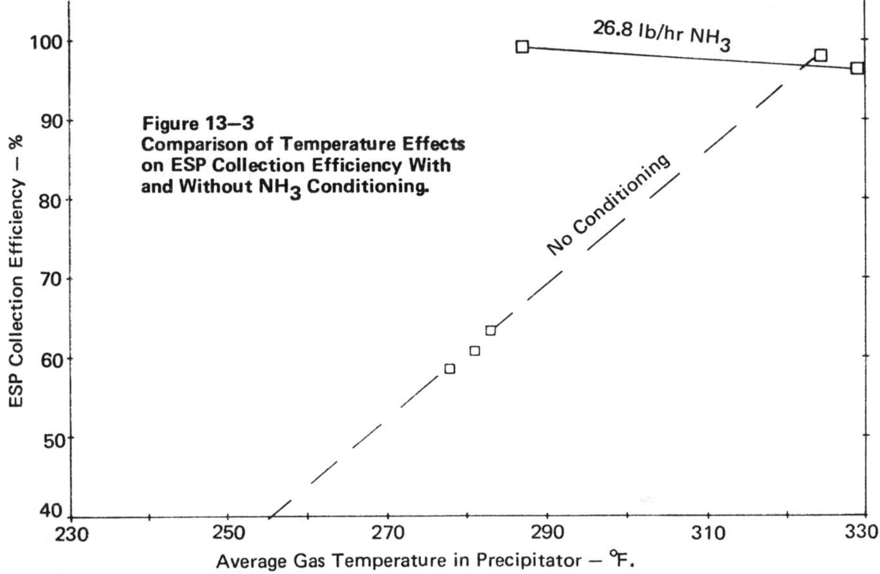

of several test measurements at the Case B plant on the effects of flue gas temperature on collection efficiency.

Case C

The fly ash precipitator at this plant would not have made the guaranteed performance if the design conditions had been met. Low resistivity ash conditions would have resulted in excessive rapper dust reentrainment at 285°F, especially with an underpowered installation. However, because the boiler had to be operated at high excess air due to an ash slagging condition (with 5.5% to 6% O_2 at the economizer outlet), the flue gas temperature entering the precipitator was in the 330 to 340°F range during most of the year. This condition nullified the potential acid dew-point characteristics; stack appearance and collector performance was acceptable. In fact, I mentioned in Chapter 3 than an increased gas flow rate could improve a precipitator performance. Here is a case where about 120% of the design flow rate produced improved power characteristics and collection efficiency when compared to performances that existed with reduced gas velocities through the precipitator.

Modifications to the boiler for reduction of the slagging problem caused a substantial drop in flue gas temperatures. Figure 13—4 shows a plan view of the precipitator units laid out

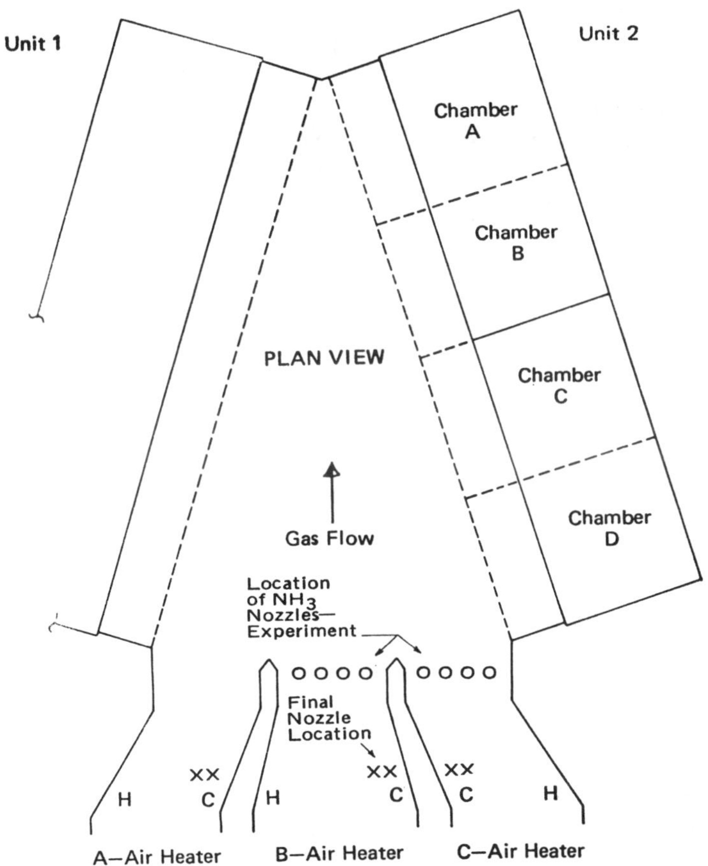

Figure 13—4 Layout of Gas System and ESP of Case C Plant Showing NH₃ Nozzle Location.

in a chevron pattern, which has been a common design as collectors have increased in size. The original rotation of Air Heater C caused severe rapping reentrainment in Chamber D where flue gas temperatures of 230 to 250°F often occurred. In very early efforts to moderate this problem before I implemented other improvements, I had the rotation of Air Heater C reversed to give the gas temperature pattern shown in Figure 13—4. This change now placed the cold flue gas zones of Air Heaters B and C together. With this modification, baffles were placed in the transition duct between the AH-B and the chevron to help moderate the span of gas temperature from this heater. Since a large pressure drop could not be taken at this location, only a 25% improvement occurred. In other words, a span of 80°F was reduced to 60°F.

Additional experimental work with ammonia injection was done with nozzles placed in existing test ports at the entry of the chevron, as noted on Figure 13—4. The object of this layout was to selectively condition the cold zone flue gases without aggravating the hotter flue gas zones. Diagnostic test results at one outlet test port are shown in Figure 13—5, with and without NH_3 injection. The results of two full-scale tests to further ascer-

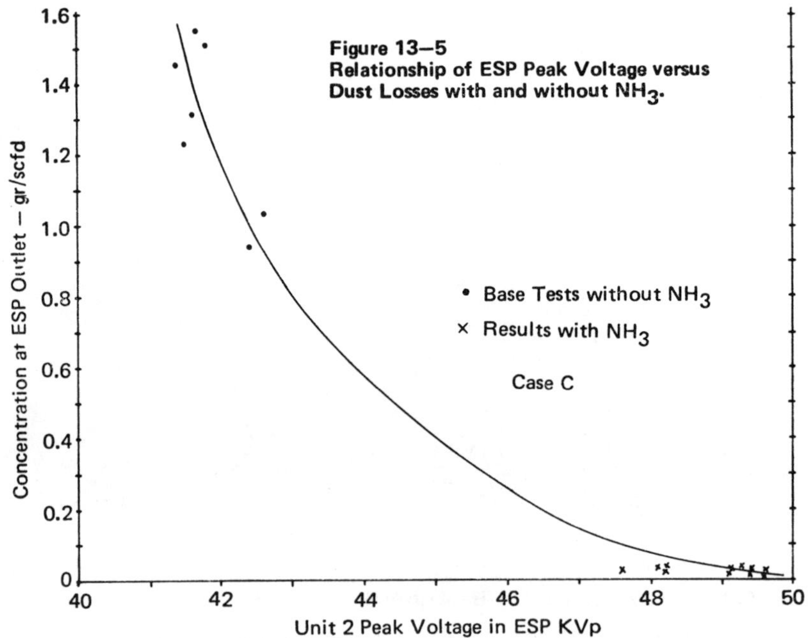

tain the effects of NH_3 on performance are shown as Tests C1 and C2 of Table 13—4. These tests were obtained at slightly reduced boiler load, but with flue gas rates still above rating. I had AH-B

Table 13—4
Test Results and Precipitator Performance Data
Type of Process — Fly Ash

PLANT IDENTIFICATION			C	C	C
Test Number			1	2	3
Steam Flow During Test K lb/hr			5160	5080	6117
At Outlet of ESP					
Gas Flow Rate Kacfm			2820	2820	2953
Gas Temperature — °F			270	270	285
O_2 — %			8.0	8.0	6.6
Moisture — %			5.2	5.5	7.2
Concentration — gr/acf			0.125	0.027	0.024
Concentration — gr/scfd			0.188	0.039	0.037
Emission — lb/10^6 BTU			0.44	0.096	0.079
ESP Efficiency — %			96.1	99.2	99.4
Test Gas Velocity — ft/sec			6.7	6.7	7.0
Test Migration Velocity — ft/sec			0.27	0.40	0.45
Test Average KVp			41	47	53
Test — watts/1000 acfm			145	140	260
ma/1000 sq ft Coll. Surf.			26	23	44
Coal as Received	Ash	%	19.30	19.30	20.67
	Sul.	%	1.73	1.73	1.76
	BTU/lb		12,100	12,100	11,891
Ash Analysis — Inlet	Al_2O_3	%			
	Fe_2O_3	%			
	SiO_2	%			
	Na_2O	%			
	SO_3	%			
	L.O.I.	%			
Conditioning Method			None	NH_3	NH_3
Amount Used (Rate)			—	67 lb/hr	25 lb/hr

biased to raise its average exit gas temperature about 47°F above that of AH-C. This technique reduced the hot end of Chamber D slightly without materially altering the characteristics of the A-B chambers. Average temperatures were 323°F and 276°F for the B and C Air Heaters respectively. A reduction in emissions of about 4½ times occurred with the NH_3 injection of 67 lb/hr. It is difficult to refer to ppm because of the method of injection.

Note that an appreciable rise in the peak voltage occurred for Test C2, while the currents only decreased slightly. Because of the large collector surface connected to each T-R set, current limit of the supply was more the rule than the exception during these tests.

An analysis of what happened in Tests C1 and C2 can be obtained from Table 13—5. The average KVp of the AB sets of Test C1 increased from 40.2 to 47.2 for Test C2, while the average

Table 13—5
Comparison of Power Characteristics of Fields of Case C
With and Without NH_3 Conditioning and Before Adding T-R Sets.

Test No.	Field	T-R Size KVA	Corona Power KW	Peak Voltage KVp	Density ma/ 1000 sq ft	Density watts/ sq ft	Totals
C1	AB1	64	29.2	45	23.3	0.69	
No	2	64	24.2	38	23.0	0.57	
NH_3	3	64	26.2	40	35.6	0.93	Fields 1-4
	4	64	24.8	38	33.8	0.88	104.4 KW
	CD1	64	28.0	47	21.8	0.66	
	2	64	23.2	40	20.6	0.55	
	3	64	27.0	43	33.5	0.96	Fields 1-4
	4	64	22.4	38	28.8	0.80	100.6 KW
C2	AB1	64	9.4	47	5.2	0.22	
With	2	64	32.9	52	24.7	0.78	
NH_3	3	64	30.4	47	36.0	1.08	Fields 1-4
	4	64	27.6	43	34.2	0.98	100.3 KW
	CD1	64	24.7	50	16.6	0.59	
	2	64	22.5	44	18.8	0.54	
	3	64	29.8	48	33.5	1.06	Fields 1-4
	4	64	21.2	40	26.0	0.75	98.2 KW

KVp of the CD sets during NH_3 injection only increased 3.5 KV over that of Test C1. Of greatest interest is the decrease in the current density of the AB sets with the use of NH_3 injection. Another point I would like to make is that the total KW input decreased about 3% from Test C1 to C2, while an increase of collection efficiency occurred from 96.1% to 99.2%. This point is still eluding many people in that more power input does not automatically mean greater collection efficiency. It is that "Effective Holding Power" that I discussed in a previous chapter that determines the bottom line, and this condition must be achieved by modifying the voltage-current relationship.

Other modifications on this precipitator included a doubling of the power supplies and correction of a gas distribution problem in the top to bottom direction. The effect of the inlet plenum, similar to that shown in Figure 11—4 of Chapter 11, was modified by closing down the holes of the middle zone of the perforated plate at the inlet face of the collector. But it was the doubling of the T-R sets as shown in Figure 13—6 that produced the greatest

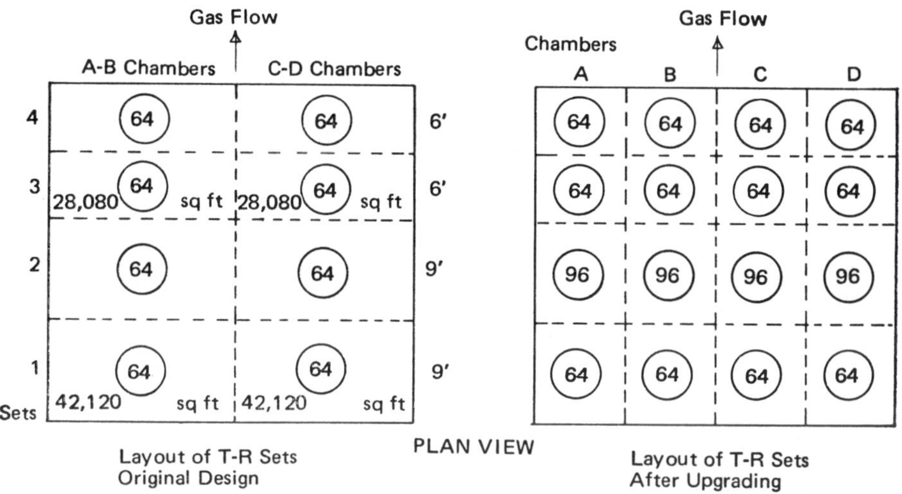

Figure 13—6 Energization of ESP Unit of Case C Before and After Modifications of Doubling of Power Supplies.

benefits. I also relocated the NH_3 injection probes to the points marked with an X on Figure 13—4. A subsequent Test C3 shown in Table 13—4 gave a collection efficiency of 99.4% at a high boiler load after the improvements even with a reduced use of ammonia. Actually, I believe that the emisson goal of 0.1 lb dust/ 10^6 BTU input could have been reached without the NH_3 addition at the gas temperatures that existed during this test. Note the higher peak voltage as well as power density that existed, primarily due to the increase in available power.

An interesting relationship of how the NH_3 effects the flue gas can be better seen in Table 13—6, where a tabulation of power characteristics are compared between Test C3 and a set of electrical readings obtained without NH_3 conditioning. Since the fields in the four chambers are energized separately, the NH_3 and temperature effects are clearly defined. For example, the individual fields of Chambers A and B with flue gas temperatures of about 260 to 270°F show high current densities without any

Table 13–6

Comparison of Power Characteristics of Individual Fields of Case C after Doubling of T-R Sets—With and Without NH$_3$ Conditioning.

Location of Nozzles Noted at Point X in Figure 13–4.

Field	T-R Size KVA	Without NH$_3$ 6000 K lb/hr Steam					With NH$_3$ (C3) 6000 K lb/hr Steam				
		Corona Power KW	Peak Voltage KVp	Density ma/ 1000 sq ft	Density watts/ sq ft	Avg KVp Total KW	Corona Power KW	Peak Voltage KVp	Density ma/ 1000 sq ft	Density watts/ sq ft	Avg KVp Total KW
A1	64	31.4	51	46	1.5		1.4	30	4	0.07	
2	96	42.5	50	64	2.0	50.5	17.6	51	26	0.8	49.2
3	64	27.9	50	68	2.0		34.4	61	71	2.4	
4	64	29.8	51	70	2.1	131.6	32.8	55	71	2.3	86.2
B1	64	30.9	54	45	1.5		7.0	47	8	0.3	
2	96	42.4	50	63	2.0	49.5	11.6	48	19	0.6	53.0
3	64	24.2	46	66	1.7		27.2	59	60	1.9	
4	64	29.2	48	63	2.1	126.7	36.4	58	71	2.6	82.2
C1	64	30.4	60	42	1.4		29.6	60	40	1.4	
2	96	39.8	51	58	1.9	53.2	OFF				56.0
3	64	28.2	52	68	2.0		30.0	55	68	2.1	
4	64	30.2	50	66	2.2	128.6	31.3	53	66	2.2	90.9
D1	64	7.4	42	10	0.4		8.2	39	11	0.4	
2	96	24.7	47	35	1.2	48.8	16.1	44	21	0.8	49.2
3	64	30.0	57	63	2.1		28.6	58	58	2.0	
4	64	25.1	49	58	1.8	87.2	33.2	56	71	2.4	86.1

NH$_3$ injection, as compared to Chamber D at a flue gas temperature of about 310 to 320°F. Note that Chamber C with a gas temperature of about 280 to 300°F shows what I would call pretty good electrical characteristics compared to Chamber A. Even though Chamber D has reduced power compared to Chamber A, its collection efficiency would be much higher because of the better ash resistivity level. The electrical readings of Test C3 shows the effect of selective NH_3 injection toward Chambers A and B. I had more NH_3 than needed for this test, which resulted in higher spark-over conditions in the first two fields of these chambers. But the overall improvement in the precipitator voltage fields still gave a high performance result.

Case D

This is another chevron applied precipitator system where the original design was shy of power for the flue gas conditions encountered. The basic design was similar to Case C, but a comprehensive attack on power, gas distribution and moderation of the temperature spread was able to overcome the high ash contents and medium sulfur levels of the coal. A doubling of the power supplies and gas mixing in the chevron that reduced a temperature differential, from 50°F to about 25°F at the face of the precipitator chambers, probably did more good than any of the other corrections implemented.

The design of the Case D precipitator is shown in Table 13—7, along with Cases E and F. Full load tests prior to the corrective steps gave collection efficiencies in the 96 to 98% range. An example of how increased power input effects performance can be seen by the curve of Figure 13—7, which is based on many diagnostic tests obtained prior to adding the extra T-R sets. These diagnostic emission tests were obtained on one chamber, where the surface area connected to a specific power supply could be halved by disconnecting the adjacent bus section.

Test D1 in Table 13—8 was one of many full scale tests that indicated that if the flue gas temperatures were kept reasonably balanced between the three air heaters, and the average temperature kept in the 280°F to 290°F range, performance goals could be met. Unfortunately, with the high ash coal that existed at this station, there was little excess collector margin. This means that continuous efforts must be maintained for the reliability of the precipitator components. If any further deterioration in coal quality or flue gas temperatures occurs, then some thought might be given to spot NH_3 injection for this type of application. But the gas distribution and temperature patterns are always capable of improvement, and I believe that 99.5% to 99.6% collection efficienies are attainable on this type of installation even with SCA's below 200.

Table 13—7
Operation and Precipitator Design Data
Type of Process — Fly Ash

PLANT IDENTIFICATION	D	E	F
Rated Steam Flow — K lb/hr	6350	4000	3925
Precipitator Design Data (original)			
No. of Units	2	2	2
No. of Fields	4	4	4
Total No. of Gas Passages	344	258	216
Size of Passage — H′ x L′	30 x 27	30 X 27	30 x 24
Spacing of Passage — inches	9	9	9
Type and Size of Discharge Electrode	Nom. 0.1″ wire	Nom. 0.1″ wire	Nom. 0.1″ wire
Total Length of Disch. Elect.—Lin. ft.	371,500	278,600	207,600
Total Collecting Surface — sq ft	556,000	417,900	311,200
No. and Size of T-R Sets	16–70 KVA	6–35 KVA 6–90 KVA	8–64 KVA 8–96 KVA
Total KVA Installed	1120	750	1280
Guarantee Efficiency — %	99.5	99.5	99.0
Gas Flow Rate — K acfm	2700	2050	1733
Rated Gas Temp. — °F	275	300	300
Gas Velocity — ft/sec	5.8	5.9	5.9
Net Aspect Ratio	0.9	0.9	0.8
Design Migration Velocity — ft/sec	0.43	0.43	0.43
watts/1000 acfm	270	238	480
watts/sq ft Coll. Surf.	1.31	1.17	2.67
SCA — sq ft/1000 acfm	206	204	179
Density — ma/1000 sq ft Coll. Surf.	29	29	64

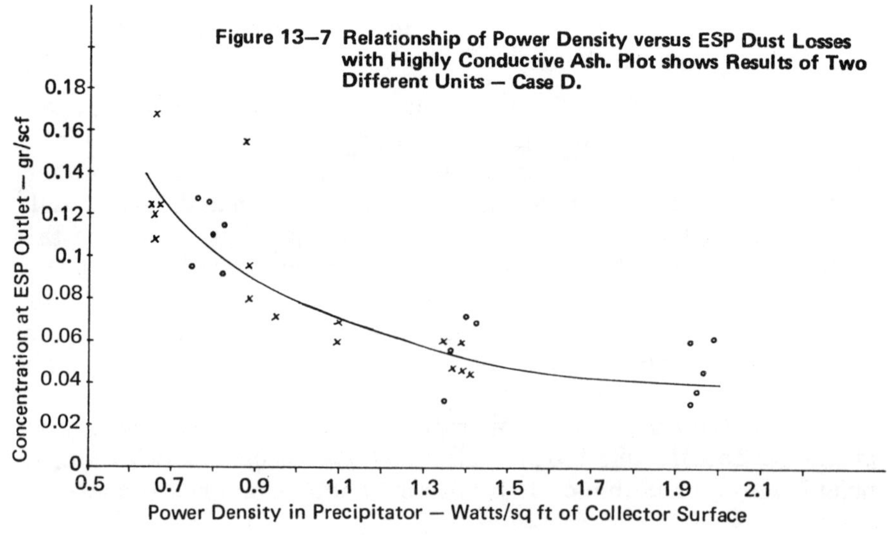

Figure 13—7 Relationship of Power Density versus ESP Dust Losses with Highly Conductive Ash. Plot shows Results of Two Different Units — Case D.

Table 13—8
Test Results and Precipitator Performance Data
Type of Process — Fly Ash

PLANT IDENTIFICATION			D	E	E	E
Test Number			1	1	2	3
Steam Flow During Test K lb/hr			6156	3450	4100	3973
At Outlet of ESP						
Gas Flow Rate — Kacfm			2800	1950	2240	2016
Gas Temperature — °F			280	282	310	295
O_2 — %			6.1	8.0	7.5	7.0
Moisture — %			7.9	7.0	6.8	6.8
Concentration — gr/acf			0.026	0.170	0.048	0.022
Concentration — gr/scfd			0.042	0.288	0.082	0.034
Emission — lb/10^6 BTU			0.078	0.63	0.170	0.083
ESP Efficiency — %			99.3	94.8	98.3	99.5
Test Gas Velocity — ft/sec			6.0	5.6	6.4	5.8
Test Migration Velocity — ft/sec			0.42	0.23	0.36	0.42
Test Average KVp			56	45	47	55
Test — watts/1000 acfm			278	184	137	273
ma/1000 sq ft Coll. Surf.			43	25	20	43
Coal as Received	Ash	%	20.04	21.58	21.21	23.58
	Sul.	%	2.03	2.70	2.78	2.12
	BTU/lb		11,886	11,821	11,896	11,594
Ash Analysis—Inlet	Al_2O_3	%	30.08			28.92
	Fe_2O_3	%	13.27			12.53
	SiO_2	%	45.78			46.14
	Na_2O	%	0.66			0.83
	SO_3	%	0.39			0.44
	L.O.I.	%	2.70	1.45	1.00	4.60
Conditioning Method			—	—	—	—
Amount Used			—	—	—	—

Case E

Here is another case where the original power supply deployment was not matched to the poor ash characteristics which were caused by the match-up of higher sulfur coal and flue gas temperatures below 300°F. The flue gas temperature spread of the rotating regenerative air heaters was not considered in the original design specifications of the precipitator, and localized pockets of ash reentrainment existed. The original layout of the precipitator had power supplies of 90 KVA capacity, energizing individual fields of 129 gas passages which spanned the width of each unit. This arrangement acted to reduce secondary voltages.

Table 13—8 shows a comparison of Tests E1 and E2 prior to any modifications. Test E1 was obtained at a reduced boiler load at the manufacturer's request in order to determine the collection efficiency at the rated gas flow. The results were quite poor because of the acid dew-point conditions. As seen by the results of Test E2 during this same evaluation period, the collection efficiency actually increased to 98.3% when a higher gas flow rate allowed the flue gas temperature to reach 310°F. However, there was no way to reach higher efficiencies without modifications that included the doubling of power supplies, gas distribution corrections and minor changes in the control of the flue gas temperature. Test E3 shows the performance results after modifications. The original guarantee level of 99.5% was met, but peak voltage fields had to be in the 55 range. These tests were obtained with distribution less than optimum and no changes were incorporated in the rapping system. I believe the precipitator was capable of additional performance gains with efforts in these areas. The key point here is that there are very few precipitators that cannot be measureably improved within the existing box size by adhering to fundamental corrections.

Case F

The precipitator of Case F did better than the performance design of 99.0% when plant personnel learned the importance of flue gas temperature relative to the medium sulfur coal being burned at the station. Some changes to the gas distribution patterns were implemented, but the major gains occurred in carefully controlling the flue gas temperature at 300°F between the two units. Sufficient power supplies and sizing was available in the original design, but there was a tendency to operate one unit colder than the other. Differences of 6 to 8 KV often occurred between the two precipitator units which had a direct effect on overall collection performance. The key lesson here is that what appeared to be relatively minor differences of 15 to 20°F can be very important in acid dew-point effects.

Case G

A number of precipitation fundamentals can be gained by a study of the industrial boiler system of Case G. The size of this boiler is typical of the pulverized coal-fired industrial systems found throughout the country. A mechanical cyclone collector was installed prior to the purchase of a series precipitator in the late 1960's. The guarantee was 99.0% overall, which called for a 96% plus precipitator performance. The design of this precipitator was specified for 1% sulfur and low ash coals, with four bus sections in series, although only two T-R sets energized the collector. The conservative design of 150,000 acfm was based on measurements which included high excess air levels. Table 13—9 shows the design data of Case G as well as that of Cases H and I.

Table 13—9
Operation and Precipitator Design Data
Type of Process — Fly Ash

PLANT IDENTIFICATION	G	H	I
Rated Steam Flow — K lb/hr	200	220	600
Precipitator Design Data (original)			
No. of Units	1	1	1
No. of Fields	2	6	2
Total No. of Gas Passages	27	31	32
Size of Passage — H' x L'	24 x 18	30 x 36	28 x 24
Spacing of Passage — inches	9	9	9
Type and Size of Discharge Electrode	Nom. 0.1" wire	Nom. 0.1" wire	Nom. 0.1" wire
Total Length of Disch. Elect.—Lin. ft.	15,532	44,640	28,672
Total Collecting Surface — sq ft	23,328	66,960	43,000
No. and Size of T-R Sets	2 x 32 KVA	3—38.5 KVA 3—51.3 KVA	1—53 KVA 1—70 KVA
Total KVA Installed	64	269	123
Guarantee Efficiency — %	Includ. Mech.Coll. 99.0	99.8	98.7
Gas Flow Rate — K acfm	150	110	232
Rated Gas Temp. — °F	505	338	325
Gas Velocity — ft/sec	5.1	2.6	5.8
Net Aspect Ratio	0.75	1.2	0.86
Design Migration Velocity — ft/sec	0.49	0.17	0.39
watts/1000 acfm	277	1590	345
watts/sq ft Coll. Surf.	1.78	2.61	1.86
SCA — sq ft/1000 acfm	155	609	185
Density — ma/1000 sq ft Coll. Surf.	43	63	45

Table 13—10 shows the results of Tests G1 and G2 obtained during a diagnostic evaluation program a number of years ago, which indicated that the precipitator could perform well on low sulfur coal. Minimal corrections were made to the inlet gas distribution pattern, but basic trouble was believed caused by the configuration of the outlet flue. High rapping reentrainment, unusual for a high resistivity ash, lent credence to a severe gas

Table 13—10
Test Results and Precipitator Performance Data
Type of Process — Fly Ash

PLANT IDENTIFICATION			G	G	G	G
Test Number			1	2	3	4
Steam Flow During Test — K lb/hr			190	195	185	190
At Outlet of ESP						
Gas Flow Rate — K acfm			140	135	139	117
Gas Temperature — °F			405	410	440	390
O_2 — %			7.5	7.5	7.8	6.3
Moisture — %			7.0	6.6	7.0	9.5
Concentration — gr/acf			0.031	0.017	0.029	0.012
Concentration — gr/scfd			0.056	0.031	0.055	0.022
Emission — $lb/10^6$ BTU			0.13	0.065	0.15	0.043
ESP Efficiency %			91.6	95.8	92.7	96.9
Test Gas Velocity — ft/sec			4.8	4.6	4.8	4.0
Test Migration Velocity — ft/sec			0.25	0.31	0.26	0.29
Test Average KVp			38	46	49	48
Test — watts/1000 acfm			75	143	76	268
ma/1000 sq ft Coll. Surf.			13	25	13	39
Coal as Received	Ash	%	8.33	7.00	9.36	10.71
	Sul.	%	1.10	0.84	1.09	1.41
	BTU/lb		13,627	13,935	13,059	13,238
Ash Analysis—Inlet	Al_2O_3	%				
	Fe_2O_3	%				
	SiO_2	%				
	Na_2O	%	0.50	1.73	0.31	
	SO_3	%				
	L.O.I.	%	12	13	—	—
Conditioning Method			—	—	—	—
Amount Used			—	—	—	—

distribution problem. A rapping cycle of one impact blow per 24 minutes was instituted for these tests instead of the normal 2½ minute cycle for the twelve top-mounted rappers. This helped reduce the emission losses during the program. Figure 13—8 shows normal rapping test results compared to dust concentrations measured with the expanded rapping cycle. Note that the 24 minute cycle produced the same emission loss for an approximate 25% reduction of power input to the precipitator.

The higher flue gas temperature was an advantage at this installation in terms of resistivity moderation. The use of the mechanical collector was a minus factor, but this factor was nullified in part by high combustible contents in the fly ash exiting the boiler. While the cyclones took out the major portion of large combustible particles, sufficient carbon particles entered the precipitator to aid the power characteristics. However, I believe the combustible segment must have also been a factor in the rapping reentrainment.

The difference in improvement between Tests G2 and G1 can be related to a change in coal, especially in the value of Na_2O. Note that the power almost doubled for Test G2 with about a 20% increase in the peak voltage.

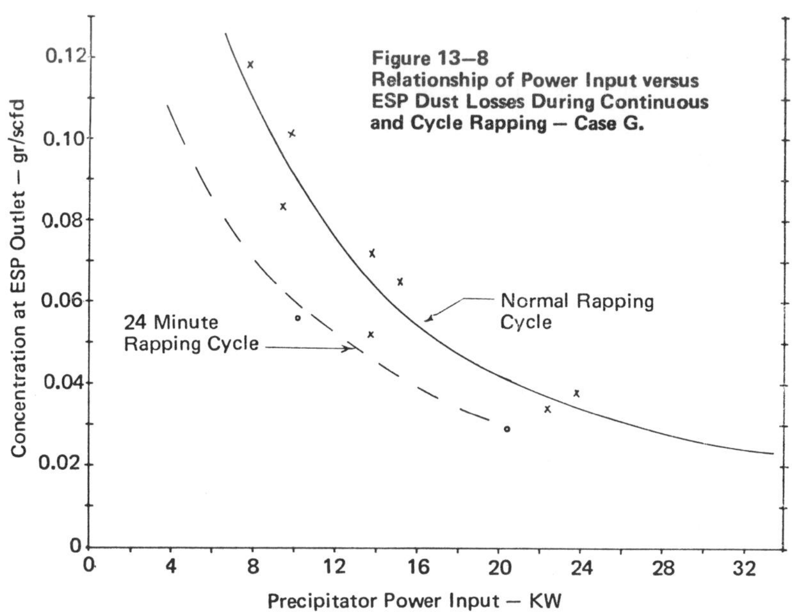

Based on this evaluation program, the following recommendations were made:

1. Eliminate air inleakage through the tubular air heater.

2. Improve combustion characteristics.

3. Work toward a more uniform coal supply with ash exhibiting satisfactory characteristics.

4. Sectionalize the precipitator with the addition of two 16 KVA power supplies for the inlet fields, with 24 and 20 mh linear reactor modifications. Use the existing 32 KVA supplies for the outlet two fields, each with extra linear reactor ballast.

5. Correct the gas distribution problem.

6. Improve reliability by installing shrouded wire electrodes, and eliminate wire oscillation by the use of 25 lb weights.

7. Install a new rapper control with a staggered frequency of rapping from inlet to outlet.

These seven (7) recommendations are no doubt familiar to you, since they are fundamental factors discussed several times through the course of this book. Most of the corrective actions in fly ash precipitators will use part or all of these basic steps in efforts to improve performances.

A later Test G3, as shown in Table 13—10, was obtained with the above items 4, 6 and 7 implemented, along with a partial correction of the gas distribution. Power input characteristics was still considered poor and the test result of 92.7% was not much better than Test G1. Prior to this test, a partial correction of the gas distribution problem occurred by locating a grid of closely spaced channels at the outlet face of the precipitator adjacent to the side discharge duct to the I.D. fan. This duct design is shown as View A of Figure 11—6 of Chapter 11. It was found that gas velocities of 10 to 11 ft/sec was channeling through several adjacent gas passages along that side of the precipitator.

Test G4 was obtained at the conclusion of the program after efforts in all 7 areas of the recommendations. Reduction of the gas flow rate occurred with the solving of leakage problems within the boiler and improvements to the burners. Completion of the grid at the outlet of the precipitator further minimized the reentrainment, so that only minor performance benefits were later observed with the expanded rapping cycle. A reduction in the flue gas temperature occurred, but the proper blending of coal counteracted any disadvantage of the cooler temperature on resistivity.

Remember that the peak of the resistivity curve would lie somewhere around 340°F. The slight pick-up of sulfur in the coal was a plus factor, but the current and power density was probably higher than I would have liked, relative to the voltage levels.

The type of installation represented by Case G is often difficult to overcome unless the precipitator is designed conservatively and every problem area explored for its maximum benefits to the performance of the collector. Too often when only one to two areas of concern are corrected, the final precipitation performance is found wanting.

Case H

This is an interesting fly ash case where a very large sized precipitator was installed for the application of low sulfur coal in a small boiler. The design of 99.8% collection efficiency was predicted on a low gas velocity of 2.6 ft/sec, but the precipitator failed to come close to expectations. One of the first values that should stand out in Table 13—9 is the high design power and current density compared to some of the other cases previously discussed. Based on the knowledge that high resistivity ash layers cannot handle large amounts of current, I believed that the presence of all this potential power had a greater effect on the problem than previously realized by the manufacturer. In other words, the power supplies were operated practically open-circuited since the 10 to 20 ma observed per field was only a small fraction of the available current rating of the T-R sets.

Part of the problem was related to the thick build-up of ash on the collector surfaces, but I could not forsee any help in rapping at higher intensities since reentrainment was already quite apparent. I was certain that a poor gas distribution was aggravating the performance, but I did not have the occasion to investigate and correct that problem. Specifically, the design of the inlet plenum allowed the larger particulate matter to segregate toward the upper half of the inlet field. In this short investigation period, I was mostly concerned with how the power input characteristics could be improved without the need of gas conditioning agents. No gas temperature stratification existed. Alignment was considered satisfactory, but the wire discharge electrodes were not taut, due to an insufficient bottle weight. While the lack of tension in the wire was not an important factor at the low power levels observed, it could result in excessive wire oscillations at higher power levels.

An evaluation study was undertaken on one of two identical precipitator units handling the flue gases from similar boilers. A diagnostic test program shown in Figure 13—9 determined that Unit B had worse electrical characteristics than Unit A, and so Unit B was chosen for modifications while Unit A was used as the control. This is a method that should usually be employed in an evaluation program. When possible, parallel precipitator units of the same boiler should be compared. Careful control of operation and coal sampling for each boiler gave me some confidence for comparing the test results between the units.

Figure 13—10 shows the electrical energization of the two units under test. The 4th field of Unit A was out of service prior to my arrival. I hooked up Unit B as shown in order to further

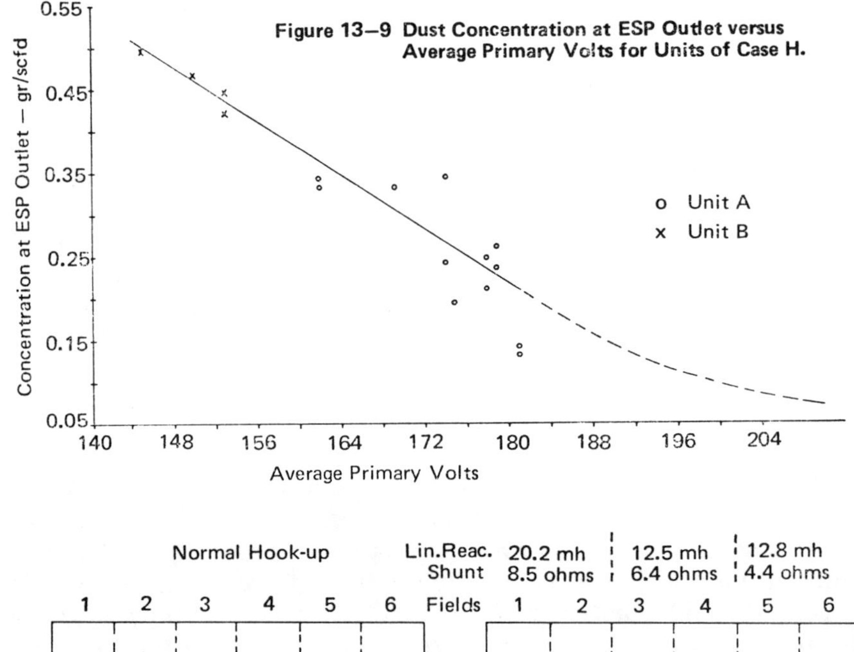

Figure 13—10   Energization of Two ESP Units at Case H Showing Modifications to Unit B During Diagnostic Program to Increase Power Input with High Resistivity Conditions.

load down the T-R sets. In addition, values of linear reactors were modified for Unit B as shown in the figure. Another modification utilized shunt resistors across the primary winding of the T-R sets of Unit B to increase the current loading of the primary circuit in an effort to effect secondary characteristics. I would like to stress several points:

1. The threshhold of spark-over appeared close to the onset of corona. In other words, electrical flash-over occurred at such low current levels that very little effective voltage fields existed.

2. At corona current levels of 10 to 20 ma, the measured conduction of the current pulse was about 15%. Current conduction approximately doubled with the circuit changes on Unit B.

3. I would have preferred smaller power supplies on the inlet fields rather than having to work with the larger existing T-R sets. For example, a 7½ KVA on the 1st 6 ft field might have improved the circuit characteristics, and so, too, with a 15 KVA supply on the 2nd 6 ft field.

4. The automatic control appeared to work against overcoming the threshhold sparking level, possibly due to the instability of circuitry at such low levels. I could increase power input by reverting to a manual mode, and adjusting the automatic mode failed to duplicate the same benefits. Since I was going through a series of diagnostic power changes, the controls of both units were left on automatic to eliminate any bias of the monitor readings.

5. The effect of the collector surface rapping on electrical characteristics was much more apparent with the reactors and shunts on Unit B than occurred without these circuit aids. The rappers were of an impact type, top mounted, and on a 4 minute cycle. Figure 13—11 shows the variation in the primary voltage envelope as the rappers were turned off and on in the 3-4 field during a two hour period. The recording voltmeter indicates that it took about 4 to 5 minutes to stabilize the T-R set primary voltage after the rappers were turned off, while the voltage reverted to its low point in about 15 minutes. I have observed this characteristic on a number of high resistivity cases, where the irregularities in the dust layer caused by rapping produced a sensitive spark-over condition. When this is observed, as in Case H, some performance gain can be achieved when all the rappers for a specific field is rapped during a short time period and then not activated for a 15 to 20 minute period. Of course, some care must be exhibited to assure sufficient cleaning during the short rap cycle. In Case H, the changes were of much smaller

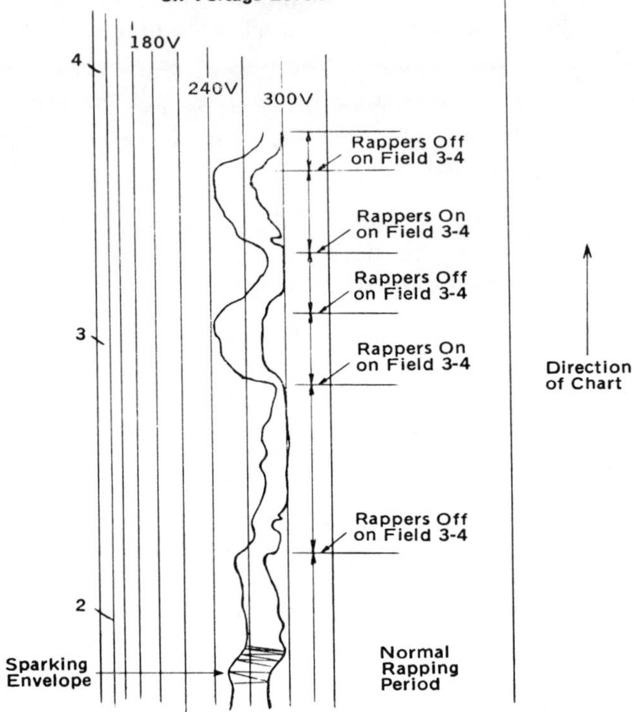

Figure 13—11
Trace of Recording Voltmeter
on Field 3—4 of Unit B-Case H—
that Shows Effect of Rappers
on Voltage Levels.

magnitude, even though I could detect some electrical power improvement without the stability of the reactors and shunts.

6. There was a tendency for the middle fields to exhibit slightly different electrical characteristics than the inlet and outlet fields. This situation is not unusual on a precipitator with many series fields where the dust properties can change during the length of the collector. For example, a set of readings at threshold spark-over on Unit A showed the following:

Field	Primary Volts	Primary Amps
1	180	3.2
2	140	0.1
3	140	0.1
5	170	2.4
6	180	3.2

Table 13—11 shows one comparison of test results for Units A and B, when the reactors and shunts were evaluated against the control conditions of Unit A. Dust concentrations were high during these tests under normal rapping conditions. But the benefits of slight improvement of power were apparent. There was no doubt in my mind that a combination of corrective steps could have achieved stack concentrations in the 0.02 to 0.03 gr/scfd range, but the final decision was to go with a gas conditioning system at this installation.

Table 13—11
Comparison of Test Results of Units A and B of Case H
Before and After Linear Reactor and Shunt Addition
on Unit B. Normal Rapping on Both Units.

Test Conditions	Unit	No. of T-R Sets	Steam Flow K lb/hr	Flue Gas Temp. °F	Coal % Sul	Coal % Ash	Corona Power Input KW	Conc. at Outlet gr/scfd
Without Reactors and Shunts	A	5	178	310	1.0	20	1.5	0.34
	B	3	175	320	1.0	20	0.4	0.42
With Reactors and Shunts on Unit B	A	5	178	310	1.0	21	1.6	0.39
	B	3	175	315	1.0	21	1.9	0.24

Case I

The use of precipitators for the cleaning of the flue gases from cyclone boilers has potentially been a difficult application. Case I, shown in Table 13—9, developed into a successful installation because of a number of corrective steps. The coal used was in the 3 to 4% S range, which allowed relatively high power inputs. However, the original two T-R sets could not supply sufficient power density to perform effective cleaning.

Fortunately, the precipitator had four (4) bus sections in series, so that two additional power supplies could be installed. Improvements in the burners reduced the combustible content down to 3 to 5%. Part of the recommendations included operation at the 25 to 30% excess air level rather than the nominal 20% range, where good combustion is often difficult to achieve. Another performance benefit was eliminating the recycle of the ash collected by the precipitator. While this may have had some merit with the carry-over of high combustible grit from the boiler, it was detrimental to the precipitator performance. Reduction of combustion problems is always preferable.

311

The tubular air heater had both excessive air inleakage and tube blockage to provide some of the starvation of combustion air. Interestingly though, the correction of the tube blockage aggravated a gas distribution problem by leading to a higher flow toward the central portion of the precipitator. Part of this corrective step involved placing a higher resistance target plate ahead of the existing perforated plate at the inlet face of the precipitator in order to break up the high velocity zone. This is one of the potential problems with a design where the gas flow from the tubular air heater immediately moves into a group of turning vanes at a close-coupled elbow, and then to the inlet plenum of the collector. The gas distribution pattern will require some monitoring over a period of time to determine the changes that can occur from tube blockage in the air heater.

A uniform rapping frequency of 5 min throughout the precipitator, with a side-mounted impact type, was modified to a 10, 20, 20, and 60 minute cycle from inlet to outlet fields. Remember that the cyclone boiler produces a much less top fly ash portion, which can allow a more drawn-out rapping cycle.

Some gas sneakage through the hoppers was apparent by viewing the cleaned off portions of the wire electrode bottle weights. A scoured area on some steel beams at the trailing wall of the inlet hopper was also observed during the internal inspection. Part of the hopper problem was believed caused by the poor gas distribution at the inlet which tried to put too much gas through too few passages and that forced part of the flow to divert to the hopper. I believe the corrections implemented with the inlet target plate probably helped this situation.

The result of this work brought stack emissions down to the 0.06 to 0.08 lb/10^6 BTU input from about 0.15 lb/10^6 BTU. Can this installation do better? I believe further work with gas distribution and gas sneakage can possibly extract another 20% reduction in emissions, but the laws of diminishing returns begin to surface.

The two new power supplies were of 36 KVA and 48 KVA sizes, placed in the first two fields. Some spark-over usually existed in the inlet 6 ft field energized by the 35 KVA supply, but power levels were generally high through the precipitator. About 500 watts/1000 acfm and 54 ma/1000 sq ft of collecting surface were common power densities after the modification.

Case J

The precipitator involved in Case J is representative of a number of collector units installed years ago for cleaning of waste gases from cement kilns. Collection efficiencies were usually guaranteed in the 95 to 98% range and while capable of achieving these goals, those installations were more often found operating below guarantee levels because of fundamental failures. The purpose of this discussion is to be concerned about past mistakes only for the sake of what can be learned from them.

As noted in Table 13—12, the installation was small, but the design was similar to the larger cement precipitators of yesteryear. Design power was low by modern standards, but as you will see, it was still more than the precipitator could handle at times. Gas distribution was less than optimum. This was common in cement precipitators, because the collector was physically lower than the back-end of the kiln, which usually resulted in a top

Table 13—12
Operation and Precipitator Design Data
Type of Process — Cement

PLANT IDENTIFICATION	J	K
Production Rate — Barrels/hr	80	460
Precipitator Design Data (original)		
No. of Units	1	3
No. of Fields	3	6
Total No. of Gas Passages	2 fields—15 1 field —12	54
Size of Passage — H' x L'	20 x 27	24 x 54
Spacing of Passage — inches	2 fields— 8 1 field —10	9
Type and Size of Discharge Electrode	3/16" twisted square wire	Nom. 0.1" wire
Total Length of Disch. Elect.—Lin. ft.	—	93,312
Total Collecting Surface — sq ft	15,120	139,968
No. and Size of T-R Sets	2—Mech.Rect. Sets	5 — 74 KVA 3 — 50
Total KVA Installed	50	520
Guarantee Efficiency — %	96	99
Gas Flow Rate — K acfm	105	350
Rated Gas Temp. — °F	700	475
Gas Velocity — ft/sec	8.8	6.0
Net Aspect Ratio	1.35	2.25
Design Migration Velocity — ft/sec	0.37	0.19
watts/1000 acfm	190	965
watts/sq ft Coll. Surf.	1.32	2.41
SCA — sq ft/1000 acfm	144	400
Density — ma/1000 sq ft Coll.Surf.	50	51

entry plenum as viewed in Figure 11—3 of Chapter 11. This wet process kiln had a slurry moisture of 33%, but air in-leakage brought the flue gas moisture content down to about 25% by volume. Most of the air leakage occurred in the dust chamber, and gas temperatures measured at the precipitator inlet was in range of 400 to 450°F.

Previous collection efficiencies were measured in the 65% range with all fields in service. Plant practice was to operate with high current levels on the power supplies in efforts to achieve higher collection. Unfortunately, the waste gas conditions was not conducive to operating at high power inputs, not unlike the high resistivity conditions of fly ash, and the spark-over found itself localized at a weak spot in the design of the electrodes. Wire electrode failures were frequent, but more importantly, the localized electrode damage soon resulted in a continuous break-down because the minimum voltage output of the T-R set exceeded the flash-over voltage inside the precipitator. Therefore, due to necked-down wires, the power supply would immediately go into a hanging-arc as soon as the set was energized. Interesting enough, this condition was not easily recognized because the noise level around the precipitator was high, the collector was well insulated which dampened the buzz sound of the continuous break-down, and the meter needles were near constant without the movement normally associated with spark-over.

The two 25 KVA mechanical rectifiers gave the following readings as found:

	Primary Volts	Primary Amps	Primary Tap
Inlet 2 fields	230	37	5
Outlet	205	39	1

Control panel meter needles were steady. Stack appearance was considered poor. During an internal inspection, approximately 30 discharge electrodes were removed because they were in a "necked-down" condition. No change was made in the gas distribution or hopper air inleakage problems. However, additional resistance grid banks were installed in series with the existing M-R resistance in order to produce a larger voltage drop. This modification was to reduce the voltage across the primary winding of the high-voltage transformer and to help break the arc condition if it occurred. The outlet field could not be energized because of a ground in the H.V. cable. But the one M-R set produced the following characteristics at light spark-over:

	Primary Volts	Primary Amps	Primary Tap
Inlet 2 fields	400	8	1

With the above power characteristics of a little over 15 watts/ 1000 acfm, the performance of the precipitator was about 87%. If the outlet field was in service, I would have expected results near the guarantee level. If the other problems were corrected, collection efficiencies of 98% or more would not be out of line.

The key point to be gained from this example is that the quantity of power is often much less important than the quality. This is another case where power was forced into a precipitator that contained a subtle design defect, that of locating a contoured surface opposite an unshrouded wire electrode. While a large radius surface will minimize high-voltage break-downs, it will still be the first place to draw a spark if the input voltage exceeds the electrode spacing and gas conditions. Uniform coatings of cement dust on the electrode surfaces was more a function of the localized spark-over than the characteristics of the dust. For example, as the voltage decreased because of the necked-down wire, the corona activity on the remainder of the wire electrodes diminished. The last point to be made is that plant personnel never understood that frequent replacement of electrodes was abnormal. Even after all these years, I still get the feeling that this point has not been stressed around the country.

Case K

The size of the cement precipitator shown for Case K in Table 13—12 would be normally considered ample to attain the 99% guarantee level. Actually 97 to 98% collection efficiencies often occurred because of several weaknesses in the system. This was a wet process kiln with a slurry feed moisture of about 36% and gas temperatures entering the precipitator of 400 to 450°F. Figure 13—12, view (A), shows a plan view of the original precipitator system.

Several modifications were made to this installation including the following:

1. Add two more power supplies and rearrange the energization of the precipitator to that shown in Figure 13—12 View (B).

2. Equalize the gas flow between chambers.

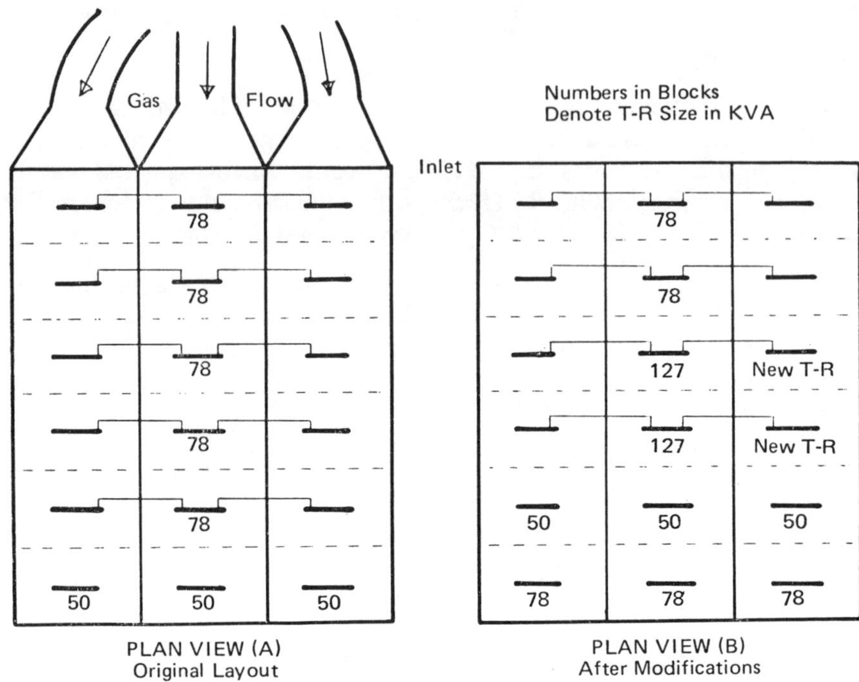

Figure 13—12   Energization of Case K Before and After Modifications.

3.   Correct a gas distribution problem from side to side, especially in the outer chambers. View (A) of the figure shows that the gas flow would tend to concentrate in the outer gas passages because of the configuration of the flues.

4.   Gas sneakage through the drag-scraper hopper area had to be minimized by horizontal shelves at the perforated diffuser plate located at the inlet of the precipitator chambers. Another input to the reentrainment of material from the hopper drag area was air inleakage through the screw conveyor system at the discharge chute. Air inleakage through roof plates and corroded wall areas were corrected.

After the above corrections, the stack appearance cleared up and collection efficiencies over 99% were measured. There were discussions of additional collection surface prior to the modifications. Often, a combination of fundamental corrective steps are required to overcome a deficient performance. There is no doubt in my mind that the improved energization was at least 50 to 70% of the reduced emissions, but corrections to the gas distribution were also a major input. Average voltages improved about 10% by the additional power supplies, since the middle to outlet fields were formerly operating current limit at nominal 200 volts on a 480 volt system.

Case L

The sinter plant precipitator represented by Case L was considered conservative compared to previous collectors observed on the sinter process. A mechanical cyclone installation preceded the precipitator, this being the normal practice. Table 13—13

Table 13—13
Operation and Precipitator Design Data
Type of Process — Steel Industry

PLANT IDENTIFICATION	L	M
Process	Sinter Plant	BOF
Production Rate	115 tons/hr	250 tons/heat
Precipitator Design Data (original)		
No. of Units	1	6
No. of Fields	3	4
Total No. of Gas Passages	50	1st — 180 Others — 220
Size of Passage — H' x L'	30 x 24	24 x 36
Spacing of Passage — inches	9	1st — 11 Others — 9
Type and Size of Discharge Electrode	Nom. 0.1" wire	1st — Barbed Others—Nom. 0.1" wire
Total Length of Disch. Elect. — Lin. ft.	48,000	240,192
Total Collecting Surface — sq ft	72,000	362,880
No. and Size of T-R Sets	1—48 KVA 2—64 KVA	3—48 KVA 3—64 KVA 6—96 KVA
Total KVA Installed	176	912
Guarantee Efficiency — %	98.5	99.7
Gas Flow Rate — K acfm	385	960
Rated Gas Temperature — °F	245	550
Gas Velocity — ft/sec	5.7	4.0
Net Aspect Ratio	0.8	1.5
Design Migration Velocity — ft/sec	0.37	0.26
watts/1000 acfm	297	618
watts/sq ft Coll. Surf.	1.59	1.63
SCA — sq ft/1000 acfm	187	378
Density — ma/1000 sq ft Coll. Surf.	38	39

shows that the 3 fields were originally energized by 3 T-R sets. This is shown in Figure 13—13, View (A). One of the first recommendations was to upgrade the power input by changes noted in Figure 13—13, View (B). The object was to obtain consistent test results below 0.02 gr/scfd, and this required addressing all facets of the installation.

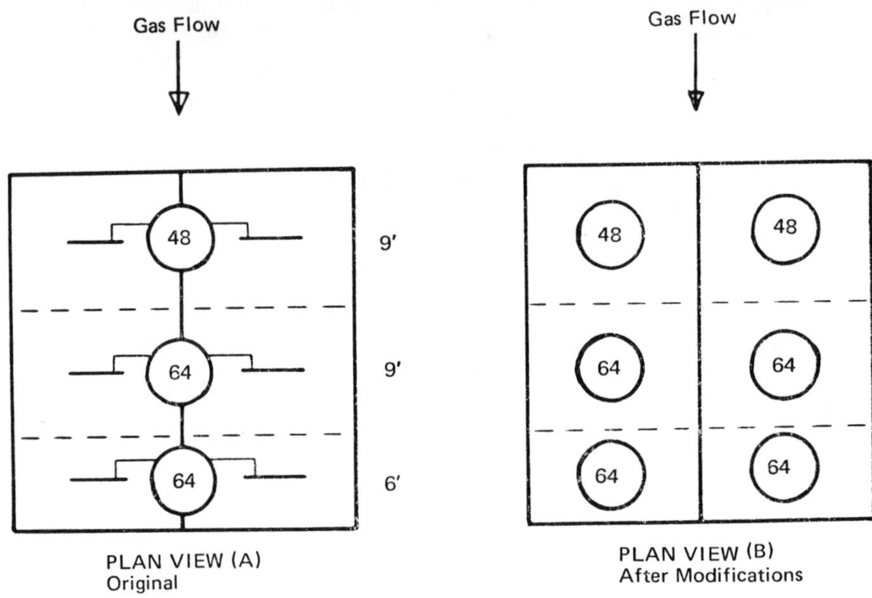

Figure 13—13   Energization of Case L ESP Before and After Modifications.

Wire electrode failures occurred because of a shroudless wire and localized spark-over zones which were caused by a close proximity of the electrode wires to new angles installed in order to stiffen up the collector surfaces. Shrouded wires corrected this problem, and in a number of locations, the wire electrodes were not installed where clearances were considered marginal. The flue gas distribution was not considered poor, but a few deflectors were placed to obtain a little more flow into the outer gas passages. The inlet plenum was an expansion type with three perforated plates in series.

Recycling of the precipitator catch back into the sinter burden was eliminated. This condition is not unlike the use of insufflation in cement kilns where the vaporization of chlorides and alkalies occur that later condense, get collected in the precipitator, and then recycle to help promote a build-up of this difficult material in the gas stream. The catch of the mechanical collector was believed a minor factor compared to the precipitator catch, and therefore was maintained in the system. However, on a critical collector I would have eliminated this portion also. The basicity of the burden was about 1.4.

Early diagnostic tests indicated that about 360 watts/1000 acfm and 62 KV_p were needed to achieve the desired performance goals. But consistency of the sinter operation was also required. Rebuilding the pallets, including new finger and seal bars, helped minimize some of the variations of the air flow through the strand caused by localized short-circuiting of air through openings.

Even the improvements of the machine itself could not overcome the variations in the raw materials that allowed a cycling of waste gas temperature from 15 to 30°F. For that reason, water sprays were still needed between the mechanical collector and the I.D. fans to smoothe out these variations. Control of the temperature entering the precipitator in the 270 to 280°F range appeared to be sufficient to keep the resistivity in a good working zone. If the blend of raw materials could be improved, the need of the water sprays might be mimized.

Test measurements have shown that the collector of Case L can perform below the 0.02 gr/scfd goal, but the variability factor must be controlled. This is one case where assigned personnel can make a difference by closely watching the process and precipitator.

Case M

The BOF process presents a unique challenge to the precipitator. The Case M design shown in Table 13—13 appeared big enough to do the anticipated cleaning, but like other multi-chamber or large installations, the problem areas tend to compound and overwhelm most plant personnel. This is understandable, and I intend to take you through one installation in an effort to sort out the problems. Most of the BOF installations have similar traits.

Along with the reliability of internal components, mostly attuned to the corrosion of collector surfaces, the main performance factors usually involve:

1. Gas distribution within chambers.

2. Gas balance between chambers.

3. Air inleakage through the screw conveyer system.

4. Conditioning of the waste gases.

5. Coordination of the fan operation to the precipitator.

These factors are not much different than those which usually exist for other processes, but the fact that the gas producing cycle only lasts about 1/3 of an hour makes evaluation of all the variables somewhat difficult. For Case M, the final results of a comprehensive program allowed the use of five (5) chambers to do effectively what six (6) chambers could not do prior to the corrections. No modifications to the power supplies occurred, although I believe that by adding small power supplies to the individual inlet fields and doubling up on the 2nd field T-R sets, the situation would have been helped. In this particular installation, each power supply energized two adjacent bus sections throughout the precipitator. Figure 13—14 shows the method of energization for the precipitator system.

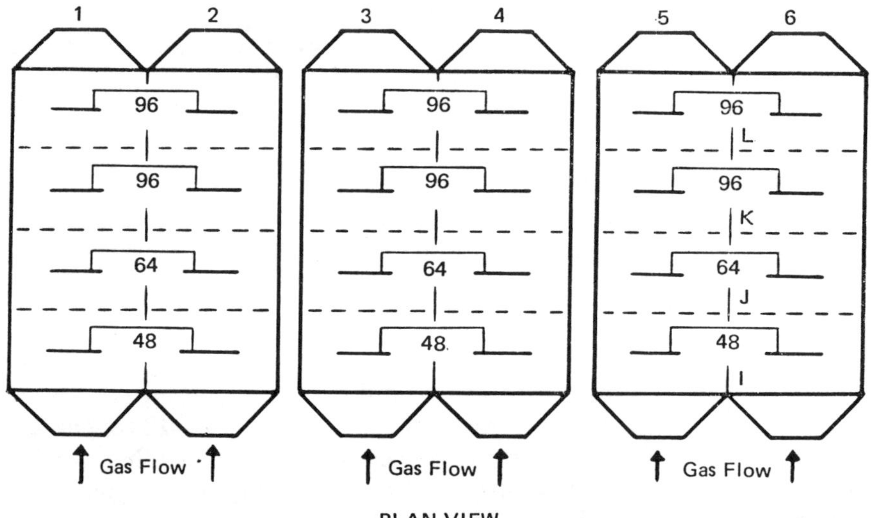

Figure 13—14   Energization of Multi-Chamber ESP Installation of Case M.

I do not mean to infer that the installation was underpowered. The use of a greater number of smaller-sized power supplies would have helped stabilize the system and minimize the effect of a disturbance that may occur in one chamber from affecting an adjacent chamber. There was sufficient power to clean the stack during the highest fume generation period that coincides with high moisture; it is in the periods at the start and conclusion of the heat, when decay of power causes a potentially worse stack appearance, that too much power capability exists. During these periods, large power supplies operating at low corona power levels present the most instability, as I had discussed in previous chapters.

Of course, for twelve (12) power supplies to have handled such a large installation, the internal integrity of the electrodes must be assured. In this case, each chamber was removed from service for a replacement of the existing wire electrodes with a shrouded type, and corrections of alignment problems were implemented. During this same time, steel sheeting was used to cover holes that occurred in the bottom of the collector surfaces of the outlet field. Actually, the corrosion was caught early enough to use this technique. The bent reenforcing steel covered the bottom 6" of the collector plate and sheet metal screws were used to secure it to the original surface. In order to minimize any localized electrical break-downs, the shroud of the new wire electrode was specified so that it extended about 1½" above the top edge of the patch plate. Where electrode failures occurred frequently in the past, the overall internal corrective work reduced the chamber outages to practically zero.

While the internal work was being conducted, a program to upgrade the automatic response of the power supplies was started. The rapid rise and decay of the power characteristics of the BOF precipitator means that a quick and reliable response of this circuitry is of prime importance, especially to minimize electrode damage during the off blow periods. Usually for 5 to 10 minutes after the conclusion of the heat, the spark-over potential is great, and the power must be backed down to conform to this change in gas conditions. Figure 13—15 shows a plot of the primary voltage of four series fields, before any internal corrective steps were taken. Note that the 2nd inlet field decayed during the middle of the blow which is indicative of an internal problem. The low voltage of the 1st two fields prior to the blow was primarily to the effect of the hot metal charge. As with most precipitators, the front half of the precipitator will generally respond rapidly to process changes. However, internal difficulties can cloud this observation. After the alignment corrections, the patterns were more uniform from inlet to outlet during the blow, but the decay at the conclusion of the blow was still pronounced on all fields.

Another characteristic of the precipitator installation is shown in Figure 13—16, where the total power input of each of

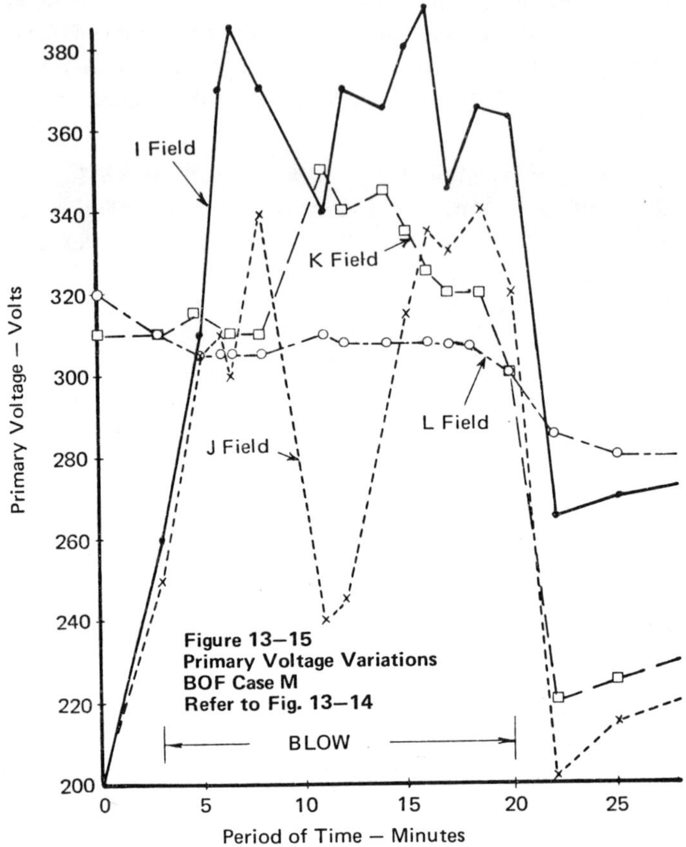

three sets of units is compared before any of the modifications. Note that the power input of Chambers 1 and 2 was much lower than the other two groups. As found out later, the internal condition of No. 1 Chamber was quite poor, dragging down the combined power. This is another reason that separate power supplies have overall advantages for the large installation.

Gas Distribution

The inlet manifold system for the six chambers had a major effect on stack appearances at the beginning and end of the blow. Figure 11—28 shows the type of distribution pattern that existed to produce higher than expected emissions. This type of distribution is more critical for the batch operation characteristics of the BOF. For example, during the high fume generation-high moisture portion of the blow, excessive gas velocities in one zone of the precipitator are not a prime factor because the high power

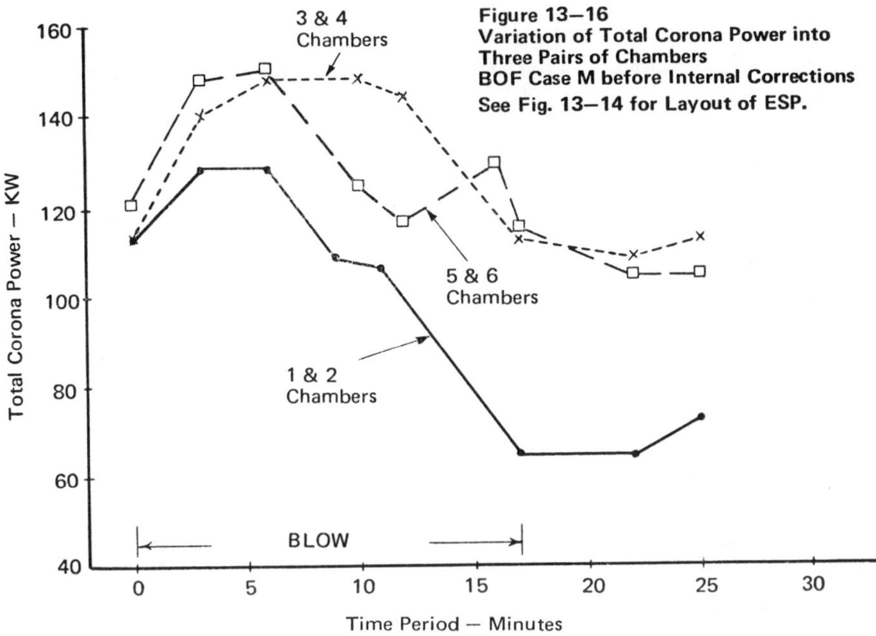

input will nullify this adverse condition. But as soon as the power at the end of the blow decays, the high velocities rapidly erode the material collected earlier, and the performance suffers for an extended period. This was especially true when the fans were maintained at high speed awaiting the analysis of a test sample to determine if a reblow was necessary.

This distribution problem was minimized by the installation of an angle grid at the inlet of the nozzle, as shown in Figure 11—29 of Chapter 11. Several plate deflectors on the discharge side of the grid helped distribute the gas flow in the vertical direction. The way the 2″ x 2″ angles were installed allowed some side to side expansion of the gas. This type of modification was relatively simple and was considered a 65 to 75% improvement over that of the original distribution pattern. That was really all that was needed to overcome the poor effects of the biased flow. About a 1½″ H_2O pressure drop was taken across the grid at normal flow rates.

Gas Balance

The balance of gas flow between chambers was more critical when a poor internal gas pattern existed. However, differences of 10 to 15% gas flow rates between chambers had to be eliminated. Part of the problem was the use of a fraction of an inch pressure drop across the nozzle of the inlet to determine the balance. The grid that was installed at the nozzle now became the measurement device and provided a reliable indicator of the flow through each chamber. Interestingly, before the system modifications were started, the stack appearance did not change much during the blow between a five or six chamber operation. In other words, the gas balance between chambers must have improved somewhat during the increased back-pressure of the system for five chamber operation.

Air Inleakage

Large amounts of excess air is part of the process, but when air is drawn into the system other than through the opening between the vessel and hood, trouble exists. The main thrust of this problem in the BOF system is the dilution effect on the waste gases, localized spark-over, corrosion potential, caking of the moisture-laden material, and reentrainment of dust from hoppers. Most BOF installations have one or more of these problem areas, and Case M was no exception.

Some inleakage air occurred through the butterfly control damper of the vessel that was not in blow. Upwards of 50,000 acfm of cool air would occur during periods of the blow, and unfortunately, most of this dilution air would travel through one or two chambers. In fact, the excessive corrosion of No. 1 Chamber was probably due to extended periods of this type of air inleakage.

The greatest problem with the BOF precipitator system has been its extensive use of screw conveyors. No doubt about it — the screw conveyor is effective and fairly reliable for this service, but the over 1000 ft of conveyor required for the system of Case M provided many areas of potential air inleakage. Air drawn into the hoppers compound the dust handling problems, promote corrosion that is often found in the lower zones of the BOF precipitator, and reentrain material from the hoppers.

The transition from the precipitator hopper screw to the main transfer screw conveyor, and the transition from the transfer screw to the storage bin, are the points of concern. Any type of valving opens up the potential for dust pluggage and maintenance breakdowns. Original valves placed between the transfer screw and the bin were removed from service because of some early difficulties. Therefore, any opening in the storage bin or associated apparatus allowed an air flow back through the system. In addition, the transfer screw conveyors had a number of access flaps installed during the early years of operation when screw pluggages were frequent, mostly because the hopper walls were not properly heated. These flap doors were difficult to seal in their existing design.

The problem of air inleakage through the screw conveyors was not overcome by the time successful stack tests were obtained. A plug seal in the screw was not successful, even though other industries such as cement and non-ferrous metals have used this technique. The bulk of the material tends to fill the transfer screw about 6 minutes after the start of the blow and remains substantially filled until about 5 minutes after the conclusion of the heat. Several techniques are possible, including double slide gates or well designed mechanical double dump valves.

Whichever approach is used, sufficient heat at critical points is mandatory. Personnel assigned to the conveyor systems must be alerted to the effects of air inleakage. Probably the major help at the screw location is to minimize the time the system is at high draft. I will talk about this later, but operating at 1″ H_2O or less suction at the precipitator, other than periods of the blow, will be most beneficial.

Another hopper modification included a recommendation not to use high levels of hopper wall vibration during the periods of maximum fume generation. It was believed better to operate the vibrators starting about 6 minutes after the conclusion of the blow and not operate them at all during the blow. Of course, optimum heat at the hoppers is recommended with this technique. Stack losses could be further reduced by cycling the screw conveyors, but the potential equipment difficulties with this approach outweighed the benefits.

Conditioning of Waste Gases

No modification was made on this installation in terms of moisture conditioning. The use of steam injection into the waste gases as soon as the O_2 began was sufficient to quiet the spark-over moderately. The gas temperature entering the precipitator usually advanced rapidly from 100°F to 300°F in a matter of a couple of minutes, and at least 1 to 2 banks of sprays were activated within the first couple minutes. Generally, five banks of sprays are used in the control of BOF waste gas temperatures. Four to five banks are required to control the gas temperature to the 500 to 600°F range at the top of the hot stack during the peak of the blow. The peak gas temperature at the precipitator inlet might reach 450°F at this time.

As discussed in the Process Chapter, moisture levels of 25% or so by volume will generally provide satisfactory power characteristics in the precipitator during the peak of the blow. Moisture levels of 12 to 15% H_2O by volume at the start of the blow appears sufficient depending to some degree on the temperature of the system. Full power is not required during the first few minutes because much less fume generation occurs. The series power supplies should react differently during these early minutes with the latter fields assuming full power in rapid fashion, possibly a $\frac{1}{2}$ minute or less. As the internal integrity of the fields improved, the ability to reach higher power levels with the same moisture levels also improved. While there may be a short reaction period at about 3 to 4 minutes into the blow, generally the power characteristics were devoid of spark-over until the end of the blow. At this point in time, a rapid decay of power characteristics caused by spark-over occurred almost simultaneously with the loss of moisture.

Several problems sometimes arise with spray conditioning. Individual vessels may have slightly different temperature control characteristics, or sprays could plug up, so that some evaluation must be made if the T-R sets do not rapidly increase to current limit. Do not be overly concerned with the inlet field acting up for a larger period of time, but at least 75% of the power supplies should be stable within several minutes. There are occasional electrical upsets during the middle of the blow, but these should be the exceptions. The characteristics of the vessel charge could effect the temperature of the waste gases and subsequent spray bank control. For example, if the carbon release is slow, temperature levels might call for less water. Modifications of the temperature control set-points may be required in this case. Of course, a danger is always present with moisture condensation at the precipitator with excessive water relative to low waste gas temperatures. This is where hopper heaters and elimination of inleakage air become mandatory.

System Gas Flows

I want to dwell on this section since an important factor in the success of a BOF precipitator, such as Case M, lies in the effective control of the gas movement through the system relative to the electrical characteristics of the collector. Stated another way, do not push a lot of gas through the system during low power conditions. Before I detail the type of operation that is most beneficial, let us key in on the problem areas of the process.

1. The hot metal charge provides a short rise in air temperature in the precipitator and coupled with no moisture, usually causes some electrical disturbances. The power levels usually climbs back to full power during the non-operating period of the cycle as the temperatures in the precipitator drop to 120 to 140°F. As soon as the hot metal pour starts, usually all the T-R sets in series will react, the inlet most of all, with a drop in power of about 10 to 30% and sporatic spark-over.

2. The first minutes of the blow may see another 5 to 10% drop of power because of spark-over if the moisture input is not sufficient. If the steam or one spray bank comes on with the initiation of the O_2, more often than not, the spark-over will quiet down from the hot metal period and the voltage-current readings will start to improve.

3. Within seconds of the end of the blow, the effect on power characteristics is pronounced. As soon as the dry air reaches the precipitator the voltage-current begins a rapid decay and stabilizes to about half the power of the blow period in 3 to 4 minutes.

4. If a reblow is required because the analysis of the test sample did not match the heat specifications, then this short resumption of 30 seconds to a minute use of the O_2 sets up another disturbance. Even though the power characteristics will start to improve as the waste gas moisture makes its effect, the transient state of the reblow makes for a difficult stack appearance.

Modifications to the Flow Rates

There is a limit to how much the precipitator's electrical characteristics can be improved during the above four critical periods. I did not mention the decarburization portion of the blow because I am not concerned with solving the problems of that period by modifying the gas flow rate through the system.

The dangers involved with the use of moisture at low gas temperatures with BOF fume are well known. Since the conditioning effect of the sprays or steam lags the dust load entering the precipitator, reduction of the gas flow rate through the system is one key method available to keep stack appearances within acceptable limits. Let's look at what can be done during the above critical periods.

1. The emission during the hot metal charge is relatively light, mostly because the concentration of material evolved from the pour is low. Actually, most of the stack discharge occurs during this period when the fan dampers are opened to full operating position to capture the fume from the pour and also prepare for the start of the O_2 blow. At the high air flow rates of this period, some additional reentrainment occurs at the onset of spark-over, even though the collector surfaces had been cleaned to a high degree during the previous down-time. Dust pick-up from ledges, manifolds, and other resting places during the low draft period also becomes another input to the discharge emission of the stack. While the disturbance is of short duration, its effect on the stack discharge is more a function of the rate of change (in the flow rate) than for any other reason.

Corrective steps can take a couple of paths. While the air flow through the system prior to the hot metal charge should be less than 15 to 20% of the flow rate of the blow, it should only take about 50% of the full system flow rate to capture the bulk of the fume and kish at the mouth of the vessel. If it takes much more air flow than this during the hot metal charge, then improvements should be implemented in the deflector hood at the crane runway, or a slower rate of pour should be utilized. For Case M, the 50% figure was satisfactory. The effects on the stack emission for a change in air movement from 20% to the percent required for the hot metal charge were minimal.

The next step is to increase the air flow through the system prior to the ignition period when the O_2 first impinges on the metal in the vessel. I recommend a relatively slow upward ramp of air flow that could automatically start by activating the fans for the hot metal pour. For example, control of the fan dampers could allow an upward ramp of 10% additional flow every minute. Since about 2 to 3 minutes usually pass from the start of the hot metal charge to ignition, then sufficient draft should be available at the time of ignition. In Case M, this type of control was not implemented, but the fans were allowed to increase draft rapidly as soon as the lance began to move. While not ideal, the time and degree of disturbance was considered satisfactory. Part of the reason for the acceptable stack results was that the gas distribution patterns were considered good compared to those usually found in BOF installations.

2. Steam injection began at the onset of the O_2 and produced immediate benefits at the start of the blow. Some red color is usually seen in the stack for a period of a couple of minutes, even in a well designed precipitator. Although the effect of this transient stage for Case M was satisfactory after all the other corrections were implemented, the first couple of minutes will usually be the most sensitive to the precipitator. Power input is not stabilized and should only be about 50 to 70% of that reached during the middle of the blow.

I believe the effects of these first couple of minutes can be minimized by matching the air flow rate to the rate required for the initial operation stage. Too often, full fan operation exists at the start of the blow when possibly only 80% is needed. The quantity of the O_2 used can also be reduced for a short period. In other words by some experimentation during this first minute or so, a substantial improvement is possible in the stack appearance, while minimal production time is lost in the overall course of a shift. The chance for excessive CO is usually small during the first couple of minutes. Ramping the fan operation from 80% to 100% air requirements can be incorporated into the controls for this initial period to allow some time for moisture stabilization.

The hotter the precipitator can be kept during the down-time of the cycle, the more moisture can be injected into the waste gases during the period immediately after ignition. This is certainly possible during periods of high production. I would especially seal off all air inleakage through hoppers, and use an abundance of heat on the hopper walls before applying this method. With this type of environment, a bank of fine-water sprays can also be initiated, along with the steam, to hasten optimum power characteristics at the onset of O_2. The above approach is only suggested as a last resort if a reduction of air flow is insufficient to abate the starting stack emission problem.

3. If the fans are not throttled down to about 50% operation, the stack appearance will show the effects of material re-entrainment from collecting surfaces at the conclusion of the blow. This condition was especially apparent for Case M before corrections to the gas distribution problem were implemented.

But regardless of the integrity of the precipitator, the rapid decay of power coupled with high gas velocities spells trouble. The draft was automatically reduced when the O_2 flow was shut off. One reason the draft is kept up in most shops is that a reblow is possible, and this procedure eliminates the cycling of the fans.

4. The reblow may be needed for 50% of the heats. With Case M, the fans can either be ramped up manually by the operator just prior to the lowering of the lance, or they can be controlled the same way as during the initial ignition. Since the precipitator is already hot, I suggest adding a spray bank to the use of steam for this period. The reblow may last less than a minute, so every technique should be employed to reduce emissions. Certainly the lowest level of air flow possible through the system should be considered. At the time of reblow, free carbon in the metal is quite low, and the relationship between the quantity of O_2, time duration of the reblow and air required through the hood can be compromised. Another technique would be to switch off the power to the rapping system at the start of the reblow. Again, the fans should be immediately throttled at the conclusion of the reblow. About 50% of the full blow draft should be sufficient from this period to the conclusion of the tap. New computerized techniques are available to reduce the frequency of reblows, and any effort in this direction is important.

The Precipitator Between Heats

I would like to stress a couple of points that may help in the BOF installation. One has to do with the use of the rapper system. Most installations rap at the same intensity and frequency throughout the cycle. I believe keying the rappers to the generation of fume and the operation of the fans has merit. I suggest using the collecting surface rappers from about 3 minutes into the blow to the conclusion of the O_2. The rappers could be kept off until the conclusion of the reblow and then operated at normal intensity and frequency until the start of the next cycle. This suggestion is based on one vessel operation where at least 20 minutes lapse between the tap and the charge of the hot metal.

Another technique that should provide benefits between heats is the ability to reduce precipitator power to low limits as soon as the fans are cut back to minimum draft. This would allow a more effective cleaning of electrode surfaces during this period, and it would conserve energy. This method should be considered especially during long hot metal delays or times of low production.

Chapter 14
Optimum Designs

Introduction

Throughout this book, I have been focusing on ways to promote the best possible operation and performance of the precipitator. Sometimes, I have probably sounded contradictory. For example, I stressed, on the one hand, that high power input is the key ingredient, but, on the other hand, I must have seemed to contradict myself by saying that too much apparent power can be bad. That is, in fact, an example of the paradox of precipitation.

It would also be foolhardy of me to say that a particular design or process change is the only right one for a specific group of collectors. Yet I know from experience that a number of better design criteria have evolved over the years, and I would like to share a number of my ideas about these criteria. Some of these design ideas seem to be contrary to recent practice in certain cases, but they are offered because my field experience has shown me that these are the directions that new designs should be taking.

At the same time, the complexity of precipitator problems must be appreciated so that any group of design factors needs to be weighed and hedged against the intangibles of the future. Optimum design is usually a cross between what would be nice if economic concerns were not pressing, and what is real. However, I believe that optimum design in precipitation should always build in long term efficiencies at low operating and maintenance costs. This approach does not add appreciably to the initial overall capital investment of a new ESP installation; in many cases it is less than 10% of the original projected cost.

Many of the items I will refer to have been discussed earlier, but I believe it is useful to summarize the most important features again, those which I consider to be the best help in reducing future troubles.

New Design

The design features I think are important include:

Gas Distribution — I stress uniformity of the flue layout at both the inlet and outlet precipitator locations. Consider the factor of dust fall-out during periods of reduced process production. Use the techniques in Chapter 11 to pick the best plenum layout and apply common sense in the installation of deflectors and turning vanes. Remember that the gas flow will head in the direction pointed.

Gas Balance — Balance the gas flow rates between the chambers of a multi-unit precipitator by utilizing pressure drops across a fixed orifice or high-pressure drop grid. Do not use the pressure drop across the precipitator chamber as the flow measurement.

Aspect Ratio — Any precipitator designed for the nominal 99.5% efficiency range should consider a net Aspect Ratio of 1.5. This suggested length to height ratio begins to mitigate against the very high collector plates or very low precipitator velocities in order to keep the economic cost of the installation reasonable.

Gas Velocity — I see disadvantages in designing below 5 ft/sec gas velocities through the precipitator. Actually, with a net A.R. of 1.5 or above, I prefer 5.5 to 6.0 ft/sec for most dry type collectors. There is only so much a low velocity precipitator can do to counteract a situation where the electrical characteristics are poor. Recall, for example, the fly ash installation of Case H in which a velocity of 2.6 ft/sec could not do the job. Another helpful guideline might take this form:

Net Aspect Ratio	Suggested Gas Velocity ft/sec
1.0	5.0
1.25	5.5
1.5	6.0
1.75	6.2
2.0	6.5

The above velocities are predicated on satisfactory internal distribution patterns and the ability to control critical variables of the process.

Sectionalization — Based on field observations, I make the following recommendations for the optimum number of fields:

Minimum Number	Efficiency Range
4 fields	99.0—99.3%
5 fields	99.4—99.6%
6 fields	99.7—99.8%
7 fields	Nominal 99.9%

Within practical cost considerations, I suggest that two bus sections be installed for every T-R set. If it is not possible to control the process variables, add two (2) additional fields to the above relationship to be very conservative.

This does not mean that a power supply has to be automatically applied for every bus section in series, even though that is generally recommended for a four (4) field or smaller precipitator. But as precipitators grow in size, the cost of the power supplies becomes a significant factor in the strategy of hedging against future performance characteristics. In other words, in the two years or so from the time of purchase to the start-up of the precipitator, there can be significant changes in fuel and raw materials. It doesn't seem wise to lock-in on all the power supplies before the process characteristics are known. This method can work with all new installations, over four (4) bus sections in series. I suggest one (1) less power supply be considered for each two (2) potential fields that number over four (4) in an installation.

Layout of the installation should allow for space to apply the T-R sets and controls at a later time. These planning procedures not only save money initially, but also allow precipitator power relationships to be refined after the fact. Physical modifications of the precipitator after start-up can be very expensive, even if an empty space is left for the addition of a field at a later time. Another reason I like to initially hold some T-R sets is that if a resistive condition occurs, too much available power exists, while if a very conductive condition occurs, then the original full deployment of normal power supplies will probably be deficient to attain the collection efficiency.

If this technique is used, then the layout of an eight (8) bus precipitator might be initially energized by six (6) power supplies as follows:

Bus Sections in Series	Length of Section	Ident. of Actual Field
1st	6 ft	A Inlet
2nd	6	B
3rd & 4th	12	C
5th & 6th	12	D
7th	6	E
8th	6	F Outlet

In this case, I would design the size of the T-R sets for the C and D fields to be the same size as the E field, which would be laid out in such a manner that they could eventually energize the future 5th and 6th bus of an expanded energization program.

In other words, many power input options are available to the designer of precipitators, and the early discussions between buyer and manufacturer should explore these departures from practice. These options can take several directions, but the object is to be flexible with the ability to put the final power where it is needed.

Length of Field — I suggest that the length of a collecting surface section be no greater than the distance it takes to reach the center line of the section with the outstretched hand. This automatically sets the length of the optimum field to 6 ft with walkways on either end. With that ability to correct all deficiencies in the field, one could subtract up to 10% of the normally designed surface area of most basic industry precipitators and be ahead of the game. Walkways should have provisions to allow inspections and corrections at the bottom, center, and upper levels of the field.

Height of Collector Plate — I appreciate the rationale for higher and higher collecting surfaces, but I believe other laws come into existence after the 30′ high surface is reached and at that point, potential troubles can become real. While I have no definite proof, I believe a performance factor should be applied toward each height group as it affects the Aspect Ratio on difficult applications. The guideline might take this form to illustrate the point:

Height of Collector Plate (ft)	Aspect Ratio Required (for 99.5% eff.)
20	1.5
24	1.6
30	1.7
36	1.8
40	1.9

After all, the higher the surface, the greater the distance to the hopper, and this condition allows for more problems with alignment, rapping, and gas distribution.

It is in the height of the collector surface where I would like to see some departure from normal design to what I call optimum design. For example, I believe that consideration should be given to a two-tier system of collector plate height in the direction of gas flow through the precipitator. The front half of the collector with 20′ high plates followed by the back half of 30′ high surfaces can be accomplished with relative ease. The benefits are many, yet the design problems should be minimal.

Type of Discharge Electrode — I have only two main suggestions for the optimum design of the discharge electrode, whether it be the weighted wire or rigid frame arrangement. First, make sure the high voltage component is enlarged sufficiently to minimize corona formation whenever a discontinuity exists on the collecting surface opposite the electrode. Secondly, provide the means to easily remove the complete discharge electrode if it fails. I have observed cases where a broken wire, either weighted or in a mast, required major efforts because of the difficulty in removing a small section. Any small cut segment left on a rigid frame could produce localized spark-over by the shear-drag of the electrode at the cut point.

Spacing of Gas Passage — Another consideration to optimum design might be in the spacing of gas passages from inlet to outlet. For example, a six (6) field precipitator could show:

Field	Size of Passage	
1 Inlet	12″	Alternate 12″
2	12″	12″
3	10″	9″
4	10″	9″
5	8″	9″
6 Outlet	8″	9″

The extra cost of design and fabricating should be reasonable if this is done routinely.

Size of Power Supply — There is an art to selecting the final surface area of a precipitator and making the decisions on the sizes of the power supplies. Usually, there is a compromise; the end result may provide a larger T-R set on the inlet field and a smaller supply on the outlet field than would be desired in actual operation. For example, the common practice of sizing supplies from inlet to outlet uses a density ratio of 0.5 to 0.75. This means that if a 50 KVA supply was applied on the inlet, you might find a 100 KVA supply or less on the outlet field. This is a major improvement over the old practice of applying uniform supplies for all fields.

If, by judging the process variables, it is deemed that there is a 90% chance that some spark-over will exist, I suggest that the design current densities be weighted much more severely from the front to back of the precipitator. This is especially true with four or more fields. While the costs of this approach may be slightly higher than for the normal practice, the benefits of matching the system's electrical characteristics are in your favor. Depending on the process, I prefer a ratio of 0.2 to 0.3 for the inlet to outlet field relationship. For example, consider a four equal-sized field precipitator energized in the normal and suggested manner.

Field	Normal Practice	Suggested Practice Case 1	Case 2
Inlet	50 KVA	25 KVA	33 KVA
2	65	50	65
3	85	100	85
Outlet	100	125	117

As the number of fields increase in number, the same philosophy can apply since there is less need for large quantities of power on the front fields. Looking at current densities of a six (6) field precipitator, we might see:

Field	Normal Practice	Suggested Practice
Inlet	25 ma/1000 sq ft	12 ma/1000 sq ft
2	25	12
3	40	40
4	40	40
5	50	63
Outlet	50	63

I make these points because 9 times out of 10 too much power has been designed into the front fields. If a highly conductive material is to be collected, then the philosophy has to be reconsidered, and in this case, larger T-R sets than normal might be desirable in the inlet fields. If the means are built into the system, such as temperature control or gas conditioning, then the need to appreciably vary the ratings of the power supplies becomes less a factor.

Linear Reactors — The use of linear reactors can provide more flexibility in an optimum design. In most marginal cases, additional impedance is needed to modify electrical characteristics, yet too much fixed impedance on occasion can limit the power output of a T-R set. Another design technique can use tapped reactors in a new installation as a means to make power input corrections after a period of operation. For example, if a 3.3 mh reactor is considered satisfactory for a 480 volt, 160 amp power supply, the installed reactor could be designed for tapped steps of 3.3, 4.3 and 5.3 mh. In fact, on any power supplies over 100 KVA capacity, I suggest providing the ability to reduce output power by half of the T-R rating through the use of a tapped reactor. For example, if a 100 KVA supply with a 1500 ma secondary output could normally use a 2.0 mh reactor, then the tapped reactor might have a normal 2.0 mh as well as 3.0 and 4.0 mh connections. Limitation of power and the ability to alter the conduction angle of secondary current could be an important tool if high resistivity condition occurs.

Control Circuitry — A key ingredient in the design of the control circuit for the power supply is to provide the means to separate a manual mode of operation from the automatic control features. This maintenance tool should only be accessible for trouble-shooting or emergency operation.

Internal Design — I have discussed a number of design features involving maintenance factors in Chapter 5. Specifically, the stabilization of the high voltage frame, insulators, hoppers and evacuation system, and rappers should be examined for access, ease of maintenance, and degree of reliability.

Retrofit Designs

When every possible improvement has been accomplished, and the performance goal of the precipitator is still elusive, then the use of additional collection surface is normally considered.

A couple of points should be stressed:

1. Nine times out of ten I would consider a series retrofit precipitator instead of a parallel one. If space and construction time permits, the best retrofit of surface occurs when the fields can be butted up to the inlet or outlet flanges. Actually. a major part of the erection step can be accomplished prior to the long outage. On a couple occasions, retrofit precipitators have been built alongside the existing collector and pulled into place during the outage.

2. If the retrofit correction is of a parallel nature, increase the time and effort spent on gas distribution and balance features.

Novel Designs

Over the years, I have observed a number of new ideas and modifications in precipitators that many groups have initiated to improve particle collection. Unfortunately, most of these concepts have not produced consistent benefits. Actually, the physical design concepts of the 1920's have stood the test of time and are still being used today with relatively minor changes. Where problems tend to occur is when designers fail to consider all aspects of high voltage phenomena on the design of individual components.

I believe in the continuing quest for improvements in precipitator components, but I'm also from Missouri when it comes to accepting new designs or concepts. That is, I believe that too often, the process and particle characteristics can nullify the best of designs, and I would like to see the most efforts for improvement be in the implementation of process knowledge to design and apply an effective collector. In other words, we already appear to have sound precipitator design concepts available, and failures occur more often from other factors.

Another direction of efforts should address itself to the conservation of energy in precipitation, since the ultra-high efficiencies are demanding larger amounts of power input.

Another area where I believe new thinking should be directed is in the improvement on existing methodology for removing collected material from hoppers. I look with favor on the trough type of hopper as long as it is placed perpendicularly to the gas flow. This type of hopper can utilize a slow moving conveyor belt, screw, or other transport means for the rear fields of large precipitators as effective ways to get around the small amounts of material collected in these zones.

Control of Process

Where the process cannot be controlled for the benefit of the precipitator, then the size of the collector must compensate for this situation. I stressed uniformity of raw materials, moisture, gas temperature, and the determination of whether a steady state condition exists rather than a variable one, as the ingredients of a successful installation. Penalties of 25 to 50% in size are sometimes required for designs that must overcome these factors.

I highly recommend consideration be given to the new philosophy in the design of the process equipment and operating techniques to take away some of the uncertainty and hazards of the precipitator performance. For example, more temperature control in the design of the gas system of the fly ash industry can reduce the precipitator size appreciably. This is a case where the money saved on the precipitator is distributed other ways. If the sulfur of the coal is reduced, it sure would be nice to bring flue gas temperatures down to the 230 to 240°F level at will. The benefits to the boiler's efficiency also cannot be overlooked with this capability. On the other hand, to be able to maintain gas temperatures above dewpoints is desirable on many occasions.

It is also possible to reconsider the design of sufficient flues for possible conditioning at a later time, or the space for a cooling spray tower which can be used as both a gas mixing and conditioning chamber. The key point here is that the precipitator should not be thought of as a separate device, but rather as a part of a total system.

Design Parameters

The range of design parameters is too wide to allow me to provide any meaningful data for all applications, but there are some key points I would like to stress:

1. Keep the collecting area for each bus section to a workable level of about 10 to 12 thousand sq ft.

2. This would place the eventual design collecting area per T-R set in the 20 to 24 thousand sq ft range. As I pointed out earlier, some compromise could occur in the original layout, but the eventual area/T-R set must be initially designed into the system.

3. The extreme inlet and outlet fields are considered more critical, and so the size of surface area and power supply should be applied carefully. I generally recommend smaller fields on both extremes with smaller T-R sets on the inlet and larger supplies than usual on the outlets. Current densities of 12 to 15 ma/1000 sq ft on the inlet and 50 to 80 ma/1000 sq ft on the outlet should not be unusual.

4. The amount of collecting area struck by one rapper is not overly critical and can range from a fraction of a percent to that of 3 to 4% of the total area. Remember that if a rapping problem exists, the process must be carefully evaluated. The means to control the frequency of the rapping cycle between fields is most important, and I like at least a ½ hour period to lapse between the inlet to outlet fields for most applications.

Where rapping problems are contemplated, the area struck on the inlet fields should be about ½ that of the outlet fields.

5. The effect of high gas temperatures will tend to promote higher current densities, and hot-side precipitators might require twice the density of cold-side collectors. Current densities of 100 to 110 ma/1000 sq ft at 700 to 750°F is not out of line.

6. Effective precipitation will usually occur with voltage gradients of 10 to 12 KV per inch of dielectric space between the discharge electrode and collecting surface. Therefore, the power supplies must be of sufficient current capacity to provide these fields. That is why the T-R sets of the latter fields are in such large ratings. You can see that as more and more particulate matter is removed from the gas stream, the dielectric space offers little resistance to current flow. This is another reason why I like smaller surface area fields on the outlet zones of the precipitator. The inlet fields usually will not have voltage deficiency problems, but the smaller fields are also useful at this location to take advantage of the rapid change in dust characteristics.

I would like to conclude this design chapter by cautioning against placing too much emphasis on strict adherence to the numbers game in precipitation. All of the above design suggestions, for example, should not apply across the board, and the purchaser and supplier must weigh many factors before the final decision on size. And size alone is less important than the shape of that size, thus, I would rather have a smaller precipitator with an ample aspect ratio more times than not. I have observed new boiler design situations where valuable days were consumed in discussions about how big to make the precipitator, and no time in looking at the system to see how much more gas temperature control could be implemented. In other words, place major emphasis on how to counter the effect of every potential process problem on the precipitator, and the size of the collector will be less critical.

Chapter 15
Specification and Bidding

Introduction

Nowhere is the art of precipitation needed more than in the specification and bidding stages. The suppliers of precipitators will judge their designs on their experiences and knowledge of the process. The buyers must look ahead to raw material and process changes that may not even be in sight yet. The variations in the physical designs of precipitators between the so-called weighted wire and rigid frame construction complicates the situation.

All of the above is not beyond handling, as long as everyone involved faces up to the fact that the precipitator application does not yet conform to all the scientific niceties (and by the way, that goes for most other air pollution equipment). The only way to minimize the potential problems is to communicate extensively about the intangibles and attempt to identify the key process inputs during the early stages. I would especially discuss the following items in the basic industries in addition to the fuel and raw material analysis and forecasts:

Utility

1. Type of boiler — dry or wet bottom
2. Number and type of air preheaters
3. Type of operation — base loaded or variable
4. Type of pulverizer and expected fineness of coal
5. Methods of gas temperature control

Cement

1. Range of slurry water content expected
2. Alkali content range
3. Type of chain or heat exchangers used
4. Methods of gas temperature control

Iron and Steel

1. Method of waste gas temperature control in BOF

2. Method of fan control and future O_2 blow rates expected

3. Basicity of burden expected in sinter strand

4. Sulfur ranges of ore expected

The above items are but a small portion of the discussions that should be held between purchaser and supplier but I believe the message is clear — every information input that has a potential effect on the eventual performance of the precipitator should be explored. A point I made before is worth repeating, that the actual design of the precipitator is less a factor in successful installations than the control of the process. But every design feature of the offered precipitator should also be explored for weak spots. It is quite possible for manufacturers to improve component design at the request of a buyer, of course, at some increased cost.

Information for precipitator specification can be obtained from the IGCI, 700 N. Fairfax Street, Alexandria, VA 22314. Other pertinent information can be obtained from articles in the Journal of the Air Pollution Control Association, V8:249 (1958), V15:256 (June 1965), and V25:362 (April 1975). Engineering firms are well equipped to supply specification knowhow, but the buyer must be responsible to supply the correct and proper input data. Make the process data all too encompassing and the precipitator size can become gigantic; miss some key factors and the best precipitator design can be found wanting. Which brings me to an area of just how I look at the specification and bidding phase of precipitation.

Concepts

It becomes exceedingly difficult to evaluate a wide spectrum of designs for the same guarantee point, especially when the cost quotations vary to the same degree. I believe the chance that the purchaser will end up with a deficient design increases with this approach.

I prefer that the purchaser submit a relatively tight set of sizing specifications to all interested manufacturers, and have all parties bid to this optimum design for a specified guarantee level. Where I have seen this done, a success story usually exists. Any supplier has the right to rebut the design or show where improvements can be implemented, but generally, if they agree with the sizing, then they should show no hesitancy in bidding.

After the successful bid, I suggest that further negotiations begin immediately between the two parties to decide on the final design criteria. At this stage, there can be trade-offs and modifications based on detailed discussions of the process forecasts and problem areas. In lieu of the above approach, the purchaser can try to discuss all details with all suppliers, which can become time consuming and disconcerting. What I am saying, in the case of a difficult application such as precipitation, is that more time should be devoted to the final details of layout and design. If too much time is spent on being confused by many designs — something may be lost in the final analysis.

Another point I would like to make on specification is that the guarantee language should recognize the difficulty of attaining the collection efficiency of the precipitator if changes in the process occurs. I do not believe it is fair to either party if a series of contingency plans are not discussed ahead of time with all the options openly explained. A study of this book certainly should make one aware of the pitfalls that can occur — and each pitfall can be countered by some modification. In many cases, contingency plans can be incorporated right into the design of the installation at relatively minor cost to that involved through a later retrofit.

The design and process problems of yesteryear have become somewhat factored into recent precipitator designs, and coupled with the push towards zero pollution, the costs and physical sizes of some of these installations begin to stagger the imagination. I believe we will all lose in the long run unless some common sense prevails in understanding where to place the emphasis and degree of control. Precipitators are not unlike other air pollution equipment; maintenance and operating costs, primarily energy, begins to work against you after certain sizes are reached. I still believe it is better to have more expenditures for the reliability of precipitator systems to help reduce these long-term burdens. Can this be done with the gigantic installations required today?

Chapter 16
References

In this short coverage of additional reading matter on the subjects discussed in this book, I intend to point out the best places to enlarge your scope of background, knowing quite well that I will probably miss many valuable references. The place to start is Dr. Harry J. White's book listed in the bibliography. This is a solid coverage of basic theory and practical background. Dr. Myron Robinson presents fundamental theory and a comprehensive biblography in his treatment. The National Technical Information Service, Springfield, Virginia 22161, makes available to the public a large number of manuals and reports on precipitators and associated subjects. Part of this distribution were Manuals of Precipitation Technology, Parts I and II, prepared by The Southern Research Institute back in 1970, in which Dr. White had a significant part. APCA has also published a number of esp papers over the years that has dealt with many phases of precipitation.

The information concerned with process effects relative to precipitation is somewhat difficult to come by; much of this data is dispensed in technical journals or in a number of communications between suppliers and purchasers, and is often known by the plant personnel who work closely with the precipitator. Some sources that can be explored.

Utility — Annual American Power Conference proceedings, recent NTIS publications, and various technical journals of the electric power industry will usually supply good background information.

Cement — The Portland Cement Association produces training and publications for the cement industry. A book by Kurt Peray and Joseph Waddell, "The Rotary Cement Kiln," Chemical Publishing Co., Inc., 1972, should be of value. There are a couple handbooks available.

Iron and Steel — The Journal of The Iron and Steel Engineers has supplied information over the years. A BOF symposium in the October 1960 issue provides some background data.

Recent years have seen more activity on the maintenance and operation techniques and problems of precipitators. The TC-I Committee of APCA embarked on maintenance surveys back in

1974, which was the cornerstone of the "Operation and Maintenance of Precipitator" Conference in April 1978, the proceedings of which can be obtained from the Air Pollution Control Association. This conference brought together a number of presenters who discussed separate subjects on the practical aspects of precipitation. The EPA has also initiated funding for booklets on maintenance techniques, one of which is entitled, "Precipitator Malfunctions in the Electric Utility Industry," January 1977, that can be obtained from NTIS of the U.S. Department of Commerce. There has been an awakening of activity in how to go about taking care of the vast number of precipitators presently in existence. The McIlvaine Co. of Northbrook, Ill. also provides a continuing service of information on precipitators.

I have freely used information from my previously published papers for inclusion in this book. These papers included: "Electrostatic Precipitator Primer," Iron and Steel Engineer, 1964; "Effective Collection of Fly Ash in Pulverized Coal-Fired," JAPCA, 1965; "Maintenance Program and Procedures to Optimize ESP's," IEEE Cement Conference, 1974; "Electrical Characteristics of ESP's," ASME Conference, 1972. I used Table 7—1 from the paper, "Corrosion Problems and Solutions for Electrostatic Precipitators," H. J. Hall and J. Katz, JAPCA, April 1976. Table 9—7 on Excess Air on CO_2 in Combustion Products was extracted in part from The North American Manufacturing Co. "Combustion Handbook."

BIBLIOGRAPHY

1. S. Whitehead and E. Benn, "Dielectric Phenomena—Electric Discharges in Gases," London, 1927.

2. J. D. Cobine, "Gaseous Conductors," Dover Publications New York, 1958.

3. H. J. White, "Role of Corona Discharge in the Electrical Precipitator Process," Electrical Engineering, Jan. 1952.

4. H. J. White, "Modern Electrical Precipitation," Industry and Engineering Chemistry, May 1955.

5. H. J. White, "Industrial Electrostatic Precipitation," Addison-Wesley, Reading, Mass.,1963.

6. M. Robinson, "Electrostatic Precipitation Chapter," Air Pollution Control I, Edited by W. Strauss, Wiley-Interscience, New York, 1971.

7. H. E. Rose and A. J. Wood, "An Introduction to Electrostatic Precipitation in Theory and Practice," Constable and Company Ltd., London, 1956.

8. K. R. Parker, "Principles and Applications of Electrical Precipitators," Chem. and Process Eng., Sept. 1963.

9. P. Muller, "Study of the Influence of Sulfuric Acid on the Dewpoint Temperature of the Flue Gas," Chem-Ing, 1959.

10. E. S. Lisle and J. D. Sensenbaugh, "The Determination of Sulfur Trioxide and Acid Dewpoint in Flue Gases," Combustion, 1965.

11. VDI Standards, Publication VDI 2112, June 1965.

Pertinent Conversion Factors Needed In Book

Length
1 cm = 0.394 inch
30.5 cm = 1.0 ft

Area
1 sq meter = 10.76 sq ft

Volume
1 cu meter = 35.31 cu ft

Mass
1 lb = 7000 grains
1 gram = 15.43 grains
1 K gram = 2.2046 lbs
1 metric ton = 2,204.6 lbs

Concentration
1 gram/cu meter = 0.437 grains/cf

Velocity
1 meter/hr = 0.0547 ft/min
1 meter/sec = 3.28 ft/sec

Flow Rate
1 cu meter/hr = 0.589 cf/min
1 cu meter/min = 35.3 cf/min
1 cf/min = 0.0283 cu meter/min

INDEX

Supplement

to the

Second Printing of

THE ART OF ELECTROSTATIC PRECIPITATION

by Jacob Katz

January 1981

PRECIPITATOR TECHNOLOGY, INC.
4525 Main Street
Munhall, Pennsylvania 15120

Table of Contents - Supplement

Supplement

Introduction

The first printing of "The Art of Electrostatic Precipitation" evidently succeeded in providing readers with a practical yet fundamental discussion of the subject. I saw no need to greatly change the original text, but comments from readers alerted me to omissions as well as to ideas which needed to be clarified. Many of the following discussions can stand alone, but in some cases references to specific text figures or tables will be made. The purpose of this supplement is to further explain the critical issues involved in precipitation as described in the text.

One critical issue has been the absence of fundamental concepts made readily available and stated simply to persons working with the precipitator. This situation has fostered misconceptions and questionable design criteria that tend to counter attempts to solve the basic problems. An understanding of these fundamental concepts will take the black magic out of precipitation.

A second critical issue is the need to identify and understand the importance of the process variables to the success of precipitation. The precipitator is normally located at the end of the process system and any one of a number of variables can effect the performance of the collector. If the physical size of the precipitator would be designed to handle all the process variables that could occur, the cost would be prohibitive. I believe, however, that it is possible to understand which variables are important and can be controlled for effective precipitation.

Another critical issue is the need of a basic understanding of the power characteristics in the precipitator and what they really mean. After all, collection of particulate matter in the precipitator is accomplished by electrical forces and practically all facets of the precipitator design and process will effect the electrical characteristics observed in the field.

The last critical issue I take up involves the day-to-day reliability of performance as it relates to the proper design and maintenance of the equipment. The key factor here is the knowledge that repetitive failures of any group of components is abnormal. The cause of the weakness must be identified and corrected early in the life of the precipitator unit.

Between the above issues, I will weave a number of comments and recommended directions that should remove some of the magic out of precipitation.

A Look At Power Input

Introduction

The original text referred to power input in a number of different ways, which confused some readers. Remember that I kept mentioning "Effective Holding Power" while saying that too much power can be as bad as too little. Lets look at some preliminary concepts.

The precipitator designer will assign an amount of surface area to each transformer-rectifier based on a number of factors. The basic size of the T-R set is meant to satisfy gas and particulate matter characteristics as well as the physical dimensions of the precipitator. For example, a fly ash application for 2% sulfur coal may call for 40 ma per 1000 sq ft of collecting surface in the inlet field of 9" spaced gas passages. If 18,000 sq ft of precipitator surface is designed into the inlet field, then a T-R set of 750 ma may be applied. A commonly used power supply might show the following rating information on the name plate:

48 KVA

400 Volt AC rating on primary winding

120 amp AC rating on primary winding

53,500 volt AC secondary winding voltage

45,000 volt DC rectifier output

750 ma DC on the secondary winding

One of several power input limits can occur for this inlet field:

1. The voltage limit of 400 on the primary winding of the T-R set can be reached before any other limitation.

2. Either the primary or secondary winding current limit could be reached.

3. Or spark-over can occur within the field limiting the available power from the T-R set.

These three limits must be understood before I go any further. First, the voltage limit is rarely observed on normal fly ash applications, but it can occur with an appreciable mismatch of the T-R set to the load requirement. That is, it can occur when a large capacity of supply is connected to a small surface area field and is combined with high concentrations of finer-sized particles.

Second, the current limit is observed more often. With a T-R set well matched to the load, the primary current of 120 amps for the above power supply should occur when approximately 750 ma is observed in the secondary circuit. However, it is possible to observe a different primary current relative to the secondary reading when the waveform appreciably deviates from the form factor of the sine-wave. Adjusting the conduction angle of the secondary current to approximately 86% by the correct application of linear reactors will generally produce a match of the primary and secondary currents. As the conduction angle decreases from 86%, the primary current will trend toward higher readings relative to the secondary current reading. The rated secondary current will sometimes be achieved before the primary current limit if conduction angles rise past the 86% point when higher levels of impedance are applied in the primary circuit. This situation can result in a sustained arc-over between the electrode systems because of the drag of the higher inductance.

This brings me to the third limit, which is a spark-over between the discharge electrode and collecting surface. When this occurs, the power supply voltage must be reduced to keep the break-downs within a reasonable level. As explained in the text, this level could range from a nominal 150 sparks/min for the inlet field to the occasional spark for the outlet fields.

However, the designer predicated his precipitator performance on a power parameter that now may not be attained because of the limitation imposed by spark-over. Basically, precipitator spark-over can occur by one of two mechanisms:

1. The impressed voltage is greater than the spacing and the physical contour or conditions between electrode surfaces will allow, regardless of the characteristic of the particulate matter. This condition if often observed with electrode misalignment where one or more of the discharge electrodes has moved too close to the collecting surface.

   Another important factor that can cause premature break-downs is the presence of severe discontinuities on the collector surface opposite the corona-emitting zones of the discharge electrode. As I explained in the text, this type of break-down tends to provide a greater electrical disturbance compared to the spark-over caused by high resistivity.

2. Spark-over caused by high resistivity levels is the most common reason for low power inputs to the precipitator. Chapter 4 discusses the concepts of resistivity, but I want

to enlarge on the key issues involved in this phenomena. The resistance of the layer of collected material on the collecting surface is the prime reason spark-over will occur. This layer will develop a voltage drop based on three factors: the resistivity value x the layer thickness x the current density. If the voltage drop is greater than the dust layer can withstand, then breakdown within the layer occurs.

This phenomenon is not unlike the breakdown of a capacitor. Reduction of the layer resistivity can be achieved by a number of methods discussed earlier, including flue gas additives and process modifications. Sufficient reduction of the layer thickness is often difficult to obtain because as resistivity increases, so does the tenacity of the particles to stick together and adhere to the collecting surface.

The third component of the voltage drop is the current density. This means that the amount of corona current attempting to pass through the dust layer must be reduced if either the resistivity value or layer thickness increases. Otherwise, spark-over can occur. For example, the 40 ma per 1000 sq ft may occur with a material resistivity of 10^{10} ohm-cm. But if the actual resistivity was 10^{11} ohm-cm, the current density might have to decrease to 25ma/1000 sq ft to keep spark-over at a reasonable level. In other words, the higher the resistivity level, the lower the current density must be to keep the precipitation process functional. It is not uncommon to see current densities below 5 ma/1000 sq ft on certain fly ash applications.

Effects of Resistivity on Power Levels

What this means in a practical sense is that an infinite number of voltage and current readings can occur in the precipitator that will not in anyway match the name plate data of the T-R sets. The important thing to remember is that higher resistivity conditions will decrease power inputs because of the spark-over limitation. Superimposed on a possible resistivity problem is the possible condition of the internal structure causing the spark-over to occur at a much lower power level. Once the dust resistivity reaches a critical point, its deposition on a sharp edge, or for that matter any kind of discontinuity on the collecting surface, will cause a localized electrical stress build-up point that will draw the spark. This is why uniformity of alignment and elimination of all internal irregularities becomes more important as the resistivity moves up from the moderate range.

Let's look at the resistivity concept in terms of a high, moderate, or low range. Although I discussed groups of numbers in Chapter 4, I now would like to separate the resistivity activity in three basic groups. The high resistivity range, where usually less than 10 ma per 1000 sq ft exists with spark-over in all fields, is usually detrimental to performance unless the precipitator is designed oversized. The moderate range of resistivity can be considered to exist when spark-over can be achieved on all fields, but near the T-R set rated current of the two outlet fields. This can be the most desirable range on most marginal precipitators where reentrainment of material back into the gas stream must be minimized. When spark-over cannot be achieved in any field of a well-designed precipitator, I would call this a well-conditioned state that might lead to a reentrainment problem depending on the voltage gradient across the space and dust thickness.

It is when there is an absence of spark-over with rated current attained on the T-R sets that there should be some concern about whether the resistivity of the dust layer has too low a magnitude. Refer to the 48 KVA T-R set on the nominal 480 volt supply to judge this condition. For fly ash, the inlet field under low resistivity conditions might show the following meter readings:

Primary Voltage — 280 volts

Secondary current — 750 ma

Spark — none

The rated primary voltage is 400 so the ratio of 280 to 400 would indicate a relatively low precipitator voltage. If an oscilloscope trace is obtained of the voltage waveform, the difference between the peak and minimum precipitator voltage could be as high as 18 kilovolts. It is likely that much of the collected material will be lost during rapping under this condition.

As I mentioned in the text under this condition, it would be possible to detect relatively slow movements in the meter readings during the rapping cycle of the collecting surfaces. However, be alert to the amount of collector surface connected to the T-R set because it can also produce a low ratio of voltage to current. For example, if 36,000 sq ft were energized by the 48 KVA supply, readings of 280 volt and 750 ma without spark-over may be observed even under reasonable resistivity conditions.

As the resistivity of the material increases, the indicated voltage will continue to increase without any change in the current reading. On our system, a primary voltage of 340 to 350 volts might occur at 750 ma with the occurrence of some spark-over. I generally consider this condition the beginning of the moderate resistivity range with incipient back-corona just starting.

A further advancement into the moderate resistivity range will decrease the amount of corona current the layer can take without excessive break-downs. As the current is reduced, the voltage must necessarily decrease in order to keep spark-over at an acceptable level. I consider the electrical readings of our inlet field to comprise the moderate zone of resistivity when they fall within these limits:

	Entry	Upper Limit
Primary Volts	350	300
Secondary ma	750	350
Spark/min	5	50

Somewhere between the above limits, the inlet field of our example will perform at optimum collection efficiency. Subsequent fields will generally show some spark-over when the inlet field is at the upper limit of the moderate zone.

As the move to higher resistivity continues, the spark-over level will continue to reduce power levels until as little as 20 to 30 ma total corona current may occur at primary voltage levels of 180 to 210 volts. For about 99% of the resistivity cases that occur in practice, spark-over will be the controlling influence and both voltage and current will follow a defined relationship downward. In the case of severe back-corona when the resistivity reaches very high levels, it is possible for current flow to rise exponentially while showing a drop in the indicated voltage.

This condition will also show a much lower minimum voltage in the precipitators as compared to the high resistivity-spark mode. For example, the difference between the peak and minimum voltage tends to flatten out, and it is possible to observe a KV_p of 36 and KV_m of 33 at the low power levels of the normal high resistivity case, while severe back corona may show minimum voltages of 20 to 22 for the same average voltage. In the case of severe back-corona, the current density across the dust layer must be kept down before the voltage begins its decline, even though this will occur at relatively low power levels. An absence of spark-over occurs for the severe back-corona state because the great number of positive ions generated from the collecting surface counteracts part of the space charge.

The bottom line of any precipitator system is its collection efficiency, so let's see what I mean by "Effective Holding Power" relative to the different resistivity levels. When low resistivity exists with low voltage conditions, it is difficult to achieve high collection efficiencies because of dust reentrainment losses. Power consumption is high because of the high current flow through the dust layer caused by a practically nil electrical resistance. Internal inspections usually show collecting surfaces devoid of any buildup. During this low resistivity, it will be difficult achieving the guaranteed efficiency at even half the design gas velocity.

Moderate resistivity will allow dust particles to bond to-gether in the dust layer by forming a charged dipole relationship not unlike those found in a magnet. The opposite polarities provide good adhesiveness at the tangent contact points of adjacent particles and even aid in holding these particles together as they become dislodged by the rapping procedure. Reentrainment losses are at a minimum level.

When the resistivity increases into the high range, the bonding can become severe and layer dislodgment by rapping is difficult. However, at some point the reentrainment portion of the total precipitator losses can be higher than in the moderate range and are caused by the reduced power levels. A sizeable part of the precipitator dust loss in high resistivity cases can also occur during high spark-over conditions because each spark blows out a small volume of dust from the layer.

When I discussed the Deutsch equation in Chapter 2, I showed that the migration velocity contained the product of two precipitator voltage gradients without regard to the corona current component. The magnitude of the current flow will partially depend on the resistivity conditions for any given voltage. For a fly ash example showing 280 volts on a primary voltmeter, a current may indicate 750 ma with low resistivity, 400 ma in the moderate range, or 150 ma in a higher resistivity range. In this example, the 400 ma condition would probably provide the better collection performance on a higher velocity precipitator. It is always advantageous to work toward higher voltage gradients and take whatever corona current results. The only exceptions would be current suppression caused by discharge electrode build-up or excessive space charge caused by high concentration levels of fine-sized particles.

Normal Versus Abnormal Power Characteristics

An understanding of the electrical readings of the precipitator must be a starting point in coping with this collector. I already stressed that the name-plate electrical values will not be observed on many fields of the precipitator, so the patterns of meter readings become an important tool of evaluation.

The key word is "uniformity" of patterns of each precipitator because gross power values used to compare one unit to another sometimes provide questionable evaluation results. Electrical pattern alterations will occur because of the following reasons:

1. Design of power supply.

2. Design of discharge electrodes.

3. Physical spacing between electrode systems.

4. Change of surface contour of discharge electrodes.

5. Surface area connected to each T-R set.

6. Space charge by changes in the number of particles within the dielectric spacing.

7. Resistivity characteristics.

8. Wave-form differences between the T-R sets.

9. Gas composition.

10. Viscosity of the gas stream in terms of pressure, density, and temperature changes.

11. Misalignment between the electrode systems.

12. Dust coating on the discharge electrode.

13. Discontinuities on the collecting surface.

14. Dust layer on the collecting surface.

Each of the above factors will make its individual mark on the electrical characteristics of a specific precipitator. Often they will combine synergistically to change the performance of the precipitator, which is why evaluating problem areas can become difficult. For purposes of this discussion, I will discuss each factor with examples of how it affects the electrical readings of a hypothetical precipitator. I am more concerned with the concepts illustrating a point than with exact numbers. The use of the primary voltmeter and secondary direct current flow meter will be used for evaluating a three (3) field precipitator with each field of 18,000 sq ft energized by a 48 KVA T-R set (the specifications were covered previously). While this arrangement would be considered a poor precipitator energization design based on concepts in the text, it offers the simplest way to explain the various factors.

Design of Power Supply

The proper match of the T-R size to the load will minimize the effect of the power supply itself on the non-uniformity of readings. For example, if the gas and dust conditions are attuned to the T-R sets, then indicated voltage will tend to decrease inlet to outlet while currents will generally rise in the same direction. This is more often true in the moderate resistivity range and the readings occur in some relatively uniform manner. If the outlet field demands much more current than the T-R set has available, then a distorted voltage/current pattern develops between the fields:

	Pattern A			Pattern B		
	Pri. Volts	Sec. Ma	Spark/ Min	Pri. Volts	Sec. Ma	Spark/ Min
Inlet	360	350	50	360	450	20
Center	330	550	20	330	700	occ
Outlet	310	750	occ	280	750	none

Note that the 50 volt difference between the center and outlet fields of pattern B is much more severe than pattern A shows for the same two fields. The fact that spark-over is also absent at the rated current of 750 ma further indicates that more power is demanded.

Other variations in power supplies that can help distort patterns of readings are meter characteristics, type of rectifier, the components of the automatic voltage control circuit, and whether or not the same primary winding taps are being used. For example, the saturable-core reactor can provide different readings than the thyristor controls.

Design of the Discharge Electrodes

Electrode design will effect the voltage-current relationship since the diameter or sharpness of the corona-emitting points or edges of the high voltage electrode will determine the starting voltage at which corona will initiate. Therefore, for a given voltage, the smaller diameter wire may show more current. Barbed wire, compared to the nominal 0.1 inch round discharge wire, might show 50% more current at reduced power levels. As the voltage levels increase toward rating, the currents begin to draw closer together.

Physical Spacing Between Electrode Systems

The wider the spacing between the discharge electrode and

collecting surface, the greater voltage will be required for any given current reading. For example:

	Pri. volt	Sec. ma
9" wide gas passage	330	750
10" wide gas passage	365	750
12" wide gas passage	440	750

As you can see, the 400 volt rated primary of the T-R set would not be sufficient to handle a 12" wide gas passage field. Normally, a greater turns-ratio transformer is applied to achieve higher secondary voltages.

Change of Surface Contour of Discharge Electrode

Localized spark-over or arcing will cause metal loss on the round or square type of discharge electrode, and the resulting surface change will eventually allow the electrical breakdowns to occur at lower and lower voltages. For example:

	Pri. volt	Sec. ma	Spark/Min
Initial spark-over	330	700	50
Minor Reduction of Wire Cross-Section	300	450	50
Severe electrode damage	200	150	50

In other words, as the discharge electrode developes sharp pits and an eventual necking down at the point of breakdown, the indicated voltage and current for that field will continually decrease in time for the same spark rate. When these conditions occur, they are symptomatic of either poor electrode design or other difficulties.

The method of spacing needles, barbs, or punched edges on the discharge electrode for the corona-emitting points will minimize the effects of the above type of damage. However, minor changes might occur in the voltage-current relationship as the electrode points become less sharp in time.

Surface Area Connected to Each T-R Set

The amount of collecting surface area connected to each T-R set can appreciably alter the voltage-current readings. This is why the current density, ma per 1000 sq ft, must be used to compare fields of various sizes. If the precipitator conditions warrant a demand of approximately 40 ma/1000 sq ft, then the match of the 48 KVA to 18,000 sq ft was a good application. But if the T-R set energized two 9,000 ft bus sections and one bus was isolated because of a grounded condition, then the indicated current flow usually decreases while the voltage will often rise for the remaining energized bus.

Remember that even at 400 ma on 9000 sq ft, the current density and performance will approximate the normal hook-up condition. This is why you must be careful to observe the characteristics of a partially energized field so that excessive break-downs or instability does not occur. It is often possible to double the current density of the 9,000 sq ft bus in the above example, but only with relatively low resistivity materials. More often, only a moderate increase in power can be obtained. With higher resistivity materials, it is possible to observe even lower voltage and current readings at spark-over when half the field is isolated.

As I mentioned earlier in this supplement, by connecting too much area to the T-R set, the rated current level may be reached before sufficient voltage gradients are achieved. Some of these conditions might show the following readings on a specific field:

Condition High Resistivity	Pri. Volts	Sec. ma	Sparks/Min
Normal Hook-Up	250	100	50
Energize ½ field	250	55	50
Double Area	250	200	50
Moderate Resistivity			
Normal Hook-Up	350	700	50
Energize ½ field	355	400	50
Double Area	320	750	None
Low Resistivity			
Normal Hook-Up	310	750	None
Energize ½ field	330	750	None
Double Area	280	750	None

Space Charge Effects

As each field of a multi-field precipitator does its work, the reduction of suspended particulate matter in the flue gas will alter the voltage-current relationship from inlet to outlet. This phenomena is best observed in the moderate resistivity range. A change in the resistivity of the material in each field can alter the patterns, but the slope of the pattern is mostly space charge oriented. For example:

	Pri. Volts	Sec. ma	Sparks/Min
Inlet	360	400	50
Center	330	550	20
Outlet	300	750	occ.

In other words, for identically sized fields, we are always looking for a stepped decrease in voltage and stepped increase in current from inlet to outlet. Deviations from this sort of linear relationship in the direction of gas flow would be considered an abnormal pattern especially in the low zone of moderate resistivity.

Resistivity Characteristics

I covered the effect on precipitator electrical readings by changes in resistivity earlier in the text and this supplement. It is difficult to identify all the variations that can be observed, but I want to stress some key concepts:

1. Patterns in voltage and current readings from inlet to outlet should form some type of recognizable pattern unless internal defects cloud the issue.

2. The high resistivity range can produce a relatively low flat voltage and current pattern, but only in the very high resistivity zones (10^{12} and above). Generally you should see an increasing pattern of current flow in the direction of gas flow.

3. The moderate range of resistivity would show the greatest magnitude of change from inlet to outlet.

4. As the resistivity becomes low enough so that all spark-over ceases from this cause, then the voltage and current readings tend to flatten out again from inlet to outlet, but at a much higher power level.

5. It is only in the moderate to high resistivity ranges that internal electrode defects will show major distortions in the voltage-current patterns between fields and adjacent cells.

6. Effects of resistivity can completely nulify the effects of the space charge on the inlet to outlet patterns.

The sets of readings on pages 72-73 in Chapter 4 show various patterns that may be observed at the different resistivity zones of a fly ash application. Let's continue this discussion to help show what constitutes a normal or abnormal pattern. The comparison of readings between fields provides the key method of detecting abnormal electrical characteristics. With equal-sized fields, the voltage pattern should follow the current patterns within reason. For example, with very high resistivities, the maximum voltage and current deviation between fields are usually less than 10%. This is also a valid limit for the low resistivity range.

When the current flow from inlet to outlet rises appreciably with dust conditions in the higher zone of the moderate resistivity range, then the voltage will also generally rise somewhat in the direction of gas flow. In this situation the voltage decrease due to the space charge reduction is overshadowed by the improving power pattern. It is not unusual to observe a level of current of the inlet field to be ½ to ¼ of that existing in the outlet field. The voltage rise, in this case from inlet to outlet, will be less percentagewise than the increase in the current value. The ratio of the voltage to current flow rise might be as much as one to three in the latter fields. In other words, it is possible to obtain 3% rise in current for every 1% rise in voltage in the outlet field, but only a 2% rise for each percentage voltage gain in the inlet field. Another key point here is that if any voltage deviates more than 10% from the average voltage and this occurs with a break in the pattern of currents, then this would be considered an abnormal pattern.

	Normal Pattern			Poor Pattern		
	Pri. Volts	Sec. ma	Spark/ Min	Pri. Volts	Sec. ma	Spark/ Min
Inlet	260	200	50	260	200	50
Center	270	400	20	200	150	20
Outlet	280	750	occ.	285	750	occ.

The center field voltage of 200 and 150 ma falls within an abnormal pattern.

Another key point is that the resistivity may change often because at each installation a family of resistivity curves usually exists depending on the variations in moisture, temperature, and other inputs of the process. These changes in resistivity will change the levels of the electrical readings, which is a normal phenomena, but the patterns at each resistivity should be fairly reproducible when conditions revert to the prior level.

Wave-form Effects on Electrical Readings

The wave-form will add another variable to the voltage-current relationship between the T-R sets of a precipitator. The ratio of actual current to the rated current will determine the peakiness of the voltage and current wave-forms. The precipitator current conduction will vary as the loading on the T-R set. Current conduction might show the following approximate percentages:

	Current Conduction
At Rated Current	86%
At 75% of rating	60%
At 50% of rating	45%
At 25% of rating	25%
At 10% of rating	15%

As the current conduction decreases, the average precipitator voltage will also decrease and a distortion in the primary/secondary currents will occur as discussed earlier. This is why the application of additional linear reactors is important to help draw out the current conduction. Unfortuately, the bulk of any gain generally occurs with existing conduction angles above 35 to 40%. In these cases, doubling the original reactor size will often add another 15 to 25% conduction time. Original reactor sizing is usually based on 50% of the ohmic value of the high voltage transformer primary winding. The more the reactor value, the lower the primary current will read relative to the secondary precipitator current.

Gas Composition Effects on Readings

The composition of most gas streams passing through industrial precipitators will have common constituents and will certainly be conducive to maintaining the ionization process. The gas composition of any given precipitator will usually effect all the fields or cells uniformly so that the patterns are controlled by other inputs. However, the voltage-current relationship between processes or even in the same process will be affected by changes in the gas flow characteristics. Disregarding spark-over, several of the effects would show:

Condition	Level	Primary Volts	Secondary ma
Water vapor	5% by vol	330	750
	15	350	750
SO_2	2.0%	320	750
	0.1% by vol	350	500
Gas Velocity	4 ft/sec	330	750
	8	350	750

A change in gas composition can alter the spark-over voltage which will vary the level of power input to the precipitator. If the moisture content of the gas stream is reduced for a particular installation, spark-over can occur at a much lower voltage. For example:

Level of Moisture/Vol.	Primary Volts	Secondary ma	Spark/ Min
15%	350	700	20
5%	320	300	50

The explanation for the reduction of water vapor drastically altering the meter readings in this case is connected to the changes in the resistivity of the particulate due to the change in humidity.

Viscosity Effects

The effect of the viscosity of the gas stream on electrical characteristics can be prominent. The level of gas temperature, aside from its effects on resistivity, will tend to distort the voltage-current relationship drastically. As the temperature rises, the ratio of current flow relative to voltage fields will also rise. For example:

	Primary Volts	Secondary ma
200°F	320	400
400°F	300	550
600°F	270	750

Another effect on the density of the gas is pressure, and its influence on the break-down voltage in the precipitator can be substantial. For example:

Absolute Pressure	Pri. Volts	Sec. ma	Min. Spark/
30.0 Hg	330	700	20
27.0	305	700	20
24.0	270	700	20

Misalignment of Electrode Systems

Misalignment between the discharge electrodes and the collecting surfaces can produce a variety of electrical readings. Uniform misalignment seldom exists between fields, so the pattern of readings is usually distorted. Misalignment will tend to produce lower voltages and current at spark-over during the higher resistivity conditions. During low resistivity conditions, the currents might trend higher relative to the voltage readings.

Coating on the Discharge Electrode or Insulators

Very seldom does the coating of material on the discharge electrodes provide major changes to the precipitator electrical readings, especially in the basic industries. Coatings might have more effect at lower power levels than when at least 50% loading levels of the T-R set have been reached. The voltage level increases substantially relative to the current flow. While this might also indicate a space charge problem in the inlet field, this observation should not occur on subsequent fields. Usually precipitators handling very fine-size particles in the high resistivity zones tend to have this problem. Designs of uniformly shaped wires are more susceptible to this condition, especially when the build-up of material also assumes a uniform shape over the contour of the wire surface. While each field might be affected if it is a gas-dust oriented problem, often this condition is observed on isolated fields because of rapper failure. Possible readings might show:

Condition	Pri. Volts	Sec. ma	Spark/ Min.
Normal Electrode	350	700	10
Coated Electrode	400	150	10

Coatings of dust on insulator surfaces will not usually generate electrical difficulty for most precipitators unless dew-point conditions exist. The key point is that when tracking or surface leakage occurs on the insulator, it should become progressingly worse in time. The voltage will decrease for elevations in the current flow in all cases.

Discontinuities on Collecting Surface

I want to stress the fact that discontinuities or irregularities on the collecting surface opposite the discharge electrodes are prime sources of distortion of electrical readings. This will be observed more in the high resistivity range. For example, a $\frac{1}{4}''$ sharp protrusion on the surface could account for a 20 to 35 volt reduction on the primary voltmeter with a substantial drop in corona-current because spark-over had limited the power level of the field. Careful internal inspections, and even attempts to locate the point of spark-over through observation ports must be employed if the pattern of readings indicate an abnormal power distortion in the field.

Effects of Dust Layer on Collecting Surface

I pointed out earlier that the impact on electrical characteristics of the dust layer thickness on the collecting surface is mostly related to the high resistivity range. The irregularities of this dust layer can be as important as the average thickness in affecting the spark-over level.

Layer thickness, however, may grow to sizable levels in the low to moderate range of resistivity without producing break-down voltages. In these cases the voltage reading will decrease relative to the current reading. Voltage drops of 20 to 30 volts are not unusual for a given current. This condition usually occurs with a failure of the rapper system on the collecting surfaces.

Design and Performance Concepts

I would like to enlarge on two main concepts of precipitator design. The simple Deutsch equation is a valid way to understand how the various critical inputs can effect the performance of the precipitator. I mentioned in the text that the exponent can be low by a factor of two because of a number of problems that the designer did not forsee. Excessive reentrainment and poor gas distribution were two of the prime reasons for the disparity between the theoretical and actual results.

Recent designs have taken the migration exponent to another $\frac{1}{2}$ power or less to correct for previous problem areas and provide additional margin for the fine-sized particles existing in the latter fields of the precipitator. What this means is that a doubling of the physical precipitator is indicated compared to what was considered a standard design of the early 1970's. Whether this is warranted is based on two things: whether or not good design concepts are applied and, second, the confidence of the user that he can exercise some control over the process gas and dust characteristics. In retrospect, the design of 8 to 12 years ago could meet their guarantees if conditions were optimized. Designs of 40 to 50% greater surface area over those of the past now appear quite reasonable if you weigh all the factors of today's environment. This still means carefully addressing the process characteristics and applying a commitment to proper operating and maintenance techniques.

Each precipitator field should be considered a separate collector unit, and for that matter, each gas passage of a field must perform well to attain the best bottom line of the overall system. For high collection efficiencies to be achieved, the inlet field must perform near design levels usually in the nominal 80% range. Theoretically, that means about 80% of the particulate matter would deposit in the front hopper. This is why the inlet field looms important in any upgrading program.

I would stress a few points:

1. Each succeeding field works on the residual of the preceeding field, but the potential collection efficiency tends to decrease in the direction of flow. Part of the reason is that collection values are harder to achieve as the magnitude of particles decrease.

2. Another reason is that the particles that are left in the gas stream in the latter half of the precipitator are tougher to collect since they usually consist of the finer-sized segment. Unless current densities above 20 ma/1000 sq ft are observed in these latter fields, their collection efficiencies can deviate substantially from design.

3. I would like to stress again that working for more effective voltage fields and maintaining the uniformity of these fields should be the cornerstones of basic design concepts. Great emphasis on gas velocity and SCA parameters is less important than being able to understand and control critical process variables.

More Operation and Maintenance Concepts

Some areas of the precipitator system cause more than 90% of all the operating and maintenance problems. I tried to cover the major points of concern and how to overcome them during earlier discussions. The following three areas must be continually emphasized and optimized beginning with the design period.

Hoppers

Unless the hoppers are being continously evacuated, trouble can occur from a number of reasons, but condensation and packing conditions are the prime factors. The pyramid hopper, common in the fly ash precipitator, is prone to have trouble at the bottom throat. The wall surface area existing for the volume of material is greatest at this point. This is compounded by the sheer weight of stored material that tends to pack the smaller volume at the throat area. This is the area of precipitation that must receive a lot of time, attention and money to minimize troubles. Stress the need for dust height monitors to signal trouble at relatively low levels rather than send an alarm after excessive build-ups occur.

Electrode Failure

I believe in time that the problem of excessive discharge electrode failures will cease to be a prominent factor in the weighted wire design. Unfortunately, too many existing precipitators still have unshrouded electrodes. The key concept is that discharge electrode failures should be infrequent in a well-designed precipitator.

The following comments are believed important for reducing the number of discharge electrode failures:

1. Do not overrap. Unless symptoms surface to indicate that dust build-up is effecting corona formation or distribution, less rapping is usually the better approach.

2. Do not replace discharge electrodes that fail in service. The cause of the failure may still be present.

3. Ascertain quickly the probable cause of failures that are observed to be repetitive at the same electrode location and to show the same scope of damage.

Rappers

From an operation and maintenance viewpoint, the problems involved with the rapper system is primarily connected to its flexibility and reliability. I am less concerned about the number of G's of force of the rappers as I am with how many will still be functioning after six months of operation. The large number of rappers used in many installations makes this component of precipitation one that requires extra maintenance concern during the early design and specification period with the manufacturer. Emphasis should be made on the need for detecting rapper difficulties in some easy manner and the ease of component replacement.

Voltage Control Circuits

The use of printed circuit boards housed in the control panel has caused some problems with dust contamination depending where the panels are located. While I mentioned the use of low-pressure dry air for a cleaning method in the text, other problems could occur with this method if the dust particles are driven into the crevices of the board components. Another technique to consider would be the use of a high-vacuum source and soft nozzle tip so as not to damage any components.

Several other points in the application of automatic control circuitry should be considered:

1. The manual control feature of the voltage supply should have a separate circuit. The ability to trouble-shoot without the influence of the automatic circuit components is desirable.

2. It is important to be able to safely adjust the controls without coming into close proximity with any portion of a live electrical circuit. This should be a primary concern with the original design as well as with the retrofit replacement of controls.

3. The 110 volt or other low voltage circuits should be clearly and safely isolated from the 440 volt or higher circuits within the control panel.

The pipe and guard system of transporting the high voltage from the T-R set to the precipitator has been widely accepted and has proven reliable in most cases. However, it has been brought to my attention that recent designs of dry high-voltage cables might again allow the placement of the T-R sets at ground level. Whether this is considered must be based on the economic factors and the information that exists on the reliability of the cables.

The Process

The importance of the process on the success of the precipitator has been emphasized at numerous times in this book. This fact has escaped many users of this equipment and variations in collection performance is still primarily laid to the fault of the precipitator.

That is one reason I have stressed the inputs of the process, both to understand which factors help and which factors can hurt the performance of the precipitator. It is an educational activity that must be continually upgraded if a company is to cope with changing gas and dust conditions.

And changing gas and dust conditions in a process are things you can always be sure will occur over a lengthy time period. There are no easy answers to all the subtle factors that occur, but some of the following concepts might help:

1. Each installation has defined foot-prints of gas and dust characteristics related to precipitator performance. For example, a cement process may show 1% O_2 at the feed-end of the kiln and 5% at the inlet face of the ESP. A fly ash installation may show 3.5% O_2 at the economizer outlet, 4.5% O_2 at the air heater outlet, and 5% O_2 at the ESP (a common abbreviation for the precipitator).

This information as well as pertinent process variables should be recorded when times are good, otherwise how can the adverse changes be easily identified when the ESP performance deteriorates.

2. Any correction program to overcome precipitator difficulties caused by the process should first return to the process for a possible solution. When the gas and dust characteristics are optimized for best precipitation, especially on cold-side applications, it is often observed that production rates or the quality of the product benefits.

3. You should always be concerned that either modifications to the process equipment or major changes in the raw materials or fuel can trigger an adverse effect on the precipitator.

4. You should also be aware of any option or possible simple hedge concept that can be designed into every system initially and which could be used to nulify adverse process changes that may occur in the future. In some cases this may only require installing a section of ductwork that could be used for a gas conditioning application. As I shall discuss later, it is generally better to design into the process the ability to control the gas temperature within as wide a range as feasible at the inlet of the ESP.

5. In-house expertise is probably the best insurance for a company to be able to detect and explain variations in precipitator performance related to the process. This requires a commitment for at least one person in the company trained to understand the factors and to set up the procedures and mechanisms to monitor both the process and the precipitator.

The Fly Ash Application

The collection of fly ash, of all the industrial applications, has enjoyed a fascinating relationship with the precipitator. Complexity of the chemical make-up of the fly ash at the low levels of moisture found in the flue gas has produced a wide range of electrical characteristics, especially over the last 30 years. When you consider that about 70 to 80% of all the new precipitator business the last few years has been for the collection of fly ash, you begin to appreciate the amount of interest in this particular area.

It is unfortunate that fly ash collection for the precipitator coincides with the material imposing the greatest assortment of problem areas. A major part of the problem lay with the wide band of fly ash electrical characteristics having to also contend with a wide band of operating factors in the boiler system. And part of the problem is man-made, primarily because the pressure to achieve extreme collection performances occurred before many of the inputs to the field problems were well understood. Just the fact that relatively small changes of flue gas temperature, often 10 to 15°F, can appreciably alter the resistivity levels of some fly ashes has been a source of trouble over the years.

Misconceptions

Superimposed on some of the above problem inputs has been a group of misconceptions that keep surfacing to complicate the effective precipitation of fly ash. While these comments also hold true for some of the other process applications, they are especially pointed toward fly ash.

The first relationship that is still being distributed and promoted is that as the content of sulfur in the coal increases, the precipitator collection efficiency is bound to improve. This concept was usually true for boiler systems designed before 1950 where gas temperatures above 310°F were often observed. During the past 15 to 20 years on many bituminous coal systems east of the Mississippi River, this relationship has been contrary to the actual performance characteristics of the fly ash precipitator.

The concept to be gained here is that there are few relationships in precipitation that can stand alone. In this case, the level of flue gas temperature in the cold-side precipitator is often as important for its effect on the ash electrical characteristics as the sulfur content of the coal. As discussed elsewhere, other constituents in the flue gas system and in the particle also contributes to the final effect on resistivity.

A second relationship that has caused some problems is tied to the exponent of the Deutsch equation which basically states that as the gas flow rate increases the performance of the precipitator must suffer, all other factors equal. While this is basically true in most situations, there happens to be a number of cases where this has not been true because the voltage fields improved with a change in the resistivity range at the higher flow. The concept here is that the quality of the flue gas stream is often more important than its magnitude.

The last misconception that always comes back to haunt certain fly ash installations is that inputs of more electrical power always results in the improved collection efficiency of the precipitator. Remember that it is the "Effective Holding Power," or the relationship between the voltage fields and the current flow, that is often more important. High levels of corona current, as described in the resistivity section, sometimes points to areas of concern.

Control of Gas Temperature

The magnitude and gradients of the flue gas temperature entering the fly ash precipitator is a tremendously important area, and yet has not been sufficiently stressed in practice. While these comments are concerned with the cold-side precipitator, a lack of gas temperature control also plays a role in the hot-side unit. A number of gas temperature comments were covered earlier in the text, but I want to emphasize two key concepts:

1. Be as concerned with the temperature gradient across the inlet face of the precipitator as with the average value.

2. Be knowledgeable of the changes of flue gas temperature with the boiler load and seasonal weather patterns.

Let's look at those two points in greater detail:

Temperature Gradient

This problem has intensified as the regenerative-rotating air heater became larger in diameter. The low speed of approximately 1 rpm produces gas temperature gradients of 70 to 90°F, and this spread is often transferred into the precipitator with only a 10 to 30°F moderation depending on the flue arrangement. If this spread occurs on the low temperature slope of a resistivity curve then the precipitator must content with a multitude of different resistivities. This reason alone has caused a number of units to perform below their guarantee, especially with coals containing 2% sulfur per weight or more.

This temperature gradient often encompassed a large area of precipitator so that possibly one T-R set had to contend with several levels of resistivity. The power supply is usually controlled by the worst zone of electrical characteristics. Moreover, the uniform application of gas conditioning systems for one level of resistivity sometimes effected the efficiency adversely in the other temperature zone.

One of several remedies can be employed to help overcome the gradient effects:

1. Adjust the average gas temperature so as to gain the best possible collection at the critical temperature zone. This technique can help a marginal performance.

2. Sectionalize with more T-R sets, so that the precipitator area in each temperature zone is energized by the best match of its own power supply. This has helped in most cases, but still leaves some problem jobs. It is interesting to observe the difficulty of holding on-to fly ash even at double power levels when acid dewpoint conditions exist in the cold temperature zones.

3. Apply appropriate gas conditioning to the selective temperature zones that require extra help. This technique has merit.

4. Nulify the temperature gradient by mixing the flue gas between the air heater and the precipitator where the flue layout allows its use. I have tried several methods, but the use of internal transfer ducts to shift part of the gases from the hot discharge side of the air heater to the cold gas zone appears the most effective. It is desirable to have at least a 36 ft straight length of flue so as to keep the angle of the transfer ducts reasonable. The shift of about 8 to 10% of the gas flow from the extreme hot side appears capable of desensitizing the effects of a cold end problem on reentrainment losses. This method might also work to transfer cold gas toward a hot zone if the resistivity is normally biased toward the higher end. As I mentioned in the text, a diffuser plate at the discharge of the transfer ducts will be needed to correct the gas distribution patterns prior to the inlet of the precipitator.

Gas Temperature Control

Many boiler systems lack the ability to maintain a fixed optimum flue gas temperature at the exit of the air heater throughout the year. This ability to control the average gas temperature at one level and fix this level within a range of 220 to 300°F can have a number of benefits. While this 80°F range appears sufficient for 95% of the eastern bituminous coal sources, I suggest that even lower gas temperatures be considered in new boiler designs for western coals.

The temperature control should exist for boiler loadings from 80 to 100% of rating. Installations located in geographical areas that experience wide swings in ambient temperature poses additional design problems. Uniform coal supplies or good blending procedures would help minimize the need for automatic temperature control, however, variable coal sources would involve some monitor and feed-back technique so as to arrive at the best gas temperature zone. This is not much different than methods used in gas conditioning.

A number of techniques are available to the designer for implementing the type and range of temperature control suggested above. Of course, a substantial amount of money must be added to the system for temperature control, but the savings in fuel costs can also be appreciable over the life of a large installation.

If the ability to operate at 250°F with certain low sulfur coals will provide acceptable performances in the precipitator then two benefits are accrued. However if 2% sulfur coal is being consumed, then 290°F may provide the best electrical characteristics for that period of time resulting in the best performance of the ESP.

I commented in the text that preheating the incoming combustion air either before or after the F.D. fan offers a good temperature control method. The use of glycol as the heat-exchange medium is commonly used on these systems. On some existing systems, insufficient heating can still occur during extreme cold weather or under boiler limitations. One company I worked with improved their system by increasing the percentage of glycol so an operating level of about 280°F existed rather than the 230°F range obtained by the original 50% mixture. Several safeguards were indicated:

1. A limit in the temperature of the mixture is the possible breakdown of the inhibitors of the glycol resulting in a fouling of the system.
2. The steam pressure must be higher than the glycol pressure in the heat exchanger.
3. It is suggested that different grades of glycol not be mixed; moreover, use the highest quality grades available.

Practically all existing boiler systems install some method for adding heat to the flue gas system. Where lower average gas temperatures are required, however, the control problem becomes difficult. As I mentioned in the text, the use of water sprays can be effective in some cases. Herb Hall mentioned one installation where additional air was pushed through the air heater and the excess air drawn off from the downstream side of the heaters to the stack.

Air Heater Leakage

The problem of inleakage of air into the flue gas side of the air heater should be monitored and corrected on a continuous basis. Leakage through the seals of the regenerative air heater will generally not be uniform so that its true magnitude must be ascertained by an O_2 traverse at the discharge of the air heater rather than rely on sparse measurements. Remember that the location of the air-inleakage can be more important than the average for its effect on electrical characteristics in one zone of the precipitator.

It is also important to keep the cold-end of a tubular air-heater as free of blockage as possible in order to keep the erosion to a minimum thus reducing the degree of air-inleakage. The higher velocities through the remaining tubes result in a cutting action especially when higher levels of combustible particles exist.

Ash Characteristics

The variability of fly ash makes the prediction of electrical characteristics difficult. Bituminous coals with reasonable sulfur contents usually supplies a high degree of confidence for the application of the fly ash precipitator. As the sulfur by weight fraction moves below 1%, then other chemical constituents increase their effect on resistivity There are even some difficult ashes that do not respond to reasonable changes in gas temperature.

Mr. H. J. Hall, during the 1980 series of the Hall–Katz Seminars, discussed some relationships that are interesting:

1. For a bituminous type ash, the Fe_2O_3 content is greater than the $CaO + MgO$.

2. In the lignitic type of ash, this relationship is reversed.

3. Poor ash characteristics might also coincide with the $CaO + MgO$ fraction above 15% by weight.

4. Low Na_2O below 0.5% by weight and Fe_2O_3 below about 4% on western coals point toward troubles.

5. When the alumina-silicate portion combines for about 80% or more of the ash analysis, the chance for high resistivity problems is increased.

6. Certain ashes appear to have the ability to coat the collecting surface with a layer of material that tends to step up the level of resistivity in a relatively short time

period. Evidently this phenomenon is tied to a depletion of sodium ions in the ash layer. This condition may exist when the power input to the precipitator decreases from a relatively high to low level exhibiting back-corona characteristics within a matter of days. This has occurred on some hot-side precipitators, but also could appear in the cold-side application. Dr. R. E. Bickelhaupt, of the Southern Research Institute, among others, have been investigating this problem area.

If there is one thing sure about fly ash it is that new variables will surface in the future. Much of this book has been devoted to practical field knowledge so that the reader can understand and cope with the large percentage of existing and future precipitators.

Reentrainment

Some additional comments in the area of the reentrainment of material are warranted. I already discussed the effect of resistivity on reentrainment, but the subject is much more complex. How far the material moves out into the gas system, where it redeposits in relation to its original position, and has it changed its physical character are but some of the unknown factors in the reentrainment syndrome.

A common description of the dust layer sliding down the collecting surface does not usually occur. The shock or tremor imparted by the rapper appears to more often dislodge some percentage of material from its resting place. If the particles have had a chance to agglomerate, the adverse effects of reentrainment are minimized. It is when the particles bounce back into the gas stream in the same condition as they were collected that troubles begin to mount. This is where proper resistivity and the timing between raps can play an important part in this interesting phase of precipitation.

All sorts of dust patterns and rapper operations have been observed over the years and I have no simple answer for rapping success across the board. In fact, I believe this area of precipitator concern is so complex I normally find it easier to upgrade performances through another route rather than spend much time on the mechanics of rapping. But make no mistake, there are certain concepts in rapping techniques that tends to either add to or minimize the reentrainment losses as observed visually or by test. These concepts include:

1. Always use the internal inspection and other visual means to help ascertain the lowest rapping intensity possible commensurate with other performance observations.

2. Always attempt to match the rapping to the dust characteristics or resistivity. For example, low resistivity requires soft rapping, the moderate range requires a harder blow, and the high resistivity zone means real trouble. Remember that hard rapping with high resistivity materials usually exhibits limited success and changing the resistivity is usually a much better way to achieve a satisfactory performance.

3. Do not feel that the inlet field must be over-rapped because it handles the bulk of material. Be aware that puffs out of the stack or other signs of reentrainment are not unique to the outlet fields. Actually, the material collected in the inlet field of some precipitators is often easier to dislodge, and excessive carry-over adds to the reentrainment potential of the following fields.

4. Rapping losses is not usually uniform across the precipitator, even discounting the effects of resistivity gradients. When reentrainment losses are observed, investigations into possible problems with gas distribution is highly recommended.

5. When reentrainment losses are severe, lengthening the time between raps on the collecting surfaces in the direction of inlet to outlet is usually recommended. As a first adjustment it might be advisable to double the rap time on succeeding fields. For example, if the inlet field rapped every 5 minutes, then the second field would be rapped every 10 minutes, the third field every 20 minutes, and so on.

6. The more the power characteristics are improved, the better the chance for reentrainment losses to diminish. Rappers should always operate across one field before the cycle moves on to the next field. As discussed in the text, excessive dust disturbances on the collected layer can lead to adverse electrical field activity in certain cases. Allowing the surface contour to smooth out slightly between raps can relieve localized stress points and reduce the spark-over potential. Usually 5 to 10 minutes time duration is needed to observe this phenomenon where it will occur.

Reentrainment does not automatically reduce when the surface area per rapper is reduced. The reverse may occur on an installation, but the signs of the reentrainment is less evident by opacity spikes because the amount of material dislodged per rap is masked by the greater quantity of the non-disturbed gas volume.

The use of the opacity spikes on a chart to ascertain the effects of rapping reentrainment is a good evaluation tool. That is why I prefer the instantaneous readout rather than the integrated time response of an opacity chart. The width of the trace band for a problem installation can be more than 15 to 20%, that is, the movement of the pen swings from the base of the band to the peak. As resistivity lowers, the swings of the pen can reach even higher levels. Isolating groups of rappers will often indicate on the chart which chamber is contributing the most to the reentrainment losses.

Gas Distribution

Whether gas distribution is effecting the precipitator performance adversely can be related to many factors. Efforts are usually worthwhile in exploring improvements in the gas distribution pattern if one or more of the following conditions exist:

1. Aspect ratios of 1.0 or less.

2. Average calculated gas velocity of more than 6 ft/sec through the precipitator.

3. Reentrainment dust losses above 50% of the total ESP emission losses.

4. Low resistivity characteristics are apparent with an absence of spark-over.

5. High resistivity characteristics are apparent with less than an average of 10 ma/1000 sq ft of collecting surface.

Some of the key concepts for keeping gas distribution problems to a minimum include:

1. The gas pattern entering the inlet plenum of the ESP should strive for uniformity, with the vectors of gas flow all perpendicular to the inlet plane of the plenum.

2. Any gas vectors striking a perforated plate, or other type of diffuser, by angles over 30° from the perpendicular tends to shift part of its flow across the plate. The sharp expansion of a nozzle plenum ahead of the precipitator provides the best chance for this condition to occur.

3. Use periodic internal inspections of the inlet plenum to ascertain effects of dust build-up or metal erosion on the gas flow patterns. Many of the dust build-up problems occur when the process goes through extended periods of low production. It is amazing how many problem installations existed because deposition of dust nullified what looked like a good model study pattern. High combustible contents in fly ash might erode relatively large areas in perforated plates or gas deflecting vanes. This will occur in the higher velocity zones, and once the wear pattern starts, the rate accelerates until the short circuiting of gas begins to effect performance.

How To Improve Performance

Programs for improving the collection performance of existing precipitators is often difficult because operating conditions always appear to be changing the scope of the problems. But the primary reason that most programs fail is that all the fundamental things that are wrong are not attacked within a reasonable time frame.

The key point here is that there are usually a number of contributors to a bad situation, and they interrelate sufficiently so as to prevent maximum performance benefits unless they are all improved to some degree. I am not saying that all the fundamental problems must be corrected simultaneously, but stopping a program after fixing one or two areas usually never achieves eventual success. The prime fundamental areas include:

1. Control over the process variables.
2. Optimum internal alignment of components.
3. Reliability of electrode components.
4. Reliability of rapper and dust evacuation system.
5. Elimination of inleakage air.
6. Elimination of temperature gradients.
7. Control over the gas temperature levels.
8. Proper sectionalization and match of T-R sets to load.
9. Optimum gas distribution patterns.
10. Reasonable control of power supplies.
11. Means to control the resistivity of the material.
12. Reasonable control over flexibility and intensity of rappers.